Advances in Intelligent Systems and Computing

Volume 419

Series editor

Janusz Kacprzyk, Polish Academy of Sciences, Warsaw, Poland
e-mail: kacprzyk@ibspan.waw.pl

About this Series

The series "Advances in Intelligent Systems and Computing" contains publications on theory, applications, and design methods of Intelligent Systems and Intelligent Computing. Virtually all disciplines such as engineering, natural sciences, computer and information science, ICT, economics, business, e-commerce, environment, healthcare, life science are covered. The list of topics spans all the areas of modern intelligent systems and computing.

The publications within "Advances in Intelligent Systems and Computing" are primarily textbooks and proceedings of important conferences, symposia and congresses. They cover significant recent developments in the field, both of a foundational and applicable character. An important characteristic feature of the series is the short publication time and world-wide distribution. This permits a rapid and broad dissemination of research results.

Advisory Board

Chairman

Nikhil R. Pal, Indian Statistical Institute, Kolkata, India
e-mail: nikhil@isical.ac.in

Members

Rafael Bello, Universidad Central "Marta Abreu" de Las Villas, Santa Clara, Cuba
e-mail: rbellop@uclv.edu.cu

Emilio S. Corchado, University of Salamanca, Salamanca, Spain
e-mail: escorchado@usal.es

Hani Hagras, University of Essex, Colchester, UK
e-mail: hani@essex.ac.uk

László T. Kóczy, Széchenyi István University, Győr, Hungary
e-mail: koczy@sze.hu

Vladik Kreinovich, University of Texas at El Paso, El Paso, USA
e-mail: vladik@utep.edu

Chin-Teng Lin, National Chiao Tung University, Hsinchu, Taiwan
e-mail: ctlin@mail.nctu.edu.tw

Jie Lu, University of Technology, Sydney, Australia
e-mail: Jie.Lu@uts.edu.au

Patricia Melin, Tijuana Institute of Technology, Tijuana, Mexico
e-mail: epmelin@hafsamx.org

Nadia Nedjah, State University of Rio de Janeiro, Rio de Janeiro, Brazil
e-mail: nadia@eng.uerj.br

Ngoc Thanh Nguyen, Wroclaw University of Technology, Wroclaw, Poland
e-mail: Ngoc-Thanh.Nguyen@pwr.edu.pl

Jun Wang, The Chinese University of Hong Kong, Shatin, Hong Kong
e-mail: jwang@mae.cuhk.edu.hk

More information about this series at http://www.springer.com/series/11156

Nelishia Pillay · Andries P. Engelbrecht
Ajith Abraham · Mathys C. du Plessis
Václav Snášel · Azah Kamilah Muda
Editors

Advances in Nature and Biologically Inspired Computing

Proceedings of the 7th World Congress on Nature and Biologically Inspired Computing (NaBIC2015) in Pietermaritzburg, South Africa, held December 01–03, 2015

 Springer

Editors
Nelishia Pillay
School of Mathematics, Statistics
  and Computer Science
University of KwaZulu-Natal
Pietermaritzburg
South Africa

Andries P. Engelbrecht
Department of Computer Science, School
  of Information Technology
University of Pretoria
Pretoria
South Africa

Ajith Abraham
Machine Intelligence Research Labs
Scientific Network for Innovation
  and Research Excellence
Auburn, WA
USA

Mathys C. du Plessis
Department of Computing Sciences
Nelson Mandela Metropolitan University
Port Elizabeth
South Africa

Václav Snášel
Faculty of Electrical Engineering
  and Computer Science
Technical University of Ostrava
Ostrava-Poruba
Czech Republic

Azah Kamilah Muda
Computational Intelligences
  and Technologies Lab, Faculty
  of Information and Communication
  Technology
Universiti Teknikal Malaysia Melaka
Durian Tunggal
Malaysia

ISSN 2194-5357   ISSN 2194-5365 (electronic)
Advances in Intelligent Systems and Computing
ISBN 978-3-319-27399-0   ISBN 978-3-319-27400-3 (eBook)
DOI 10.1007/978-3-319-27400-3

Library of Congress Control Number: 2015956133

This Springer imprint is published by SpringerNature
The registered company is Springer International Publishing AG Switzerland

Preface

Welcome to the 2015 Seventh World Congress on Nature and Biologically Inspired Computing (NaBIC'15)! The Congress is organized to discuss the state of the art as well as to address various issues with respect to building up of computers in man's image. The theme for NaBIC 2015 is "Nurturing Intelligent Computing Towards Advancement of Machine Intelligence." The congress will provide an excellent opportunity for scientists, academicians, and engineers to present and discuss the latest scientific results and methods. The conference includes a keynote address and contributed papers.

The papers were solicited in the following areas:

Artificial Neural Networks
Biodegradability Prediction
Cellular Automata
Evolutionary Algorithms
Swarm Intelligence
Emergent Systems
Artificial Life
Lindenmayer Systems
Digital Organisms
Artificial Immune Systems
Membrane Computing
Simulated Annealing
Communication Networks and Protocols
Computing with Words
Common Sense Computing
Cognitive Modeling and Architecture
Connectionism
Metaheuristics

Hybrid Approaches
Quantum Computing
Nano Computing
Industrial Applications of Nature and Biologically Inspired Computing

The aim is to bring together worldwide leading researchers, developers, practitioners, and educators interested in advancing the state of the art in Nature and Biologically Inspired Computing for exchanging knowledge that encompasses a broad range of disciplines among various distinct communities. It is hoped that researchers will bring new prospects for collaboration across disciplines and gain ideas facilitating novel breakthrough.

The conference will provide an exceptional platform to researchers to meet and discuss the utmost solutions, scientific results, and methods in solving intriguing problems with people who are actively involved in these evergreen fields.

NaBIC 2015 received contributions from more than 26 countries. Each paper was sent to at least five reviewers from our International Program Committee for a standard peer-review evaluation, and based on the recommendations 39 papers were included in the proceedings. NaBIC 2015 is technically supported by the IEEE SMC Technical Committee on Soft Computing. Many people have collaborated and worked hard to make NaBIC 2015 a success. First and foremost, we would like to thank all the authors for submitting their papers to the conference, and for their presentations and discussions during the conference. Our thanks to the Program Committee members and reviewers who carried out the most difficult work of carefully evaluating the submitted papers. We also thank the Springer Series on Advances in Intelligent Systems and Computing Editorial Team: Prof. Dr. Janusz Kacprzyk, Dr. Thomas Ditzinger, and Mr. Holger Schaepe for the wonderful support to publish this volume so quickly.

We wish all NaBIC 2015 delegates an exciting meeting and a pleasant stay in Pietermaritzburg, South Africa! Enjoy the Congress!

Nelishia Pillay
Andries P. Engelbrecht
Ajith Abraham
Mathys C. du Plessis
Václav Snášel

Contents

A Diverse Meta Learning Ensemble Technique to Handle Imbalanced Microarray Dataset 1
Sujata Dash

A Self-configuring Multi-strategy Multimodal Genetic Algorithm 15
Evgenii Sopov

An Ensemble of Neuro-Fuzzy Model for Assessing Risk in Cloud Computing Environment 27
Nada Ahmed, Varun Kumar Ojha and Ajith Abraham

k-MM: A Hybrid Clustering Algorithm Based on k-Means and k-Medoids 37
Habiba Drias, Nadjib Fodil Cherif and Amine Kechid

A Semantic Reasoning Method Towards Ontological Model for Automated Learning Analysis 49
Kingsley Okoye, Abdel-Rahman H. Tawil, Usman Naeem and Elyes Lamine

Cooperation Evolution in Structured Populations by Using Discrete PSO Algorithm 61
Xiaoyang Wang, Lei Zhang, Xiaorong Du and Yunlin Sun

Memetic and Opposition-Based Learning Genetic Algorithms for Sorting Unsigned Genomes by Translocations 73
Lucas A. da Silveira, José L. Soncco-Álvarez, Thaynara A. de Lima and Mauricio Ayala-Rincón

**Aesthetic Differential Evolution Algorithm for Solving
Computationally Expensive Optimization Problems**............... 87
Ajeet Singh Poonia, Tarun Kumar Sharma, Shweta Sharma
and Jitendra Rajpurohit

**Automatic Discovery and Recommendation for Telecommunication
Package Using Particle Swarm Optimization**..................... 97
Shanshan Liu, Bo Yang, Lin Wang and Ajith Abraham

**EEG Signals of Motor Imagery Classification Using Adaptive
Neuro-Fuzzy Inference System**.............................. 105
Shereen A. El-aal, Rabie A. Ramadan and Neveen I. Ghali

**Modeling Insurance Fraud Detection Using Imbalanced
Data Classification**.. 117
Amira Kamil Ibrahim Hassan and Ajith Abraham

**Using Standard Components in Evolutionary Robotics
to Produce an Inexpensive Robot Arm**........................ 129
Michael W. Louwrens, Mathys C. du Plessis and Jean H. Greyling

Sinuosity Coefficients for Leaf Shape Characterisation............ 141
Jules R. Kala, Serestina Viriri and Deshendran Moodley

**A Study of Genetic Programming and Grammatical Evolution
for Automatic Object-Oriented Programming: A Focus
on the List Data Structure**................................. 151
Kevin Igwe and Nelishia Pillay

**Evolving Heuristic Based Game Playing Strategies
for Checkers Incorporating Reinforcement Learning**.............. 165
Clive Frankland and Nelishia Pillay

**PARA-Antibodies: An Immunological Model for Clonal
Expansion Based on Bacteriophages and Plasmids**............... 179
Mark Heydenrych and Elizabeth Marie Ehlers

Bioinspired Tabu Search for Geographic Partitioning............. 189
María Beatríz Bernábe-Loranca, Rogelio González Velazquez,
Martín Estrada Analco, Jorge Ruíz-Vanoye, Alejandro Fuentes Penna
and Abraham Sánchez

A Hyper-Heuristic Approach to Solving the Ski-Lodge Problem..... 201
Ahmed Hassan and Nelishia Pillay

Real-Time Vehicle Emission Monitoring and Location Tracking Framework. 211
Eyob Shiferaw Abera, Ayalew Belay and Ajith Abraham

Applying Design Science Research to Design and Evaluate Real-Time Road Traffic State Estimation Framework. 223
Ayalew Belay Habtie, Ajith Abraham and Dida Midekso

Application of Biologically Inspired Methods to Improve Adaptive Ensemble Learning. 235
Gabriela Grmanová, Viera Rozinajová, Anna Bou Ezzedine,
Mária Lucká, Peter Lacko, Marek Lóderer, Petra Vrablecová
and Peter Laurinec

A Variable Neighbourhood Search for the Workforce Scheduling and Routing Problem. 247
Rodrigo Lankaites Pinheiro, Dario Landa-Silva and Jason Atkin

Modelling Image Processing with Discrete First-Order Swarms. 261
Leif Bergerhoff and Joachim Weickert

Training Pattern Classifiers with Physiological Cepstral Features to Recognise Human Emotion 271
Abdultaofeek Abayomi, Oludayo O. Olugbara,
Emmanuel Adetiba and Delene Heukelman

Identification of Pathogenic Viruses Using Genomic Cepstral Coefficients with Radial Basis Function Neural Network. 281
Emmanuel Adetiba, Oludayo O. Olugbara and Tunmike B. Taiwo

An Aphid Inspired Evolutionary Algorithm 293
Michael Cilliers and Duncan Coulter

A Neural Network Model for Road Traffic Flow Estimation 305
Ayalew Belay Habtie, Ajith Abraham and Dida Midekso

Optimization of a Static VAR Compensation Parameters Using PBIL. 315
Dereck Dombo and Komla Agbenyo Folly

ImmunoOptiDrone—Towards Re-Factoring an Evolutionary Drone Control Model for Use in Immunological Optimization Problems. 327
Kevin Downs and Duncan Coulter

A Generative Hyper-Heuristic for Deriving Heuristics
for Classical Artificial Intelligence Problems   337
Nelishia Pillay

A Priority Rate-Based Routing Protocol for Wireless
Multimedia Sensor Networks   347
Loini Tshiningayamwe, Guy-Alain Lusilao-Zodi and Mqhele E. Dlodlo

Newcastle Disease Virus Clustering Based on Swarm
Rapid Centroid Estimation   359
Fatma Helmy Ismail, Ahmed Fouad Ali, Saleh Esmat
and Aboul Ella Hassanien

Design of Nature Inspired Broadband Microstrip Patch
Antenna for Satellite Communication   369
Pushpendra Singh, Kanad Ray and Sanyog Rawat

Using Headless Chicken Crossover for Local Guide Selection
When Solving Dynamic Multi-objective Optimization   381
Mardé Helbig and Andries P. Engelbrecht

Updating the Global Best and Archive Solutions of the Dynamic
Vector-Evaluated PSO Algorithm Using ϵ-dominance   393
Mardé Helbig

Detection of Zero Day Exploits Using Real-Time Social
Media Streams   405
Dennis Kergl, Robert Roedler and Gabi Dreo Rodosek

Profile Matching Across Online Social Networks
Based on Geo-Tags   417
Robert Roedler, Dennis Kergl and Gabi Dreo Rodosek

Detection of Criminally Convicted Email Users by Behavioral
Dissimilarity   429
Maqsood Mahmud, Pranavkumar Pathak, Vaishalibahen Pathak
and Zahra Afridi

Cognitive Radio Networks: A Social Network Perspective   441
Efe F. Orumwense, Thomas J. Afullo and Viranjay M. Srivastava

Author Index   451

A Diverse Meta Learning Ensemble Technique to Handle Imbalanced Microarray Dataset

Sujata Dash

Abstract One of the challenging issues in bioinformatics field is that, microarray datasets are imbalance in nature i.e., the majority class dominates the minority class making it difficult for the conventional classifiers to achieve accurate and useful predictions. However, some studies have addressed this issue merely by focusing on binary–class problems. In this article, an ensemble framework is proposed for multiclass imbalance classification problem that combines a meta learning algorithm 'decorate' with a sampling technique to deal with the problem in microarray datasets. The meta-learning algorithm builds diverse ensembles of classifiers constructing artificial samples and the sampling technique introduces bias to achieve uniform class distribution to reduce misclassification error. Experimental results on the two highly imbalanced multiclass microarray cancer datasets indicate that the technique applied provides significant improvement in comparison to other conventional ensembles.

Keywords Imbalanced dataset · Sampling method · Decorate · Meta-learning · Multiclass datasets · Ensemble technique

1 Introduction

Microarray datasets contain gene expression profiles of tens of thousands of genes and has been widely used to predict the functions of gene, help to find gene regulatory system, find subtypes of a particular tumor, help in drug discovery and also useful for classifying cancers [1]. From the above mentioned applications, cancer classification research has gained a momentum in research community all

S. Dash (✉)
North Orissa University, Baripada, Odisha, India
e-mail: sujata238dash@gmail.com; sujata_dash@yahoo.com

© Springer International Publishing Switzerland 2016
N. Pillay et al. (eds.), *Advances in Nature and Biologically Inspired Computing*,
Advances in Intelligent Systems and Computing 419,
DOI 10.1007/978-3-319-27400-3_1

1

around the world. In addition to this microarray data suffers some problem such as high dimension, high noise, small sample size, high redundancy and class imbalance problem. The skewed class distribution or when samples of one class outnumber the samples of other class happens in a dataset, which affects the classification performance of minority class to a great extent that is called class imbalance problem. This problem also has an effect on the overall classification performance of the dataset. Moreover, the current research works have taken up this issue in the context of cancer classification using microarray cancer dataset [2–8]. But, most of the existing work has done with binary class imbalanced problem and it has been observed that the conventional learning algorithms which tries to minimize classification errors results in poor prediction performance. However, addressing the imbalanced problem in multiclass microarray cancer datasets is more difficult than binary class imbalance problems [6, 9].

One of the important issue which is very closely associated with imbalance problems include biasness that is involved in majority class which increases the number of false negatives and as a result reduced classification performance. This issue severely affects the performance of most of the classifiers by classifying the instances of the majority class available in the dataset. To overcome this problem, some standard sampling techniques such as under sampling (removing instances of the majority class in a direct or random manner), oversampling [9] (adding instances to the minority class in a direct or random manner), sampling by synthetically generating instances namely, 'Synthetic Minority Oversampling Technique' (SMOTE) [10] have been used. The sampling methods change the distribution of the dataset by adding or removing the instances in order to arrive at a balanced class problem.

Furthermore, in algorithmic level, the ensemble methods [11] also have been used to increase the predictive performance of imbalanced microarray datasets. They use to integrate the outcome of multiple classifiers learned on different datasets to reduce the prediction errors so that new samples can be predicted accurately. The overfitting problem can be avoided in this case by providing varied datasets to each classifier. Many ensemble methods such as bagging, boosting and AdaBoost have been paired with sampling techniques by many studies to handle the issue of class imbalance datasets. Generally, these learning algorithms work on the entire feature space as they aim to modify the samples which are being selected for the training sets. As a matter of fact, sampling techniques and algorithmic methods are not sufficient enough to handle high-dimensional class imbalance problems.

Forman [12] has analyzed highly imbalanced text classification problems and stated that feature selection increases the performance than classification algorithm and also helps in curbing the overfitting problem.

Table 1 Details of the datasets

Dataset	Number of samples	Number of classes	Number of genes	Imbalance ratio	Diagnostic task
Brain Tumor1	90	5	5920	15.00	5 human brain tumor types
11 Tumors	174	11	12533	4.50	11 various human tumor types

In the present paper, it is assumed that the reduced feature of the dataset might help the re-sampling method to acquire more refined information which not only helps to increase the diversity among the base learners of the ensemble but also to the bias introduced through the sampling method. A widely used filter method namely, "ReliefF" algorithm is employed to enhance the capability of the proposed framework to handle the class imbalance problem. The functionality of meta-classifier Decorate [13] is extended by integrating the sampling technique [11] into its domain and then the model is tested on reduced multi-class highly imbalanced microarray datasets with different degrees of class imbalance shown in Table 1. The performance of the proposed ensemble method is compared with conventional ensemble techniques like bagging, boosting, random subspace and AdaBoost paired with sampling technique. The statistically tested experimental result has established that the sampling method combined with Decorate Meta learning is more efficient than the other combinations of sampling and ensemble learners. The paper is organized as follows. Section 2 explains the methodology used in this study in detail and Sect. 3 introduce the multi class datasets. The experimental settings and the metrics used for evaluating performance are described in Sect. 4. Section 5 describes experimental results and discussions. Section 6 concludes the paper followed by references.

2 Methodology

2.1 Feature Selection

The 'curse of dimensionality' and 'overfitting' problems are prevalent in high dimension and small sample size microarray imbalance datasets which create lot of computational challenges for finding significant features [14].

Relief is one of the popular and widely accepted feature selection metric based on the nearest neighbor principle devised by Kira and Rendell [15]. This algorithm is very effective as it does not assume that attributes are independent from each other [14, 15]. ReliefF [16] is an advance version of Relief algorithm that deals with multi class problems, incomplete and noisy data. Therefore, the advantage of this algorithm over the others are, it interacts with other features, attain local dependencies and low bias which is lacking by most of the other filter algorithms.

2.2 Sampling Techniques

Class imbalance [9] is a common problem in cancer microarray dataset and it occurs when instances of negative/major class or classes is significantly more than the positive/minor class or classes. The class or classes with scanty instances are always a problem of interest. Examples are plenty in cancer microarray datasets like cancer patients are less than non-cancer patients. The conventional learning algorithms are more lenient towards predicting major classes because general rules are more preferred than the specific rules required for predicting the samples from minority class. This class imbalance problem also persists in multi-class microarray datasets.

The traditional data level methods referred to as re-sampling [11] used for balancing the classes of the datasets either by under-sampling the majority class or by over-sampling the minority class until equal representation of the classes occur. The re-sampling technique depends on many aspects of the imbalanced dataset especially the ratio between major and minor samples and the characteristics of the classifier. The major drawback of this technique is: under-sampling which eliminates useful data, whereas over-sampling increases the size of the minority class artificially, consequently encounters the problem of over-fitting. SMOTE (Synthetic Minority Oversampling Technique) [8] is one of the popular oversampling techniques which introduce synthetic samples to the minority class by randomly joining all of the nearest neighbors of k-minority class. A multiple re-sampling method [17] adaptively selects appropriate re-sampling rate to facilitate achieving balanced dataset. Similarly, different types of under-sampling techniques are introduced by many researchers. Random majority under sampling (RUS) [8] is one of the common under-sampling techniques which remove samples from the majority class randomly. However, the problem with RUS is that it might lose potential information while randomly removing samples from the datasets. In past few years, ensemble of learning algorithms in combination with sampling techniques [18] have been increasingly used by the researchers as an alternative solution to the class imbalance problem especially for imbalance microarray cancer datasets. In this study, a re-sampling method which creates random subsample of the datasets using sampling with replacement approach has been applied on reduced datasets.

2.3 DECORATE (Diverse Ensemble Creation by Oppositional Relabeling of Artificial Training Examples): A Meta-Learning Ensemble Algorithm

DECORATE [13] is a meta-learner generates diverse ensembles of classifiers iteratively by creating artificial training instances. The ensemble is initialized with the base classifier which is trained on the training data provided with the ensemble. In the successive iteration, classifiers are trained on the original dataset along with some artificial data which is generated from the data distribution. The number of artificial samples to be generated is specified as the fraction of the size of the training dataset. The labels that are assigned to the artificial samples are selected from a probability distribution which is inversely proportional to the prediction of the current ensemble. The artificially labeled training dataset is considered as the diversity data of the ensemble. Then a new classifier is trained using both original and diversity training dataset and becomes a member of the committee provided it reduces the training error of the ensemble otherwise will be discarded from the ensemble and the iteration of the algorithm continues to find another classifier as the member of the committee. The iteration is repeated as long as the desired size of the classifier in the DECORATE ensemble [13] is not reached or not exceeds the number of iterations that is set.

Moreover, the prediction of individual classifier's are combined together to predict the outcome of DECORATE ensemble. The following method is used to classify a new sample x.

Let Ci: be the *i*th base classifier of the ensemble C*

PCiy(x): probability associated with instance x belonging to class y for base classifier Ci. Then the probability of membership of the instance x in class y in ensemble C* is:

$$P_y(x) = \frac{\sum_{C_i \in C^*} P_{C_{i,y}}(x)}{C^*} \tag{1}$$

where Py(x) is the probability that x belongs to class y and

$$C^*(x) = argmax_{y \in Y} P_y(x), \tag{2}$$

such that Y is the set of classes. The following pseudocode [19] gives an idea, how DECORATE builds a diverse ensemble.

Pseudo code description of DECORATE algorithm

BaseLearn - base learning algorithm

T - set of m training examples with labels $<(x_1, y_1), ..., (x_m, y_m)>$ with labels $y_j \in Y$

C_{size} - ensemble size

I_{max} - maximum number of iterations to build the ensemble

R_{size} - number of artificial examples; a proportion of m examples

$$err(C) = \frac{\Sigma_{x_j \in T:[C(x_j) \neq y_j]} 1}{m}$$

Train Decorate Classifier:

   trials = 1

   Initialize ensemble $C^* = \{BaseLearn(T)\}$

   Compute ensemble error $\varepsilon = err(C^*)$

   while $|C^*| < C_{size}$ and *trials* $< I_{max}$ {

      Generate $R_{size} * |T|$ training examples, R

      Label examples in R from the inverse of C^*'s probability of class labels, using equation (1)

      $T' = T \cup R$, new training set

      $C' = BaseLearn(T')$

      $C^{*'} = C^* \cup \{C'\}$

      Compute new ensemble error $\varepsilon' = err(C^{*'})$

      if $\varepsilon' \leq \varepsilon$

         $\varepsilon = \varepsilon'$

         $C^* = C^{*'}$

      trials++

   }

The advantage of using this meta-ensemble [19] is it performs consistently than the base classifier. The experiments show that it attains higher accuracy than traditional ensembles like Bagging, Boosting and Random Subspace on small training samples. The aim of studying diverse ensemble is to understand the concept of bias-variance decomposition and the related ambiguity decomposition.

In this paper, re-sampling technique has been used to handle skewed nature and other properties of the multi-class imbalanced microarray cancer datasets paired with meta-learning ensemble DECORATE.

3 Datasets

Two skewed imbalance multiclass cancer microarray datasets are considered to verify the effectiveness of the proposed ensemble framework. It deals with 5–11 classes, 90–174 samples, 5920–12533 genes and imbalance ratios (IRs) of 4.50 and 15.00 respectively. Information of all these two benchmark datasets is recorded in Table 1 and they are referred from http://www.gems-system.org/.Multi-class datasets are converted to imbalance two-class problems in which union of one or more classes became the positive class and the union of one or more of the remaining classes was labeled as the negative class. The ratio of positive and negative class provides the imbalance ratios (IRs) of the datasets.

4 Experimental Settings and Performance Evaluation Metrics

4.1 Evaluation Metrics

For a balanced dataset, the overall classification accuracy is used as a metric to estimate the quality of the classifier. On the other hand, for an imbalanced problem, this metric does not estimate the quality of the classifier appropriately whereas Confusion matrix is considered as an appropriate performance evaluator [12] of the classifier.

The performance evaluation metrics such as F-measure, G-mean, Precision, Recall and Accuracy are more suitable for binary class imbalance problem rather than multi-class problem. Therefore, some transformations are required to be applied to all these metrics to make it suitable for multi-class imbalanced problems. The accuracies of all the classes are computed by a metric called G-mean which is evaluated using Eq. (3). It computes the geometric mean of all the classes of the problem. Similarly, F-measure of all the classes can be transformed to F-score [20] and can be calculated by using the following formula:

$$G-\text{men} = \left(\prod_{i=1}^{C} \text{Acc}_i \right)^{1/C}, \tag{3}$$

$$F-\text{score} = \frac{\sum_{i=1}^{C} F-\text{measure}_i}{C}, \tag{4}$$

$$F-\text{measure}_i = \frac{2 \times \text{Precision}_i \times \text{Recall}_i}{\text{Precision}_i + \text{Recall}_i}, \tag{5}$$

Finally, the accuracy of the multi-class problem can be computated as follows:

$$\text{Acc} = \sum_{i=1}^{C} (Acc_i \times P_i), \tag{6}$$

where, i represent the percentage of samples in ith class. On the other hand TN rate has a negative impact on F-score [20] which can be overcome by Cohen's Kappa statistic. Kappa statistic was introduced by Cohen [21] as a statistical measure to evaluate the percentage of agreement between the diversity and accuracy of measurement. It also measures the reliability of classification using the following formula:

$$\kappa = \frac{\text{observed agreement} - \text{chance agreement}}{1 - \text{chance agreement}} \tag{7}$$

However, the value of the statistic ranges from 0 to 1 and value 1 indicates a better agreement than value 0. In some cases it varies from −1 to 1. In this study, four different metrics, namely, Accuracy, F-score, Area under Curve (AUC) of Receiver Operating characteristics (ROC) and Kappa statistic (CKC) are chosen to assess the performance of the multi-class imbalance problem.

4.2 Experimental Setup

An experimental setup is explained here to find appropriate solution for multi-class imbalanced problem employing the proposed re-sample based ensemble DEC-ORATE [13]. The performance is assessed with two highly imbalanced benchmark microarray datasets described in Table 1. Both the datasets before being used by the method was standardized with mean 0 and standard deviation 1 and then feature selection mechanism was applied on the pre-processed datasets. The size of the feature space influences the classification accuracy and prevent overfitting problem, so, to address this, fixed size feature subspace i.e., 30 best ranked features are selected by reducing the dimension of the datasets by ReliefF [16] filter method. Then, the application of re-sampling [18] technique produces random subsamples of the datasets by oversampling the minority class to address the imbalance characteristic of the given problem. Furthermore, the Re-sampling technique biases the class by introducing 1 to maintain a uniform distribution. The size of the randomly generated subsample specified as a percentage of the size of the original datasets i.e., 100, 200, 300, 400 and 500 % s to overcome the problem of imbalance. To perform the experiment with DECORATE, the maximum size of the ensemble and the maximum number of iterations is set to 15 and 50 respectively. The numbers of artificially generated samples are varying from 0 to 1 but it produces a significant result when it varies from 0.5 to 1 but affects adversely when less than 0.5.

Therefore, the size is set to 1, i.e., the number of artificially generated samples is equal to the training sample size. Then 10-fold cross-validation was used to find the performance of each learning algorithm and the average value of F-score, Accuracy and AUC were calculated from the above 10 iterations. The performance of DECORATE was compared with four competitive relevant ensemble algorithms including the base classifier namely, re-sampled-Bagging, re-sampled-AdaBoost, re-sampled-Random Subspace and re-sampled-J48, using J48 as the base learner in all the ensembles.

5 Results and Discussion

The classification results of the experiment conducted for two highly imbalanced cancer microarray datasets, namely, Brain-Tumor1 and 11-tumor using the proposed re-sampling based ensemble algorithm is recorded in Table 2. The results are highlighted in three different ways to indicate the best in bold, second best in italic and the worst one in underline. Table 1 shows that, Brain Tumor1 is highly

Table 2 Results of different ensemble methods using 10-fold CV

Datasets	IR	Re-sampling (%)	Methods	Accuracy	F-score	AUC	Kappa-statistic
Brain-Tumor1	15.0	100	Decorate	**94.444**	**0.945**	**0.994**	**0.933**
			Bagging	90.00	0.989	0.991	0.879
			AdaBoost	*93.333*	*0.933*	*0.997*	*0.9195*
			Random Subspace	92.222	0.921	0.997	0.9061
			J48	90.00	0.898	0.991	0.879
		200	Decorate	**93.333**	**0.933**	**0.996**	**0.9249**
			Bagging	88.333	0.883	0.981	0.8689
			AdaBoost	*92.5*	*0.925*	*0.987*	*0.9155*
			Random Subspace	90.83	0.908	0.986	0.8967
			J48	88.333	0.883	0.958	0.8685
		300	Decorate	**97.222**	**0.972**	**1.0**	**0.9687**
			Bagging	94.444	0.944	0.990	0.9373
			AdaBoost	*96.667*	*0.967*	*0.997*	*0.9624*
			Random Subspace	94.444	0.943	0.993	0.9373
			J48	93.333	0.932	0.983	0.9248
		400	Decorate	**99.167**	**0.992**	**1.0**	**0.9906**
			Bagging	93.333	0.932	0.992	0.9247
			AdaBoost	*97.083*	*0.971*	*0.995*	*0.9671*
			Random Subspace	97.083	0.971	0.995	0.9671
			J48	95.833	0.958	0.991	0.953

(continued)

Table 2 (continued)

Datasets	IR	Re-sampling (%)	Methods	Accuracy	F-score	AUC	Kappa-statistic
		500	Decorate	**99.667**	**0.997**	**1.0**	**0.9962**
			Bagging	96.667	0.966	0.997	0.9624
			AdaBoost	*97.667*	*0.976*	*0.998*	*0.9737*
			Random Subspace	*97.667*	*0.976*	*0.998*	*0.9737*
			J48	97.00	0.969	0.993	0.9662
11-Tumor	4.50	100	Decorate	**91.954**	**0.918**	**0.987**	**0.9108**
			Bagging	82.184	0.813	0.977	0.8021
			AdaBoost	*86.782*	*0.867*	*0.988*	*0.8536*
			Random Subspace	83.908	0.833	0.975	0.8212
			J48	78.736	0.791	0.929	0.7645
		200	Decorate	**96.839**	**0.968**	**0.999**	**0.9651**
			Bagging	92.816	0.928	0.993	0.9208
			AdaBoost	*96.264*	*0.963*	*0.999*	*0.9588*
			Random Subspace	94.540	0.945	0.996	0.9398
			J48	91.954	0.919	0.980	0.9113
		300	Decorate	**99.042**	**0.990**	**1.0**	**0.9899**
			Bagging	96.935	0.969	0.999	0.9662
			AdaBoost	*98.351*	*0.989*	*1.0*	*0.9873*
			Random Subspace	97.509	0.975	1.0	0.9726
			J48	96.935	0.969	0.992	0.9662
		400	Decorate	**99.569**	**0.996**	**1.0**	**0.9953**
			Bagging	98.419	0.984	0.998	0.9826
			AdaBoost	*99.282*	*0.993*	*1.0*	*0.9921*
			Random Subspace	98.851	0.988	0.999	0.9873
			J48	96.983	0.970	0.995	0.9668
		500	Decorate	**99.770**	**0.998**	**1.0**	**0.9975**
			Bagging	98.735	0.987	1.0	0.9861
			AdaBoost	*99.652*	*0.997*	*1.0*	*0.9962*
			Random Subspace	99.310	0.993	1.0	0.9924
			J48	98.735	0.987	0.998	0.9861

imbalanced in nature than the other one. It can be observed from Table 2 that, the proposed re-sampling based Decorate ensemble has shown a remarkable performance in comparison to other ensembles including the base classifier. The re-sampling technique created the uniformly distributed random subsamples at 100, 200, 300, 400 and 500 % of the size of the original datasets. The weighted average values of the metrics, F-score, Accuracy, AUC and Kappa statistic can be observed from Table 2 for both the datasets. The average values of Accuracy reported by DECORATE for Brain-Tumor1 and 11-Tumor are quite significant in comparison to other four algorithms. The value of Accuracy is consistently increasing with the increasing size of the datasets from 100 to 500 % only for DECORATE ensemble. The results reported by other four ensembles are not quite impressive comparing to DECORATE. Therefore, it is quite apparent from Table 2 that DECORATE is sensitive to both types of multi-class imbalance problems which have high and low imbalance ratio. Again, from Table 2, it can be observed that DECORATE always attains the highest average values of F-score for both the datasets. It surpasses all the remaining methods by an impressive margin for all the different sizes of the subsample. It is also seen from the Table 2 that F-score value obtained by AdaBoost is closer to DECORATE. However, the performance of DECORATE still exceeds all the four algorithms with a certain degree and becomes the best one.

Similarly, when compared with the average value of AUC of other algorithms, DECORATE algorithm was found to approach the value 1 for all the different percentage of the original size of the datasets. It is also observed from Table 2 that when size of the dataset increases from 300 to 500 % of the original size, value of AUC of DECORATE becomes 1 for both the datasets. Another interesting observation is that other algorithms have attained the value 1 for 11-Tumor dataset along with DECORATE which has lower imbalance ratio than Brain-Tumor1. Therefore, it indicates that the randomly selected positive sub-samples of different data sizes ranked higher than the negative subsamples. Lastly, it is observed from the statistical measure that the proposed method demonstrated a significant improvement over other algorithms for the imbalanced problems. The value of Kappa statistic for different subsample size for both the imbalanced problems shows a better percentage of agreement between the diversity and accuracy of measurement in DECORATE ensemble in comparison to other four algorithms. In DECORATE; the reliability of classification is very high because the computed value of Kappa is very close to 1 for both the imbalanced problems. Table 2 shows all the above results for the analysis of the experiment and concludes that proposed ensemble is sensitive to multi-class imbalanced problem.

6 Conclusion

In this article, a sampling based ensemble learning framework was employed to find appropriate solution for multi-class imbalanced classification problem using microarray cancer datasets. The proposed framework has addressed three important aspects of the given multi-class imbalanced problem. The first and most important aspect is the selection of ensemble algorithm to handle multi-class imbalanced cancer microarray datasets. DECORATE, an advanced meta learning algorithm was chosen which generates diverse ensembles of classifiers iteratively by creating artificial training instances so that a perfect balance of relationship can be established between accuracy and diversity of the base classifiers. The second important aspect of the problem is the dimension of the feature space and feature subspace which has a direct influence on the classification performance. To address this, an efficient filter mechanism is applied to the imbalance datasets to select significant features which removes the overfitting problem at the time of classification. The third and last important aspect which is addressed in this paper is the imbalance class distribution. To obtain a uniformly distributed subsample of the sample space, a novel sampling technique that oversamples the minority/positive class is used to overcome the imbalance nature of the datasets. The future extension of this paper will vary the dimension of the feature subspace and will study its influence on the classification performance of the ensemble algorithm using many multi-class datasets.

References

1. Ghorai, S., Mukherjee, A., Sengupta, S., Dutta, P.K.: Cancer classification from gene expression data by NPPC ensemble. IEEE/ACM Trans. Comput. Biol. Bioinf. **8**(3), 659–671 (2011)
2. Yin, Q.Y., Zhang, J.S., Zhang, C.X., Ji, N.N.: A novel selective ensemble algorithm for imbalanced data classification based on exploratory understanding. Math. Probl. Eng. **2014**, 1–14, article ID 358942 (2014)
3. Kamal, A.H.M., Zhu, X., Narayanan, R.: Gene selection for microarray expression data with imbalanced sample distributions. In: Proceedings of the International Joint Conference on Bioinformatics, Systems Biology and Intelligent Computing (IJCBS '09), pp. 3–9. Shanghai, China (2009)
4. Blagus, R., Lusa, L.: Class prediction for high-dimensional class-imbalanced data. BMC Bioinf. **11**, article 523 (2010)
5. Wasikowski, M., Chen, X.-W.: Combating the small sample class imbalance problem using feature selection. IEEE Trans. Knowl. Data Eng. **22**(10), 1388–1400 (2010)
6. Yu, H., Hong, S., Yang, X., Ni, J., Dan, Y., Qin, B.: Recognition of multiple imbalanced cancer types based on DNA microarray data using ensemble classifier. BioMed Res. Int. **2013**, 1–13 (2013)
7. Lin, W.J., Chen, J.J.: Class-imbalanced classifiers for high dimensional data. Briefings Bioinf. **14**(1), 13–26 (2013)
8. Blagus, R., Lusa, L.: Evaluation of SMOTE for high dimensional class-imbalanced microarray data. In: Proceedings of the 11th International Conference on Machine Learning and Applications, pp. 89–94. Boca Raton, Fla, USA (2012)
9. Wang, S., Yao, X.: Multiclass imbalance problems: analysis and potential solutions. IEEE Trans. Syst. Man. Cybern. B **42**(4), 1119–1130 (2012)
10. Chawla, N.V., Cieslak, D.A., Hall, L.O., Joshi, A.: Automatically countering imbalance and its empirical relationship to cost. Data Min. Knowl. Disc. **17**(2), 225–252 (2008)
11. Lee, P.H.: Resampling methods improve the predictive power of modeling in class-imbalanced datasets. Int. J. Environ. Res. Public Health **11**, 9776–9789 (2014)
12. Forman, G.: An extensive empirical study of feature selection metrics for text classification. J. Mach. Learn. Res. **3**, 1289–1305 (2003)
13. Melville, P., Mooney, R.J.: Constructing diverse classifier ensembles using artificial training examples. In: Eighteenth International Joint Conference on Artificial Intelligence, pp. 505–510 (2003)

14. Dash, S., Dash, A.: A correlation based multilayer perceptron algorithm for cancer classification with gene-expression dataset. In: Proceedings of the International Conference on Hybrid Intelligent Systems (HIS), published in IEEE Xplore, 978-1-4799-7633-1/14, Kuwait (2014)
15. Kira, K., Rendell, L.: The feature selection problem: Traditional methods and new algorithms. In: Proceedings of the 9th International Conference on Machine Learning, pp. 249–256 (1992)
16. Robnik-Sikonja, M., Kononenko, I.: Theoretical and empirical analysis of ReliefF and RreliefF. Mach. Learn. **53**, 23–69 (2003)
17. Estabrooks, Andrew, Jo, Taeho, Japkowicz, Nathalie: Multiple resampling method for learning from imbalanced data sets. Comput. Intell. **20**(1), 18–36 (2004)
18. Galar, M., Fernandez, A., Barrenechea, E., Bastince, H., Herrera, F.: A review on ensembles for the class imbalance problem: bagging, boosting and hybrid-based approaches. IEEE Trans, Syst. Man Cybern. Part C: Appl. Rev. **42**(4), 463–484 (2012)
19. Melville, P., Mooney, R.: Constructing diverse classifier ensembles using artificial training examples. In: Proceedings of the IJCAI. pp. 505–510. Acapulco, Mexico, August (2003)
20. Ramaswamy, S., Tamayo, P., Rifkin, R. et al.: Multiclass cancer diagnosis using tumor gene expression signatures. In: Proceedings of the National Academy of Sciences of the United States of America, vol. 98, no. 26, pp. 15149–15154 (2001)
21. Cohen, J.: A coefficient of agreement for nominal scales. Educ. Psychol. Meas. (1960)

A Self-configuring Multi-strategy Multimodal Genetic Algorithm

Evgenii Sopov

Abstract In recent years many efficient nature-inspired techniques (based on evolutionary strategies, particle swarm optimization, differential evolution and others) have been proposed for real-valued multimodal optimization (MMO) problems. Unfortunately, there is a lack of efficient approaches for problems with binary representation. Existing techniques are usually based on general ideas of niching. Moreover, there exists the problem of choosing a suitable algorithm and fine tuning it for a certain problem. In this study, an approach based on a metaheuristic for designing multi-strategy genetic algorithm is proposed. The approach controls the interactions of many MMO techniques (different genetic algorithms) and leads to the self-configuring solving of problems with a priori unknown structure. The results of numerical experiments for benchmark problems from the CEC competition on MMO are presented. The proposed approach has demonstrated efficiency better than standard niching techniques and comparable to advanced algorithms. The main feature of the approach is that it does not require the participation of the human-expert, because it operates in an automated, self-configuring way.

Keywords Multimodal optimization · Genetic algorithm · Metaheuristic · Ensemble · Self-configuration · Niching

1 Introduction

Many real-world problems have more than one optimal solution, or there exists only one global optimum and several local optima in the feasible solution space. Such problems are called multimodal. The goal of multimodal optimization (MMO) is to find all optima (global and local) or a representative subset of all optima.

E. Sopov (✉)
Siberian State Aerospace University, Krasnoyarsk, Russia
e-mail: evgenysopov@gmail.com

© Springer International Publishing Switzerland 2016
N. Pillay et al. (eds.), *Advances in Nature and Biologically Inspired Computing*,
Advances in Intelligent Systems and Computing 419,
DOI 10.1007/978-3-319-27400-3_2

15

Evolutionary and genetic algorithms (EAs and GAs) demonstrate good performance for many complex optimization problems. At the same time, traditional EAs and GAs have a tendency to converge to the best-found optimum losing population diversity.

In recent years MMO have become more popular, and many efficient nature-inspired MMO techniques were proposed. Almost all search algorithms are based on maintaining the population diversity, but differ in how the search space is explored and how optima basins are located and identified over a landscape. The majority of algorithms and the best results are obtained for real-valued MMO problems [1]. The main reason is the better understanding of landscape features in the continuous search space.

Unfortunately many real-world MMO problems are usually considered as black-box optimization problems and are still a challenge for MMO techniques. Moreover, many real-world problems contain variables of many different types, including integer, rank, binary and others. In this case, usually binary representation is used. Unfortunately, there is a lack of efficient approaches for problems with binary representation. Existing techniques are usually based on general ideas of niching and fitness sharing.

In this study, a novel approach based on a metaheuristic for designing multi-strategy MMO GA is proposed. Its main idea is to create an ensemble of many MMO techniques and adaptively control their interactions. Such an approach would lead to the self-configuring solving of problems with a priori unknown structure.

The rest of the paper is organized as follows. Section 2 describes related work. Section 3 describes the proposed approach. In Sect. 4 the results of numerical experiments are discussed. In the Conclusion the results and further research are discussed.

2 Significant Related Work

The MMO, in general, can have at least 3 goals: to find a single global optimum over the multimodal landscape only; to find all global optima; to find all optima (global and local) or a representative subset of all optima [2]. It is obvious that the second and the third goals are more interesting from both a theoretical and a practical point of view. Over the past decade interest for this field has increased. The recent approaches are focused on the goal of exploring the search space and finding many optima to the problem. There exist many different MMO techniques. Good surveys can be found in [1, 3, 4].

Many real-world problems contain variables of many different types, including integer, rank, binary and others. In this case, usually binary representation is used. Binary and binarized MMO problems are usually solved using the GA based on general techniques. Also special techniques are applied, but some of their features can be lost in the binary space. Unfortunately, many efficient nature-inspired MMO algorithms have no binary version and cannot be easily converted to binary representation.

As we can see from many studies, there is no universal approach that is efficient for all MMO problems. Many researches design hybrid algorithms, which are generally based on a combination of search algorithms and some heuristic for niching improvement. For example, here are four top-ranked algorithms from the CEC'13 competition on MMO: Niching the CMA-ES via Nearest-Better Clustering (NEA2), A Dynamic Archive Niching Differential Evolution (dADE/nrand/1), CMA-ES with simple archive (CMA-ES) and Niching Variable Mesh Optimization algorithm (N-VMO) [5].

Another way is combining many basic MMO algorithms to run in parallel, migrate individuals and combine the results. In [6] an island model is applied, where islands are iteratively revised according to the genetic likeness of individuals. In [7] four MMO niching algorithms run in parallel to produce offspring, which are collected in a pool to produce a replacement step. In [8] the same scheme is realized using the clearing procedure.

The conception of designing MMO algorithms in the form of an ensemble seems to be perspective. A metaheuristic that includes many different MMO approaches (different search strategies) can deal with many different MMO problems. And such a metaheuristic can be self-configuring due to the adaptive control of the interaction of single algorithms during the problem solving.

In [9] a self-configuring multi-strategy genetic algorithm in the form of a hybrid of the island model, competitive and cooperative coevolution was proposed. The approach has demonstrated good results with respect to multi-objective and non-stationary optimization.

3   Proposed Methodology

In the field of statistics and machine learning, ensemble methods are used to improve decision making. This concept can be also used in the field of EA. The main idea is to include different search strategies in the ensemble and to design effective control of algorithm interaction. Our hypothesis is that different EAs are able to deal with different features of the optimization problem, and the probability of all algorithms failing with the same challenge in the optimization process is low. Moreover, the interaction of algorithms can provide the ensemble with new options for optimization, which are absent in stand-alone algorithms.

The general structure of the self-configuring multi-strategy GA proposed in [9] is called Self*GA (the star sign corresponds to the certain optimization problem) and it is presented in Fig. 1.

The Self*GA technique eliminates the necessity to define an appropriate search strategy for the problem as the choice of the best algorithm is performed automatically and adaptively during the run.

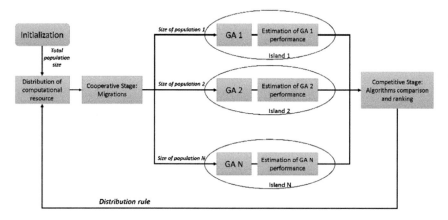

Fig. 1 The Self*GA structure

Now we will discuss the design of a Self*GA for MMO problems that can be named SelfMMOGA.

At the first step, we need to define the set of individual algorithms included in the SelfMMOGA. In this study we use six basic techniques, which are well-studied and discussed, and they can be used with binary representation with no modification [1, 10]. Algorithms and their specific parameters are presented in Table 1. All values for radiuses and distances in Table 1 are in the Hamming metric for binary problems and in the Euclidean metric for continuous problems. The motivation of choosing certain algorithms is that if the SelfMMOGA performs well with basic techniques, we can develop the approach with more complex algorithms in further works.

The adaptation period is a parameter of the SelfMMOGA. Moreover, the value depends on the limitation of the computational resource (total number of fitness evaluations).

The key point of any coevolutionary scheme is the performance evaluation of a single algorithm. For MMO problems performance metrics should estimate how many optima were found and how the population is distributed over the search

Table 1 Algorithms include in the SelfMMOGA

Algorithm	Approach	Parameters
Alg1	Clearing	Clearing radius, capacity of a niche
Alg2	Sharing	Niche radius, α
Alg3	Clustering	Number of clusters, min distance to centroid, max distance to centroid
Alg4	Restricted tournament selection (RTS)	Window size
Alg5	Deterministic crowding	–
Alg6	Probabilistic crowding	–

space. Unfortunately, good performance measures exist only for benchmark MMO problems, which contain knowledge of the optima. Performance measures for black-box MMO problems are still being discussed. Some good recommendations can be found in [11]. In this study, the following criteria are used.

The first measure is called Basin Ratio (BR). The BR calculates the number of covered basins, which have been discovered by the population. It does not require knowledge of optima, but an approximation of basins is used.

$$BR(pop) = \frac{l}{k},$$

$$l = \sum_{i=1}^{k} min \left\{ 1, \sum_{\substack{x \in pop \\ x \neq z_i}} b(x, z_i) \right\}, b(x, z) = \begin{cases} 1, & if \ x \in basin(z) \\ 0, & otherwise \end{cases} \tag{1}$$

where pop is the population, k is the number of identified basins by the total population, l is the indicator of basin coverage by a single algorithm, $b(x, z)$ is a function that indicates if an individual is in basin z.

To use the metric (1), we need to define how to identify basins in the search space and how to construct the function $b(x, z)$.

For continuous MMO problems, basins can be identified using different clustering procedures like Jarvis-Patrick, the nearest-best and others [12]. In this study, for MMO problems with binary representation we use the following approach. We use the total population (the union of populations of all individual algorithms in the SelfMMOGA). For each solution, we consider a predefined number of its nearest neighbours (with respect to the Hamming distance). If the fitness of the solution is better, it is denoted as a local optima and the centre of the basin. The number of neighbours is a tunable parameter. For a real-world problem, it can be set from some practical point of view.

The function $b(x, z)$ can be easily evaluated by defining if individual x is in a predefined radius of basin centre z. The radius is a tunable parameter. In this study, we define it as total population size divided by the number of previously identified basins.

The second measure is called Sum of Distances to Nearest Neighbour (SDNN). The SDNN penalizes the clustering of solutions. This indicator does not require knowledge of optima and basins. The SDNN can be calculated as

$$SDNN(pop) = \sum_{i=1}^{popsize} d_{nn}(x_i, pop),$$

$$d_{nn}(x_i, pop) = \min_{y \in pop \setminus \{x_i\}} \{dist(x_i, y)\}, \tag{2}$$

where d_{nn} is the distance to the nearest neighbour, $dist$ is the Hamming distance.

Finally, we combine the *BR* and the *SDNN* in an integrated criterion *K*:

$$K = \alpha \cdot BR(pop) + (1 - \alpha) \cdot \overline{SDNN}(pop), \tag{3}$$

where $\overline{SDNN}$ is a normalized value of *SDNN*, α defines weights of the BR and the SDNN in the sum ($\alpha \in [0, 1]$).

Next, we need to design a scheme for the redistribution of computational resources. New population sizes are defined for each algorithm. In this study, all algorithms give to the "winner" algorithm a certain percentage of their population size, but each algorithm has a minimum guaranteed resource that is not distributed. The guaranteed resource can be defined by the population size or by problem features.

At the cooperative stage, in many coevolutionary schemes, all individual algorithms begin each new adaptation period with the same starting points (such a migration scheme is called "the best displaces the worst"). For MMO problems, the best solutions are defined by discovered basins in the search space. As we already have evaluated the approximation of basins (*Z*), the solutions from *Z* are introduced in all populations replacing the most similar individuals.

4   Experimental Results

To estimate the approach performance we have used the following list of benchmark problems. Six binary MMO problems are from [7]. These test functions are based on the unitation functions, and they are massively multimodal and deceptive. Eight real-valued MMO problems are from CEC'2013 Special Session and Competition on Niching Methods for Multimodal Function Optimization [13].

We have denoted the functions as in the source papers. Some details of the problems are presented in Table 2.

The following criteria for estimating the performance of the SelfMMOGA over the benchmark problems are used for continuous problems:

- Peak Ratio (*PR*) measures the percentage of all optima found by the algorithm (Eq. 4).

$$PR = \frac{|\{q \in Q | d_{nn}(q, pop) \leq \varepsilon\}|}{k}, \tag{4}$$

where $Q = \{q_1, q_2, \ldots, q_k\}$ is a set of known optima, ε is accuracy level.

- Success Rate (*SR*) measures the percentage of successful runs (a successful run is defined as a run where all optima were found) out of all runs.

Table 2 Test suite

Problem	Number of desirable optima	Problem dimensionality	Problem	Number of desirable optima	Problem dimensionality*
binaryF11	32 global	30	cecF1	2 global + 3 local	9, 12, 15, 19, 22
binaryF12	32 global	30	cecF2	5 global	4, 7, 10, 14,17
binaryF13	27 global	24	cecF3	1 global + 4 local	4, 7, 10, 14,17
binaryF14	32 global	30	cecF4	4 global	14, 22, 28, 34, 42
binaryF15	32 global	30	cecF5	2 global + 2 local	11, 17, 24, 31, 37
binaryF16	32 global	30	cecF6	18 global + 742 local	16, 22, 30, 36, 42
			cecF7	36 global	14, 20, 28, 34, 40
			cecF8	12 global	8, 14, 20, 28, 34

*Real-valued problems have been binarized using the standard binary encoding with 5 accuracy levels

The maximum number of function evaluation and the accuracy level for the *PR* evaluation are the same as in CEC completion rules [13]. The number of independent runs of the algorithm is 50.

In the case of binary problems, we have substituted the *SR* measure with Peak Distance (*PD*). The *PD* indicator calculates the average distance of known optima to the nearest individuals in the population [11].

$$PD = \frac{1}{k} \sum_{i=1}^{k} d_{nn}(q_i, pop) \qquad (5)$$

To demonstrate the control of algorithm interaction in the SelfMMOGA, we have chosen an arbitrary run of the algorithm on the cecF1 problem and have visualized the distribution of the computational resource (see Fig. 2). The total population size is 200 and the minimal guaranteed amount of the computational recourse is 10. The maximum number of generations is 200 and the size of the adaptation period is 10, thus the horizontal axis contains numeration of 20 periods.

As we can see, there is no algorithm that wins all the time. At the first two periods, Sharing (Alg2) and Clearing (Alg1) had better performance. The highest amount of the resource was won by Clustering (Alg3) at the 10th period. At the final stages, Deterministic Crowding (Alg5) showed better performance.

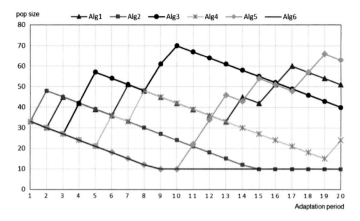

Fig. 2 Example of the SelfMMOGA run

The results of estimating the performance of the SelfMMOGA with the pack of binary problems are presented in Table 3. The table contains the values of the *PR*, the *SR* and the *PD* averaged over 50 independent runs. We also have compared the results with Ensemble of niching algorithms (ENA) proposed in [7]. There is only the *SR* value for the ENA.

The setting for the SelfMMOGA are:

- Maximum number of function evaluation is 50000 (as for the ENA);
- Total population size is 200 (the ENA uses 500);
- Adaptation period is 10 generations (25 times);
- All specific parameters of individual algorithms are self-tunable using the concept from [14].

We have also compared the results with the average of 6 stand-alone algorithms (Alg1–Alg6). The average value can be viewed as the average performance of a randomly chosen algorithm. Such an estimate is very useful for black-box optimization problems, because we have no information about problem features and, consequently, about what algorithms to use. If the performance of the SelfMMOGA

Table 3 Results for binary problems

Problem	Average of 6 stand-alone algorithms			SelfMMOGA			ENA
	PR	SR	PD	PR	SR	PD	SR
binaryF11	0.91	0.89	2.30	1.00	1.00	0.00	1.00
binaryF12	0.96	0.95	1.38	1.00	1.00	0.00	1.00
binaryF13	0.95	0.93	2.34	1.00	1.00	0.00	1.00
binaryF14	0.89	0.91	2.37	1.00	1.00	0.00	1.00
binaryF15	0.84	0.82	2.61	1.00	1.00	0.00	1.00
binaryF16	0.78	0.79	3.08	1.00	1.00	0.00	0.99

is better that the average of its component, we can conclude that on average the choice of the SelfMMOGA will be better. As we can see from Table 3, the Self-MMOGA always outperforms the average of its stand-alone component algorithms for binary problems. Moreover, for some problems no stand-alone algorithm has a SR value equal to 1, but the SelfMMOGA does.

The results of estimating the performance of the SelfMMOGA with the pack of continuous problems are presented in Tables 4 and 5. Table 4 shows results averaged over all problems. Table 5 contains ranks of algorithms by separate criteria.

All problems and settings are as in the rules of the CEC'13 competition on MMO. For each problem there are 5 levels of accuracy of finding optima ($\varepsilon = \{1e-01, 1e-02, \ldots, 1e-05\}$). Thus, each problem have been binarized 5 times. The dimensionalities of binarized problems are presented in Table 2.

We have also compared the results of the SelfMMOGA runs with some efficient techniques from the competition. The techniques are DE/nrand/1/bin and Crowding DE/rand/1/bin [13], N-VMO [15], dADE/nrand/1 [16], and PNA-NSGAII [17].

The settings for the SelfMMOGA are:

- Maximum number of function evaluation is 50000 (for cecF1–cecF5) and 200000 (for cecF6–cecF8);
- Total population size is 200;
- Adaptation period is 10 generations 25 times (for cecF1–cecF5) and 25 generations 40 times (cecF6–cecF8);
- All specific parameters of individual algorithms are self-tunable.

As we can see from Tables 4 and 5, the SelfMMOGA shows results comparable with popular and well-studied techniques. It yields to dADE/nrand/1 and N-VMO, but we should note that these algorithms are specially designed for continuous MMO problems, and have taken 2nd and 4th places, respectively, in the CEC competition. At the same time, the SelfMMOGA has very close average values to the best two algorithms, and outperforms PNA-NSGAII, CrowdingDE and DE, which have taken 7th, 8th and 9th places respectively.

In this study, we have included only basic MMO search techniques in the SelfMMOGA. Nevertheless, it performs well due to the effect of collective decision making in the ensemble. The key feature of the approach is that it operates in an automated, self-configuring way. Thus, the SelfMMOGA can be a good alternative for complex black-box MMO problems.

Table 4 Average PR and SR for each algorithm

ε	SelfMMOGA		DE/nrand/1/bin		cDE/rand/1/bin		N-VMO		dADE/nrand/1		PNA-NSGAII	
	PR	SR	PR	SR	PR	SR	PR	SR	PR	SR	PR	SR
1e-01	0.962	0.885	0.850	0.750	0.963	0.875	1.000	1.000	0.998	0.938	0.945	0.875
1e-02	0.953	0.845	0.848	0.750	0.929	0.810	1.000	1.000	0.993	0.828	0.910	0.750
1e-03	0.943	0.773	0.848	0.748	0.847	0.718	0.986	0.813	0.984	0.788	0.906	0.748
1e-04	0.907	0.737	0.846	0.750	0.729	0.623	0.946	0.750	0.972	0.740	0.896	0.745
1e-05	0.816	0.662	0.792	0.750	0.642	0.505	0.847	0.708	0.835	0.628	0.811	0.678
Average	**0.916**	**0.780**	**0.837**	**0.750**	**0.822**	**0.706**	**0.956**	**0.854**	**0.956**	**0.784**	**0.893**	**0.759**

Table 5 Algorithms ranking over cecF1-cecF8 problems

Rank by PR criterion	Algorithm	Rank by SR criterion	Algorithm
1	N-VMO and dADE/nrand/1	1	N-VMO
2	SelfMMOGA	2	dADE/nrand/1
3	PNA-NSGAII	3	SelfMMOGA
4	DE/nrand/1/bin	4	PNA-NSGAII
5	cDE/rand/1/bin	5	DE/nrand/1/bin
–	–	6	cDE/rand/1/bin

5 Conclusions

In this study, a novel genetic algorithm (called SelfMMOGA) for multimodal optimization is proposed. It is based on self-configuring metaheuristic, which involves many different search strategies in the process of MMO problem solving and adaptively control their interactions.

The SelfMMOGA allows complex MMO problems to be dealt with, which are the black-box optimization problems (a priori information about the objective and its features are absents or cannot be introduces in the search process). The algorithm uses binary representation for solutions, thus it can be implemented for many real-world problems with variables of arbitrary (and mixed) types.

We have included 6 basic MMO techniques in the SelfMMOGA realization to demonstrate that it performs well even with simple core algorithms. We have estimated the SelfMMOGA performance with a set of binary benchmark MMO problems and continuous benchmark MMO problems from CEC'2013 Special Session and Competition on Niching Methods for Multimodal Function Optimization. The proposed approach has demonstrated a performance comparable with other well-studied techniques. The proposed approach does not require the participation of the human-expert, because it operates in an automated, self-configuring way.

In further works, we will investigate the SelfMMOGA using more advanced component techniques.

Acknowledgements The research was supported by President of the Russian Federation grant (MK-3285.2015.9). The author expresses his gratitude to Mr. Ashley Whitfield for his efforts to improve the text of this article.

References

1. Das, S., Maity, S., Qub, B.-Y., Suganthan, P.N.: Real-parameter evolutionary multimodal optimization: a survey of the state-of-the art. Swarm Evol. Comput. **1**, 71–88 (2011)
2. Preuss, M.: Tutorial on multimodal optimization. In: The 13th International Conference on Parallel Problem Solving from Nature, PPSN 2014, Ljubljana, Slovenia (2014)

3. Liu, Y., Ling, X., Shi, Zh., Lv, M., Fang. J., Zhang, L.: A survey on particle swarm optimization algorithms for multimodal function optimization. J. Softw. **6**(12), 2449–2455 (2011)
4. Deb, K., Saha, A.: Finding multiple solutions for multimodal optimization problems using a multi-objective evolutionary approach. In: Proceedings of the 12th Annual Conference on Genetic and Evolutionary Computation, GECCO 2010, pp. 447–454 (2010)
5. Li, X., Engelbrecht, A., Epitropakis, M.: Results of the 2013 IEEE CEC competition on niching methods for multimodal optimization. In: Report presented at 2013 IEEE Congress on Evolutionary Computation Competition on: Niching Methods for Multimodal Optimization (2013)
6. Bessaou, M., Petrowski, A., Siarry, P.: Island model cooperating with speciation for multimodal optimization. Parallel Problem Solving from Nature PPSN VI, Lecture Notes in Computer Science, vol. 1917. pp. 437–446 (2000)
7. Yu, E.L., Suganthan, P.N.: Ensemble of niching algorithms. Inf. Sci. **180**(15), 2815–2833 (2010)
8. Qu, B., Liang, J., Suganthan P.N., Chen, T.: Ensemble of Clearing Differential Evolution for Multi-modal Optimization. Advances in Swarm Intelligence Lecture Notes in Computer Science, vol. 7331. pp. 350–357 (2012)
9. Sopov, E.: A Self-configuring Metaheuristic for Control of Multi-Strategy Evolutionary Search. ICSI-CCI 2015, Part III, LNCS 9142. pp. 29–37 (2015)
10. Singh, G., Deb, K.: Comparison of multi-modal optimization algorithms based on evolutionary algorithms. In: Proceedings of the Genetic and Evolutionary Computation Conference, Seattle, pp. 1305–1312 (2006)
11. Preuss, M., Wessing, S.: Measuring multimodal optimization solution sets with a view to multiobjective techniques. In: EVOLVE—A Bridge between Probability, Set Oriented Numerics, and Evolutionary Computation IV. AISC, vol. 227, pp. 123–137. Springer, Heidelberg (2013)
12. Preuss, M., Stoean, C., Stoean, R.: Niching foundations: basin identification on fixed-property generated landscapes. In: Proceedings of the 13th Annual Conference on Genetic and Evolutionary Computation, GECCO 2011. pp. 837–844 (2011)
13. Li, X., Engelbrecht, A., Epitropakis, M.G.: Benchmark functions for CEC'2013 special session and competition on niching methods for multimodal function optimization. Evolutionary Computation, Machine Learning Group, RMIT University, Melbourne, Australia. Technical Report (2013)
14. Semenkin, E.S., Semenkina, M.E.: Self-configuring Genetic Algorithm with Modified Uniform Crossover Operator. Advances in Swarm Intelligence. Lecture Notes in Computer Science, vol. 7331. Springer, Berlin Heidelberg. pp. 414–421 (2012)
15. Molina, D., Puris, A., Bello, R., Herrera, F.: Variable mesh optimization for the 2013 CEC special session niching methods for multimodal optimization. In: Proceedings of the 2013 IEEE Congress on Evolutionary Computation (CEC'13), pp. 87–94 (2013)
16. Epitropakis, M.G., Li, X., Burke, E.K.: A dynamic archive niching differential evolution algorithm for multimodal optimization. In: Proceeding of 2013 IEEE Congress on Evolutionary Computation (CEC'13), pp. 79–86 (2013)
17. Bandaru, S., Deb, K.: A parameterless-niching-assisted bi-objective approach to multimodal optimization. In: Proceedings of 2013 IEEE Congress on Evolutionary Computation (CEC'13), pp. 95–102 (2013)

An Ensemble of Neuro-Fuzzy Model for Assessing Risk in Cloud Computing Environment

Nada Ahmed, Varun Kumar Ojha and Ajith Abraham

Abstract Cloud computing is one of the hottest technologies in IT field. It provides computational resources as general utilities that can be leased and released by users in an on-demand fashion. Companies around the globe are showing high interest in adopting cloud computing technology but cloud computing adaptation comes with greater risks that need to be assessed. In this research, an ensemble of adaptive neuro-fuzzy inference system (ANFIS) is proposed to assess risk factors in cloud computing environment. In the proposed framework, various membership functions were used to construct ANFIS model and finally, an ensemble of ANFIS models was constructed using an evolutionary algorithm. Empirical results indicate a high performance of the proposed models for assessing risk in the cloud environment.

Keywords Cloud computing · Risk assessment · Adaptive neuro-fuzzy inference system · Feature selection · Ensemble · Evolutionary algorithm

N. Ahmed (✉)
Faculty of Computer Science and Information Technology, Sudan University of Science, Technology, Khartoum, Sudan
e-mail: naessa@pnu.edu.sa

N. Ahmed
College of Computer and Information Sciences, Princess Nourah bint Abdul Rahaman University, Riyadh, KSA

A. Abraham
Machine Intelligence Research Labs (MIR Labs), Washington, WA, USA
e-mail: ajith.abraham@ieee.org

V.K. Ojha · A. Abraham
IT4Innovations, VSB - Technical University of Ostrava, Ostrava, Czech Republic
e-mail: varun.kumar.ojha@vsb.cz

© Springer International Publishing Switzerland 2016
N. Pillay et al. (eds.), *Advances in Nature and Biologically Inspired Computing*,
Advances in Intelligent Systems and Computing 419,
DOI 10.1007/978-3-319-27400-3_3

1 Introduction

The emergence of cloud computing represents a fundamental change in the way information technology service is invented, deployed, developed, maintained, scaled, updated, and paid [1]. Cloud computing provides on-demand service such as processing and storage. Consumers use these services as they need and pay only for what is used [2–4]. In recent years, there was obvious migration to cloud computing with end users who access remote servers via network to manage a growing number of personal data such as photographs, music files, bookmarks, and much more [5]. The use of cloud computing services can cause great risks to consumers. Before consumers start using cloud computing services they must confirm whether the product satisfies their needs and understand the risks involved in using those services [6].

Risk has been defined as "the chance that someone or something that is evaluated will be adversely affected by the hazard". Risk assessment is an essential tool for safe tying policies in any organization. Risk formulation is essential in estimating the level of the risk on the basis of likelihood of a risk mapped against estimated negative impact. The risk level assessment is a complex subject that is usually covered in uncertainty and ambiguity. Risk level is often assessed using linguistic terms. In that case, fuzzy set analysis is an efficient tool because of its ability to deal with the problems having uncertainty and ambiguity. The imprecision may be dealt by using fuzzy rules in a fuzzy inference system. Adaptive neuro-fuzzy inference system (ANFIS) is a fuzzy inference system implemented using neural network learning methods [7]. The aim of this research is to construct a reliable and effective ensemble of adaptive neuro fuzzy inference systems (En-ANFIS) model for risk assessment in cloud computing environment.

This paper is organized as follows: Sect. 2 gives a summary of literature related to risk assessment for adoption to cloud service. In Sect. 3, we describe the basics of ANFIS and the ANFIS ensemble learning. Methodology used to develop the ANFIS ensemble model is presented in Sect. 4. Section 5 presents the results obtained in this work followed by Conclusions in Sect. 6.

2 Related Works

European Network and Information Security Agency (ENISA) identified 35 risks and classified them into four categories: policy and organizational, technical, legal, and other scenarios not specific to cloud computing. A group of experts determined the likelihood of each risk and their business impact [8]. Risk management standard by International Organization for Standardization (ISO) published a generic standard called ISO 31000 [9], for risk management. Later, ISO and International Electrotechnical Commission (IEC) complemented ISO 31000 by a joint publication ISO/IEC 31010 [10] about generic standards in risk assessment techniques. COBIT is another generic framework in IT, introduced by Information Technology

Governance Institute and the Information Systems Audit and Control Association (ISACA) in 1996 that provides a common language to communicate results, goals, and objectives of business. An enterprise risk management recommendations was provided in the last versions of COBIT in 2013 [11]. Cloud adoption risk assessment model (CARAM) complement the various recommendation from European Network and Information Security Agency (ENISA), and Cloud Security Alliance (CSA) for a complete risk assessment framework. CARAM help in assessing various risks to business, privacy, and security that cloud customers has to face when moving to use cloud computing environment [12].

3 Adaptive Neuro-Fuzzy Inference System and Ensemble

System modeling based on mathematical and statistical methods is not suitable to represent data that requires knowledge closer to human-like thinking. By contrast, a fuzzy inference system (FIS) [15] can be viewed as a real-time expert system used to model and utilize human experience by employing fuzzy *if—then* rules [7, 13, 14], Fuzzy *if—then* rules are an expression of the form "*if x is A then y is B*", where *x* is an input (antecedent) and *y* is corresponding output (consequent). A and B are fuzzy sets [15] representing antecedent and consequent respectively. Fuzzy *if —then* rules are employed to capture imprecise modes of reasoning that represents a basic role in human-like decision making [14]. Takagi and Sugeno [16, 17] proposed a form of fuzzy if—then rules that has fuzzy sets involved only in the premise part. The core part of fuzzy inference system was represented by fuzzy *If— Then* rules. Fuzzy inference system is primarily applied to cases that are either difficult to precisely model or ambiguous to describe [18, 19]. The fuzzy inference system is the foundation of adaptive neuro-fuzzy system (ANFIS).

3.1 Adaptive Neuro-Fuzzy Inference System (ANFIS)

Adaptive neuro-fuzzy inference system was first introduced by Jang [13] and [20]. ANFIS is a multilayer feed forward network that uses neural network learning algorithms and fuzzy reasoning to map input characteristics into input membership functions (MFs), input MFs to a set of *if—then* rules, rules to a set of output characteristics, output characteristics to output MFs, and output MFs to a single-valued output [14, 18, 21–24].

Figure 1 shows architecture of ANFIS consisting of five layers used for training Sugeno-type FIS through learning and adaptation. It shows structure of a system with (m = 2) inputs x and y [in general we may say $(x_1, x_2, ..., x_m)$] each wiith (*n*) membership functions (MFs), a fuzzy rule base of *R* rules, and a single output *z*. The number of nodes in the first layer is the product of the number of inputs *m* and MFs *n* for each input, i.e., $N = m \times n$. In layers 2 to 4, the number of nodes is

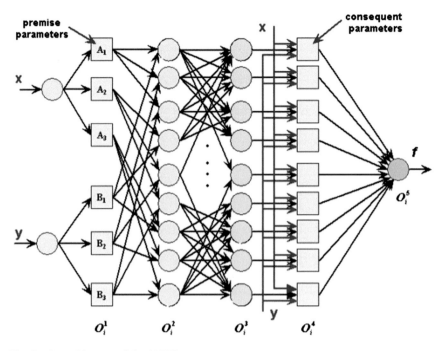

Fig. 1 The architecture of the ANFIS

equal to the number of rules R in the fuzzy rule base. ANFIS implements a Takagi Sugeno Kang (TSK) fuzzy inference system. ANFIS structure comprising of 5 layers is briefly described below:

- Layer 1: each node in this layer is an adaptive node with a node function $O_{1i} = \mu Ai\ (x)$ or $O_{1i} = \mu Bi\ (y)$, where A_i and B_i are linguistic lable (such as "small" or large"). Each node generates membership grades O_{1i} of a fuzzy sets A $(A_1, A_2, ., \text{ or } B_1, B_2.)$ using membership functions.
- Layer 2: the nodes in this layer are fixed nodes. The output O_{2i} in this layer are the product of all input signals represents the firing strength of a rule, usually AND operator is applied to compute incoming signal.
- Layer 3: average nodes, the nodes in this layer are also fixed nodes. The output of this layer is called normalized firing strengths.
- Layer 4: consequent nodes, all nodes in this layer are adaptive node, whose output is the product of the normalized firing and a first-order polynomial.
- Layer 5: this layer has a single node and it computes the overall output by summing all the signals.

ANFIS parameters are trained/updated by using a hybrid learning algorithm, where a least-squares method is used in forward pass (offline learning) manner to identify consequent linear parameters iteratively by minimizing error between actual state and desired state of an adaptive network, while the antecedent parameters (membership functions) are assumed are kept fixed. A gradient descent method is then employed in backward pass manner to tune premise parameters, by back propagating error rate from the output layer towards the input layers, while consequent parameters remain fixed [21, 25].

3.2 Ensemble of Adaptive Neuro Fuzzy Inference System

Generally an individual model may or may not offer the best solution. ANFIS employ only a single fuzzy inference system and an ensemble of ANFIS is investigated by the combination of *M* networks. In-fact, generalization may not be achieved by using a single model. In an ensemble model, a combined decision of many predictors gives us a generalized solution [26]. For a regression problem, ensemble decisions are obtained by averaging the decisions of candidate predictors. However, an average decision lacks in providing the due credit to the best predictors in an ensemble system. Therefore, a weighted average is an alternative, where each predictor in an ensemble system are pre-assigned a weight according to their accuracy/credibility. Now, we have option to assign weight according to the predictor's accuracy, however, in this way, we will lose insights of predictor's decision. Hence, we used an evolutionary algorithm method for computing the weights of each predictors [27] in the proposed ensemble of ANFIS system. An evolutionary algorithm applies the principles of evolution found in nature to the problem of finding an optimal solution to a Solver problem. In an evolutionary method, we start by initializing random weights to the predictors from a range −1 to 1. During the evolutionary process, a predictor may acquire negative and positive weight according to its credibility in the ensemble decision. The fitness of the ensemble system having *k* predictors was computed as:

$$RMSE^{F'}(w_1, w_2, \ldots, w_k) = \sqrt{\frac{1}{N}\sum_i^N \left(\left(\sum_j^k w_j x_{ij}\right) - y_i\right)^2}, \qquad (1)$$

where x_{ij}, is *i*th decision of *j*th predictor and y_i denote target output in learning set that consists of total of *N* samples. In the present work, we use an evolutionary algorithm [28] for searching weights $w_1, w_2, \ldots, w_5$ of predictors.

4   Experimental Methodology

4.1   Datasets

We conducted an online survey to determine and finalize the various risk factors. In the survey, we requested the participants to categorize risk factors to three levels (*Important, Neutral*, and *Not important*) according to their effect on cloud computing. 35 international experts responded to the survey from different countries and all of them agreed that the previously defined factors are important. Next, we mapped each risk factor to a qualitative scale and finally we formulated expert rules and used statistical methods to simulate data based on the expert rules. The collected dataset contains 18 input attributes and comprises of 1951 instances. Input attributes were labeled as data transfer (DT), insufficient due diligence (IDD), regulatory compliance (RC), business continuity and service availability (BCSA), third party management (TPM), interoperability and portability (IP), data loss (DL), insecure application programming (IAP), data location and Investigative Support (DLIS), recovery (RY), resource exhaustion (RE), service level agreement (SLA), authentication and access control (AAC), shared environment (SEnv), data breaches (DB), data segregation (DS), virtualization vulnerabilities (VV) and data integrity (DI). Risk Factors and numerical ranges are illustrated in Table 1. We used following feature selection methods to extract subset instances from the original dataset. Best-first: This algorithm search in the space of set as a graph called "feature selection lattice", then applies one of the standard graph searching algorithms [29]. Random search: First, it randomly selects subset and then continues in two different ways. One of them is to follow sequential search, the second one is to continue randomly and generate next subset randomly [30]. Feature selection reduced the number of features to 3 attributes (IDD, DL, and DLIS) we labeled as First Dataset, and the second dataset with 4 attributes (RC, DL, DLIS, and VV). In the dataset, we also used different random percentage splits: 60–40 % (A); 70 –30 % (B); 80–20 % (C); 90–10 % (D) to train and test the different algorithms [31, 32]. We used ANFIS algorithm to estimate the risk.

Table 1 Risk factors and their range values

Risk factor	Range value	Risk factor	Range value
DI	0–3	R	1–3
IDD	1–3	RE	0–2
RC	0–1	SLA	0–3
BCSA	1–3	AAC	0–3
TPM	0–2	SEnv	1–3
IP	0–1	DB	0–2
DL	0–3	DS	0–1
IAP	0–1	VV	1–3
DLIS	0–3	DI	0–2

4.2 Development of ANFIS Model

To generate Sugeno fuzzy inference systems (FIS), 5 types of membership functions were used: For each membership function two and three input membership functions were used for each variable. The following MF's were considered in our experiments:

- Triangular membership function (*Tri*)
- Trapezoidal membership function (*Trap*)
- Generalized bell membership function (*Gbell*): $\mu_{Ai} = 1/(1 + (((x - c_i)/a_i)^2)^{b_i})$
- Gaussian membership function (*Guass*): $\mu_{Ai}(x) = \exp(-((x - c_i)^2/a_i))$
- A two-sided version of Gaussian membership function (*Guass2*)

Performance of the constructed model was assessed using correlation coefficient (*r*) and root mean square error (RMSE). ANFIS architecture was generated using a grid-partitioning algorithm and the algorithms were run for 100 epochs for all experiments.

4.3 Ensemble Method

As reported earlier, there are two datasets and four sets of train and test data. The experiments repeated using two membership functions and three membership functions for each dataset. With each membership function, we obtained five outputs. Our objective was to obtain best membership functions (with two or three membership function) and the best set of train/test data (A, B, C, D). To fulfill the mentioned objective, we need to obtain the best combination of weights of predictors. We used the evolutionary algorithm with population size 20, crossover: 0.8, mutation 0.2.

5 Results and Discussions

5.1 ANFIS Results Discussion

The ANFIS models are compared based on their performance in test sets. We observed that the ANFIS algorithm was accurate and can effectively predict the values of risk. Due to space constriants, only the results for the first data set are presented in this paper. Tables 2 and 3 represent the results of first data with 2 and 3 MFs, where all the values of RMSE are smaller and correlation coefficients are close to 1. As evident from Tables 2 and 3, the triangular membership function gives the lowest error with dataset A and 2 MFs.

Table 2 RMSE computed over datasets with 2 membership functions

MF shape	No of MF's	RMSE			
		A	B	C	D
Tri	2	**2.882e−06**	3.285e−06	4.022e−06	5.870e−06
Trap	2	**1.112e−05**	1.320e−05	1.937e−05	3.636e−05
Gbell	2	**1.874e−05**	2.294e−05	2.787e−05	3.669e−05
Gauss	2	**1.296e−05**	1.628e−05	2.440e−05	3.466e−05
Gauss2	2	**1.748e−05**	2.115e−05	2.739e−05	3.868e−05

Table 3 RMSE computed over datasets with 3 membership functions

MF shape	No. of MF's	RMSE			
		A	B	C	D
Tri	3	**4.118e−06**	4.886e−06	5.789e−06	7.961e−06
Trap	3	**1.985e−05**	2.577e−05	3.320e−05	3.320e−05
Gbell	3	**1.871e−05**	2.115e−05	2.882e−05	3.991e−05
Gauss	3	**1.976e−05**	2.108e−05	2.840e−05	3.666e−05
Gauss2	3	**1.977e−05**	2.153e−05	2.826e−05	3.145e−05

5.2 Ensemble Results

Tables 4 and 5 provide a comprehensive overview of results and obtained weights. The results demonstrate that the ensemble of ANFIS can be successfully applied to establish the risk assessment model that could provide accurate and reliable risk

Table 4 Weights and RMSE of ensemble systems with 2 membership functions

Dataset	Predictors weights					Weighted ensemble (RMSE)
	Tri	*Trap*	*Gbell*	*Gauss*	*Gauss2*	
D	0.626	0.187	0.049	0.162	−0.023	9.43936E−06
B	0.485	0.030	0.240	0.178	0.067	1.21276E−05
C	0.331	0.3175	0.071	0.258	0.023	6.69358E−06
A	**0.760**	**0.082**	**0.057**	**−0.021**	**0.122**	**1.92408E−06**

Table 5 Weights and RMSE of ensemble systems with 3 membership functions

Dataset	Predictors weights					Weighted ensemble (RMSE)
	Tri	*Trap*	*Gbell*	*Gauss*	*Gauss2*	
D	0.534	0.058	0.018	0.114	0.276	1.27589E−05
B	0.519	0.011	−0.008	0.352	0.126	1.09499E−05
C	0.788	0.239	0.248	−0.153	−0.122	8.64922E−06
A	0.383	0.217	0.114	0.070	0.216	5.21580E−06

level prediction and minimize risk in cloud computing environment. Dataset A with two membership functions offered the best performance in terms of accuracy (RMSE). We performed our experiment using five predictors. We found that the average performance with Triangular MF was the best among the five predictors.

6 Conclusions

We proposed an ensemble of adaptive neuro-fuzzy inference system (En-ANFIS) to construct a model to predict risk level in a cloud-computing environment. The experimental results show that the ensemble of ANFIS can be applied effectively without excessively compromising the performance and ensemble provides much better performance than an single ANFIS model. Results obtained with the proposed architecture of ensemble of ANFIS minimizes the prediction error against the results obtained by methods as reported in by Ahmed et al. [31, 32].

7 Acknowledgment

This work was supported by the IPROCOM Marie Curie initial training network, funded through the People Program (Marie Curie Actions) of the European Union's Seventh Framework Program FP7/2007-2013/under REA grant agreement No. 316555.

References

1. Avram, M.-G.: Advantages and challenges of adopting cloud computing from an enterprise perspective. Procedia Technol. **12**, 529–534 (2014)
2. Paquette, S., Jaeger, P.T., Wilson, S.C.: Identifying the security risks associated with governmental use of cloud computing. Gov. Inf. Q. **27**(3), 245–253 (2010)
3. Carroll, M., Van Der Merwe, A., Kotze, P.: Secure cloud computing: benefits, risks and controls. In: Information Security South Africa (ISSA), IEEE (2011)
4. Sun, D., Chang, G., Sun, L., Wang, X.: Surveying and analyzing security, privacy and trust issues in cloud computing environments. Procedia Eng. **15**, 2852–2856 (2011)
5. Zissis, D., Dimitrios, L.: Addressing cloud computing security issues. Future Gener. comput. Syst. **28**(3), 583–592 (2012)
6. Chandran, S.P., Mridula, A.: Cloud computing: analysing the risks involved in cloud computing environments. In: Proceedings of Natural Sciences and Engineering, pp. 2–4 (2010)
7. Fragiadakis, N.G., Tsoukalas, V.D., Papazoglou, V.J.: An adaptive neuro-fuzzy inference system (anfis) model for assessing occupational risk in the shipbuilding industry. Saf. Sci. **63**, 226–235 (2014)
8. Catteddu, D.: Cloud Computing: benefits, risks and recommendations for information security. In: Web Application Security, pp. 17–17. Springer (2010)

9. Purdy, G.: ISO 31000: 2009—setting a new standard for risk management. Risk Anal. **30**(6), 881–886 (2010)
10. Commission, I.E.: IEC/ISO 31010: 2009. Risk management-risk assessment techniques (2009)
11. Association, I.S.A.C.: COBIT 5: A Business Framework for the Governance and Management of Enterprise IT. 2012: ISACA
12. Cayirci, E., Garaga, A., Santana, A., Roudier, Y.: A cloud adoption risk assessment model. In: Proceedings of the 2014 IEEE/ACM 7th International Conference on Utility and Cloud Computing, IEEE Computer Society (2014)
13. Jang, J.-S., ANFIS: adaptive-network-based fuzzy inference system. IEEE Trans. Syst. Man Cybern. **23**(3), 665–685 (1993)
14. Abraham, A.: Rule-Based Expert Systems. Handbook of Measuring System Design (2005)
15. Zadeh, L.A.: Fuzzy sets. Inf. Control **8**(3), 338–353 (1965)
16. Takagi, T., Sugeno, M.: Fuzzy identification of systems and its applications to modeling and control. IEEE Trans. Syst. Man Cybern. **1**, 116–132 (1985)
17. Takagi, T., Michio, S.: Derivation of fuzzy control rules from human operator's control actions. In: Proceedings of the IFAC Symposium on Fuzzy Information, Knowledge Representation and Decision Analysis (1983)
18. Khoshnevisan, B., Rafiee, S., Omid, M., Mousazadeh, H.: Development of an intelligent system based on ANFIS for predicting wheat grain yield on the basis of energy inputs. Inf. Process. Agric. **1**(1), 14–22 (2014)
19. Yang, Z., Liu, Y., Li, C.: Interpolation of missing wind data based on ANFIS. Renewable Energy **36**(3), 993–998 (2011)
20. Yüksel, S.B., Alpaslan, Y.: Modelling uniform temperature effects of symmetric parabolic haunched beams using adaptive neuro fuzzy inference systems (ANFIS). In: Metaheuristics and Engineering, p. 83
21. Chang, F.-J., Chang, Y.-T.: Adaptive neuro-fuzzy inference system for prediction of water level in reservoir. Adv. Water Res. **29**(1), 1–10 (2006)
22. Wang, Y.-M., Elhag, T.M.S.: An adaptive neuro-fuzzy inference system for bridge risk assessment. Expert Syst. Appl. **34**(4), 3099–3106 (2008)
23. Rezaei, E., Karami, A., Yousefi, T., Mahmoudinezhad, S.: Modeling the free convection heat transfer in a partitioned cavity using ANFIS. Int. Commun. Heat Mass Transf. **39**(3), 470–475 (2012)
24. Hayati, M., Rashidi, A.M., Rezaei, A.: Prediction of grain size of nanocrystalline nickel coatings using adaptive neuro-fuzzy inference system. Solid State Sci. **13**(1), 163–167 (2011)
25. Lima, C.A.M., Coelho, A.L.V., Von Zuben, F.J.: Fuzzy systems design via ensembles of ANFIS. In: Proceedings of the IEEE International Conference on Fuzzy Systems, 2002. FUZZ-IEEE'02. IEEE (2002)
26. Breiman, L.: Bagging predictors. Mach. Learn. **24**(2), 123–140 (1996)
27. Ojha, V.K.J., Abraham, A., Snášel, V.: Dimensionality reduction, and function approximation of poly (lactic-co-glycolic acid) micro-and nanoparticle dissolution rate. Int. J. Nanomed. **10**, 1119 (2015)
28. Goldberg, D.E.H., John, H.: Genetic algorithms and machine learning. Mach. Learn. **3**(2), 95–99 (1988)
29. Jain, A., Zongker, D.: Feature selection: evaluation, application, and small sample performance. IEEE Trans. Pattern Anal. Mach. Intell. **19**(2), 153–158 (1997)
30. Liu, H. Yu, L.: Toward integrating feature selection algorithms for classification and clustering. IEEE Trans. Knowl. Data Eng. **17**(4), 491–502 (2005)
31. Ahmed, N.A., Abraham A.: Modeling cloud computing risk assessment using ensemble methods. In: Pattern Analysis, Intelligent Security and the Internet of Things, p. 261–274. Springer (2015)
32. Ahmed, N.A., Abraham A,. Modeling cloud computing risk assessment using machine learning. In: Afro-European Conference for Industrial Advancement, Springer (2015)

k-MM: A Hybrid Clustering Algorithm Based on k-Means and k-Medoids

Habiba Drias, Nadjib Fodil Cherif and Amine Kechid

Abstract k-means and k-medoids have been the most popular clustering algorithms based on partitioning for many decades. When using heuristics such as Lloyd's algorithm, k-means is easy to implement and can be applied on large data sets. However, it presents drawbacks like the inefficiency of the used metric, the difficulty of the choice of the input k and the premature convergence. In contrast, k-medoids takes more time to come up with a clustering but ensures a better quality of the result. Moreover, it is more robust to noise and outliers. In this article, we design a hybrid algorithm, namely k-MM to take advantage of both algorithms. We experimented k-MM and we show that, when compared to k-means and k-medoids, it is very efficient and effective. We present also an application to image clustering and show that k-MM has the ability to discover clusters faster and more effectively than a recent work of the literature.

Keywords Data mining · Clustering based on partitioning · K-means · PAM · Hybrid algorithm · Image clustering application

1 Introduction

Nowadays, clustering has become increasingly important because it has shown its usefulness in several fields like industry, business and image processing. A plethora of clustering algorithms have been proved effective but unfortunately, they also present drawbacks. The leading idea of this study is to explore some of these algorithms and try to improve them. We focus on the most common clustering algorithms based on partitioning that are k-means and k-medoids. We undertook a detailed review of these algorithms in order to bring judicious enhancements. We therefore developed a hybrid algorithm called k-MM so as, to preserve the benefits

H. Drias (✉) · N.F. Cherif · A. Kechid
Computer Science Department, LRIA, USTHB, Algiers, Algeria
e-mail: hdrias@usthsb.dz

© Springer International Publishing Switzerland 2016
N. Pillay et al. (eds.), *Advances in Nature and Biologically Inspired Computing*,
Advances in Intelligent Systems and Computing 419,
DOI 10.1007/978-3-319-27400-3_4

while minimizing the disadvantages of the mentioned algorithms. The idea is to start constructing relatively good clusters using k-means then to improve the result by launching k-medoids. But as k-means can yield non object centroids, we integrate a process between them to get the closest objects from these centroids. According to this framework, we gain the effectiveness of k-medoids because the latter terminates the algorithm. The efficiency will be intermediary between that of k-means and that of k-medoids because the new algorithm adds instructions after k-means and starts k-medoids with relatively good centroids computed by k-means.

Intensive experiments were performed on the Breast Cancer data set of the well-known UCI benchmark used by the data mining community, for k-means, PAM and k-MM to test the efficiency and effectiveness of the latter. The achieved results were in favour of k-MM.

In a second stage, k-MM was applied to image clustering in order to validate its usefulness in concrete domains. Using the COIL-100 benchmark and the same measures, we show that it is more efficient and more effective than a recent method found in the literature.

The article is organized as follows. Section 2 presents the most known partitioning algorithms and recent studies that are the most relevant to our work. Section 3 gives more details about k-means and k-medoids in order to introduce our proposal. Section 4 describes k-MM, the new clustering algorithm. Section 5 provides an experimental validation of k-MM and an application to image clustering to show its contribution to concrete domains.

2 Related Works

The clustering optimization problem is NP-hard [1] and despite this fact, several efficient heuristics were designed and used widely. Among the known clustering algorithms are k-means [2, 3], k-medoids [4, 5], CLARA [6] and CLARANS [7]. Given a set S of n objects, the problem is to partition the n objects into k (k <= n) subsets $S = \{S_1, S_2, ..., S_k\}$ so as, to group similar elements together in one cluster and separate the dissimilar ones in different clusters. A distance function has to be defined in order to model the notion of similarity between objects.

k-means is perhaps the most popular clustering algorithm because it is simple to implement and can be even applied to large data sets. Initially, the algorithm chooses randomly k objects representing the centres of the clusters. Then all the remaining objects are assigned to the nearest cluster respectively. A new centroid for each cluster is calculated as the means of the group members and the process is repeated until getting stable centroids. This algorithm runs rapidly however, it converges to a local optimum and therefore it compromises its effectiveness. On the other hand, it needs to specify in advance k, the number of clusters. Considering the centroid as the means of the objects belonging to the same cluster may be a constraint for data types other than numerical. Moreover, k-means is unable to handle noise and outliers.

k-medoids also called PAM (**P**artitioning **A**round **M**edoids) presents the same algorithmic framework as k-means. The major difference resides in the representation of the centroid, which is the most central object of the cluster called medoid. Of course, it takes more time to calculate medoids, leading to a longer running time, but it has the ability to control noise and outliers. PAM is then considered more robust for small data sets but not interesting for scalability.

CLARA (**C**lustering **Lar**ge **A**pplications) brings an enhancement of PAM so as, to handle scalability. Instead of applying the algorithm on the entire data set, it runs on data samples with small sizes and selects the best clustering afterwards. It then reduces considerably the algorithm complexity but at the same time, it decreases the quality of the clustering as it is based on samples.

CLARANS (a **Cl**ustering **A**lgorithm based on **RAN**domized **S**earch) is based on CLARA and PAM. It performs a randomized search in the whole search space that is, the set of all possible solutions, a solution being a set of objects representing medoids. It searches for a good solution in a local graph representing the neighbourhood of a solution. Then it changes the region by considering another arbitrary solution and repeats the search. It keeps the best solution at each iteration.

During the last decade, many efforts have been deployed to enhance the efficiency and effectiveness of these classic well-known methods. In [8], the authors proposed an interesting algorithm from the effectiveness point of view but not for scalability. On the other hand, another algorithm was developed and presented in [9] with the aim to shorten the runtime. Unfortunately, it was designed in detriment of a large memory space.

In [10], the authors propose a subtle technique to reduce the time taken to calculate the medoids, while keeping the same algorithmic structure of PAM. This way, they contribute to improve the efficiency of PAM but they did not come up with a new algorithm.

Hybrid centroid-medoid [11] is a heuristic developed to improve *k*-Means and more precisely its convergence to local optimum. After running a certain number of k-means iterations, the algorithm changes the centroid by a medoid calculated from a subset drawn randomly from the cluster. The algorithm performs better than k-means in terms of convergence but no comparison was provided with k-medoids.

In our proposal, we bring an enhancement to both k-means and k-medoids. The idea we suggest is to gather the majority of benefits from these algorithms such as implementation simplicity, efficiency, effectiveness, scalability and insensitivity to noise and outliers in order to perform better than all the above methods. Concretely, we develop k-MM, an algorithm based on k-means and k-medoids. Unlike centroid-medoids, where medoids operations are introduced in k-means, we propose a simple framework that includes both algorithms entirely. This way, a comparison with the existing algorithms k-means and PAM, can be undertaken.

3   k-Means Versus k-Medoids

This section is devoted to the presentation and the analysis of k-means and
k-medoids as well as to the comparison between both algorithms, which are
complementary.

3.1   k-Means Algorithm

k-means combines statistics and unsupervised machine learning. It starts by
drawing at random k objects representing centres of clusters. Then each object is
inserted in the cluster whose centre is the nearest. The centres of the clusters are
updated and replaced by the means of the clusters objects. The process is iterated
until reaching the stability of the centres. k-means is outlined in Algorithm 1.

```
Algorithm 1. k-means
Input: k the number of clusters;
       D a set of n objects;
Output: a set of k clusters;
begin
1. Select k initial centres arbitrarily from D;
2. repeat until no changes in clusters centers
   2.1. for each object
      2.1.1. calculate its distance from the center of
             each cluster;
      2.1.2. assign the object to the cluster with the
             nearest center;
   2.2. compute the means of each cluster and update
        its center;
end.
```

It is clear that the computational complexity of the algorithm is $O(nkt)$ where t is
the number of iterations. The latter is in practice low and the algorithm is considered as linear. Consequently, k-means is able to treat large data sets. Another
benefit of the algorithm is the simplicity of its implementation, this is the reason it
has been used widely. However, k-means suffers from several drawbacks. The
choice of the number k at random can bias the results. The representative of the
cluster, which is the means of the objects inside the cluster yields premature convergence and makes the algorithm sensitive to noise and outliers. It presents also a
constraint to non-numerical data such as categorical and ordinal data.

3.2   k-Medoids Algorithm

The k-medoids structure is the same as that of k-means. The representative of a cluster is an object calculated as the most centrally located in the cluster. The distance between two objects is computed and the object that has the minimal average dissimilarity to all the other objects is selected to be the center. k-medoids is outlined in Algorithm 2. Instruction 2.1 is a loop of n iterations of simple operations. Instruction 2.2 is a loop of k iterations including two operations. The first one costs $O(n - k)$ and the second $O((n - k)^2)$ for the worst case. The complexity of k-medoids is then $O(k(n - k)^2)$. Therefore, PAM is not scalable but it is more robust to noise and outliers than k-means because as the medoid considers the minimal distance that separates it from the other objects, it will be distant from noise and outliers. The representation of the centres by medoids avoids the premature convergence and hence generates relatively effective outcomes.

```
Algorithm 2. k-medoids
Input: k the number of clusters;
       D a set of n objects;
Output: a set of k clusters;
begin
1. Select k initial medoids arbitrarily from D;
2. repeat until no changes in clusters medoids
   2.1. for each object
      2.1.1. calculate its distance from the medoid of
             each cluster;
      2.1.2. assign the object to the cluster with the
             nearest medoid;
   2.2. for each medoid m do
      2.2.1. min := sum of the distances from m to
             the other objects;
      2.2.2. for each non-medoid object o do
         2.2.2.1. calculate the sum of the distances
                  From o to the other objects;
         2.2.2.2. if the sum is less than min
                  then swap m and o and update min;
end.
```

4   k-MM: A New Clustering Algorithm

In this section, we propose a new algorithm called k-MM, which is based on the hybridization of k-means and k-medoids algorithms as follows:

1. Call k-means on the initial data set. We get then a set of clusters with means centers (fictive or not).
2. Then replace each centroid by the closest object in each cluster by calculating the distance between the centroid and each object of the cluster and considering the object that has the minimum distance to the centroid.
3. Call k-medoids on the new centers representing initial medoids.

Let us recall that k-means centroids can be real objects or fictive objects that are not objects but points computed as the means of the cluster members. To measure the similarity between different objects, we need to define a distance between them. We consider for our algorithm, not only numerical data but also binary, categorical, ordinal and even mixed types of data. For more details, see [12] for the calculation of the distance of two objects with these types of data. k-MM is summarized in Algorithm 3.

Algorithm 3. k-MM
Input: *k* the number of clusters;
      D a set of *n* objects;
Output: a set of *k* clusters;
begin
1. call k-means;
2. calculate the distance that separates each centroid
   from the other objects of the cluster;
      2.1. replace each centroid by the object that
      has the minimal distance;
3. call k-medoids on the new centroids;
end;

The worst case complexity in theory is the same as that of k-medoids because the complexity of instruction 3 is greater than those of step 1 and 2, which are respectively near $O(n)$. In practice, it is intermediary between those of k-means and k-medoids and it is close intuitively to that of k-means because the third instruction runs with fewer iterations. k-means yields good initial centres relatively to random ones. Consequently, k-MM is able to handle large data sets. We also conclude that it is more effective than k-means because the results of k-means are refined. They are at least as good as those of k-medoids because the latter is called at the end of k-MM. The only disadvantage that remains is the initialization with a random number *k*, which bias less the results because of the call of k-means and k-medoids consecutively.

Let us now compare k-MM to centroid-medoids [11], which proceeds as follows. First, it executes for an empirical number of iterations *r*, k-means algorithm. Then it changes each centroid randomly chosen from a set of candidates by a medoid, if the latter is of better quality. The whole process is repeated until no further change occurs. The Centroid-medoids framework is shown in Algorithm 4.

Algorithm 4. Centroid-medoids
Input: *k* the number of clusters;
       D a set of *n* objects;
Output: a set of *k* clusters;
begin
1. call k-means for *r* iterations;
2. replace the centroids by medoids calculated from a
   randomly chosen subset of objects, if better;
3.pursue with k-means until convergence;
end;

From this description, we notice at first that the strength of k-MM is its simplicity of design. The centroid-medoids algorithm indeed improves k-means because it integrates some medoids operations inside its loop. However, this design cannot allow the comparison with PAM. Theoretically, k-MM is more robust because it is more scalable than and at least as effective as PAM.

5 Experimental Validation

To demonstrate the effectiveness of k-MM, we conducted several experiments in order to compare it to k-means and PAM. We used the Normal Mutual Information (NMI) to measure the degree of closeness of the result of the classification to the one that should be in reality [11]. The mathematical formula for calculating NMI is:

$$NMI = -2* \frac{\sum \sum n_{h,l} \log\left(\frac{n_{h,l}*n}{n_h*n_l}\right)}{\sum n_h \log\left(\frac{n_h}{n}\right) + \sum n_l \log\left(\frac{n_l}{n}\right)} \tag{1}$$

where:

- n is the total number of objects.
- n_h is the theoretical number of objects in cluster h.
- n_l is the actual number of objects in the cluster given by the algorithm.
- $n_{h,l}$ is the theoretical number of objects in cluster h that exist in cluster l.

The results are significant when the NMI is close to 1. We also used the inertia to measure the effectiveness of the algorithm. We considered two kinds of inertia:

- The intraclass inertia, which represents the distance between a centroid and the cluster elements, so it is a value to minimize.
- The interclass inertia, which measures the distance between the centres of the clusters. An interesting result is expressed by a high value, so it is a value to maximize.

5.1 Experimental Environment

The algorithms k-means, PAM and k-MM we presented, were implemented in Java and run on a machine with a processor Intel (R) Core (TM) i5-3230 M, 2.60 GHz CPU and a RAM of 4.00 GB under Windows 8.1 64-bit.

5.2 Experimental Results

We first tested k-means, PAM and k-MM on the Breast Cancer Data set [13] provided by the UCI Archive. We compared the performances of the three algorithms considering the criteria of execution time, interclass inertia, intraclass and NMI. Figure 1 depicts the experimental results for the inertia for the three algorithms. We see that k-MM has the smallest intraclass inertia and the largest interclass inertia.

Figure 2 shows the results of the NMI measure for the three algorithms on the Breast Cancer Data Set. We observe that k-MM has the best NMI, which is equal to that of PAM. Figure 3 illustrates the execution time for the three algorithms. We remark that, the execution time for k-MM is intermediary between that of k-means, which is the lowest and that of PAM, which is the highest.

According to this set of experiments, we can conclude that for Breast Cancer data set, k-MM yields the best results.

We also tested k-means, PAM and k-MM on The COIL-100 Benchmark [14], which is a collection of 100 colour images with 72 shots each. The total number of pictures is then equal to 7200. Figure 4 shows the computed inertia for the three algorithms. We observe that k-MM and PAM yield the best results that is, the smallest intra-class inertia and the largest interclass inertia.

Figure 5 shows the comparison of the NMI for the three algorithms on COIL-100, knowing that the exact clusters are supplied with the benchmark. We stopped the experiments at 5000 pictures because of a memory overflow due to a performance limit of our machine. We can see that k-MM has a NMI equal to that of PAM when the number of images is less than 2000 and a NMI higher than the

Fig. 1 Comparing inertia for k-means, PAM and k-MM on Breast Data Set

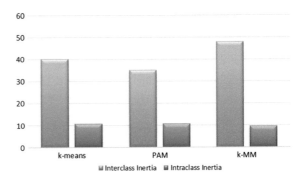

Fig. 2 Comparing NMI for k-means, PAM and k-MM on Breast Data Set

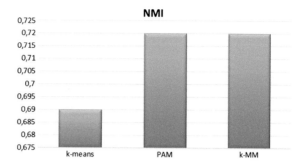

Fig. 3 Comparing runtime for k-means, PAM and k-MM on Breast Data Set

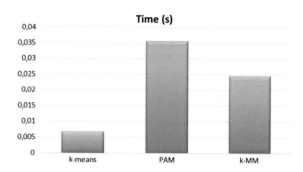

Fig. 4 Comparing inertia for k-means, PAM and k-MM on COIL-100

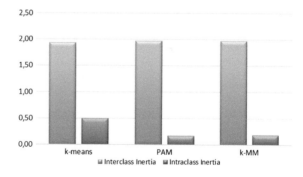

Fig. 5 Comparing NMI for k-means, PAM and k-MM on COIL-100

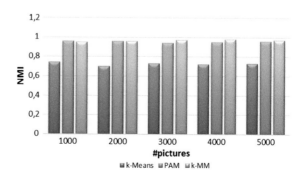

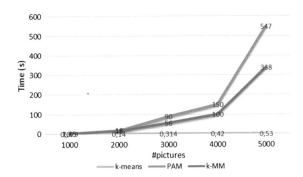

Fig. 6 Comparing runtime for k-means, PAM and k-MM on COIL-100

two other algorithms when the number of images is greater than 2000. We notice that data with more than 2000 objects are considered as large data set according to [7]. Our algorithm has then the best NMI for COIL-100 large data subsets.

Figure 6 exhibits the running time for the three algorithms on COIL-100. We remark that k-MM has an execution time longer than that of k-means and lower than that of k-medoids. It therefore meets our expectations by being faster than k-medoids. For the tested benchmark, k-MM has been proved effective and efficient in getting the best of both algorithms, rapid time of k-means and precision of k-medoids.

Figures 7 and 8 depict respectively the runtime and the NMI for k-MM and centroid-medoid, run on 1440 images grouped into 20 clusters. We stopped executing k-MM at 1440 images because the centroid-medoids results are provided with this number of pictures. For both criteria, k-MM outperforms centroid-medoids.

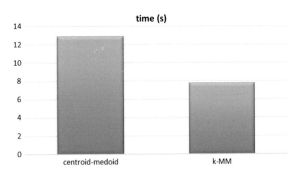

Fig. 7 Comparing running time for centroid-medoids and k-MM

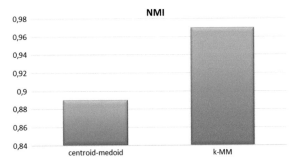

Fig. 8 Comparing NMI for centroid-medoids and k-MM

6   Conclusion

In this work, we proposed a hybridization between two complementary clustering algorithms that are k-means and k-medoids to achieve greater efficiency and effectiveness. The experimental validation of this new algorithm called k-MM was tested on the Breast Cancer data set of the UCI archive and on a Columbia Object Image Library benchmark, which contains 100 different images with 72 shots each, that is on 7200 pictures in total. The results of k-MM outperform both those of k-means and PAM. Also under the same data sets, it performs better than a recent work of the literature.

Even if k-MM presents a large spectrum of advantages, we continue undertaking experiments to prove its superiority on other clustering algorithms such as CLARA and CLARANS. Second, we plan to investigate other types of clustering algorithms such as hierarchical, density-based, grid-based and model-based in order to reproduce our main idea on these methods.

References

1. Mouratidis, K., Papadias, D., Papadimitriou, S.: Medoid queries in large spatial databases. In: Advances in Spatial and Temporal Databases. Lecture Notes in Computer Science, vol. 3633, pp. 55–72. Springer, Berlin (2005)
2. Hartigan, J.A., Wong, M.A.: Algorithm AS 136: A K-Means clustering algorithm. J. Roy. Stat. Soc.: Ser. C (Appl. Stat.) **28**(1), 100–108 (1979)
3. MacQueen, J.B.: Some methods for classification and analysis of multivariate observations. In: Proceedings of 5-th Berkeley Symposium on Mathematical Statistics and Probability, vol. 1, pp. 281–297. University of California Press, Berkeley (1967)
4. Kaufman, L., Rousseeuw, P.J.: Clustering Large Data Sets (With Discussion). In Pattern Recognition in Practice II, pp. 425–437. North-Holland, Amsterdam (1986)
5. Kaufman, L., Rousseeuw, P.J.: Clustering by means of Medoids. In: Dodge, Y. (ed.) Statistical Data Analysis Based on the L_1–Norm and Related Methods, pp. 405–416. Elsevier Berlin, North-Holland (1987)
6. Kaufman, L., Rousseeuw, P.J.: Finding Groups in Data: An Introduction to Cluster Analysis. Wiley (1990)
7. Ng, R., Han, J.: Efficient and effective clustering method for spatial data mining. IEEE Trans. Knowl. Data Eng. **14**(5), 1003–1016 (2002)
8. Zhang, Q., Couloigner, I.: A new and efficient k-Medoid algorithm for spatial clustering. In Computational Science and Its Applications. Lecture Notes in Computer Science, vol. 3482, pp. 207–224. Springer, Berlin (2005)
9. Park, H.S., Jun, C.-H.: A Simple and fast algorithm for k-Medoids clustering. Expert Syst. Appl. **36**(2, Part 2) 3336–3341 (2009)
10. Paterlini, A.A., Nascimento, M.A., Caetano Traina Jr., C.: Using pivots to speed-up k-Medoids clustering. J. Inf. Data Manage. **2**(2), 221–236 (2011)
11. Grira, N., Houle, M.E.: Best of both: a hybridized centroid-medoid clustering heuristic. In: ICML, Volume 227 of ACM International Conference Proceeding Series, pp. 313–320. ACM (2007)
12. Han, J., Kamber, M.: Data mining: concepts and techniques 3rd edn. Morgan Kaufmann (2011)

13. Breast Cancer Data Set, Wisconsin University, https://archive.ics.uci.edu/ml/datasets/Breast Cancer Wisconsin
14. Nayar, S.K., Murase, H.: Columbia object image library: Coil-100. Technical Report CUCS-006-96, Department of Computer Science, Columbia University (1996)

A Semantic Reasoning Method Towards Ontological Model for Automated Learning Analysis

Kingsley Okoye, Abdel-Rahman H. Tawil, Usman Naeem
and Elyes Lamine

Abstract Semantic reasoning can help solve the problem of regulating the evolving and static measures of knowledge at theoretical and technological levels. The technique has been proven to enhance the capability of process models by making inferences, retaining and applying what they have learned as well as discovery of new processes. The work in this paper propose a semantic rule-based approach directed towards discovering learners interaction patterns within a learning knowledge base, and then respond by making decision based on adaptive rules centred on captured user profiles. The method applies semantic rules and description logic queries to build ontology model capable of automatically computing the various learning activities within a Learning Knowledge-Base, and to check the consistency of learning object/data types. The approach is grounded on inductive and deductive logic descriptions that allows the use of a Reasoner to check that all definitions within the learning model are consistent and can also recognise which concepts that fit within each defined class. Inductive reasoning is practically applied in order to discover sets of inferred learner categories, while deductive approach is used to prove and enhance the discovered rules and logic expressions. Thus, this work applies effective reasoning methods to make inferences over a Learning Process Knowledge-Base that leads to automated discovery of learning patterns/behaviour.

K. Okoye (✉) · A.-R.H. Tawil · U. Naeem
School of Architecture Computing & Engineering, University of East London, London, UK
e-mail: u0926644@uel.ac.uk

A.-R.H. Tawil
e-mail: A.R.Tawil@uel.ac.uk

U. Naeem
e-mail: U.Naeem@uel.ac.uk

E. Lamine
Université de Toulouse, Mines-Albi, CGI, Campus Jarlard, Albi Cedex 09, France
e-mail: Elyes.Lamine@mines-albi.fr

© Springer International Publishing Switzerland 2016
N. Pillay et al. (eds.), *Advances in Nature and Biologically Inspired Computing*,
Advances in Intelligent Systems and Computing 419,
DOI 10.1007/978-3-319-27400-3_5

49

Keywords Process model · Learning process · Ontology · Semantic reasoning

1 Introduction

Ontology is one of the scientifically proven technique that is used to model different kinds and structure of objects, events and processes as they happen in reality. The concept can be layered on top of existing information asset to provide a more conceptual analysis of real time processes capable of providing real world answers that are closer to human understanding [1]. Ontology presents to the data science world the capability of using semantics to classify instances to explain the dependent variables in terms of independent ones; which is a great way to compliment the way we look at processes. The concept of semantic annotations and reasoning makes it possible to match same ideas as well as use the coherence and structure itself to inform and answer questions about relationships learning objects share within information knowledge base. The various units/activities within a learning process model can be related to exactly one case and assigned a case identifier [2] which results in automatic creation of workflow processes [3] and can help to maintain the resulting hierarchy correctly. This approach is made possible by using the semantic annotation scheme to represent sets of various entities, properties and classes within the learning knowledge base and create inferences capable of providing new knowledge or a richer set of intelligence within the model.

In this paper, we use ontology web rule languages and schema to discover sets of relationships that can be found within a learning process knowledge base. As a result, suitable learning paths are determined by means of semantic reasoning, which is then used to address the problem of extracting useful patterns from captured data source to provision of knowledge.

The rest of the paper is structured as follows; in Sect. 2, we present a description of learning process model and how we apply the representations for learning activities to draw conclusions and make predictions based on analysis of captured learning data. Section 3 shows the proposed semantic process model, describing in detail its ontological representation and reasoning using Ontology Web languages. The prototype implementation and preliminary outcomes was presented in Sect. 4, and in Sect. 5, we analyse and discuss appropriate related work in this area of study. Finally, Sect. 6 concludes the paper and points out directions for future research.

2 Ontological Description of Process Model

Ontological description of process model is based on computer logic programming [4], and has been related to the natural process of human thinking. Lumpe [5] mentions that Inductive intelligent is the process of reasoning from the particular to the general which involves the observation of particular events or data. The approach associates new content with prior knowledge; which can lead to unrelated data being discovered, examined and further grouped and labelled in order to draw conclusions or make predictions based on the analysis of the data. Following the works in [6, 7], the ability to analyse information and create concepts is fundamental to ontological reasoning and can be applied towards automation of learning processes, as we define below;

Step 1: Examine the Learning Knowledge base to define unrelated entities.
Step 2: Group entities with common attributes and provide descriptive labels to objects or data property.
Step 3: Identify relationships in order to generalise, predict and extract patterns from the existing properties.
Step 4: Apply discovered patterns to a new and different context to demonstrate understanding.
Step 5: Check that all entities within the discovered classes are true and at least falls within the universal restriction of validity by definition, and that there are no inconsistency of data or repeatable contradicting discovery.

The purpose of the definition is to use the concept of ontological model to describe and understand learning process reality based on captured knowledge or historic data, and the ability to provide a link between learning objects or data types.

3 Concept Matching and Association of Variables

Association Rule Learning aims at finding rules that can be used to predict the value of some response variables that has been identified as being important just like decision systems [8] but without focusing on a particular response variable. The rule aims at creating rules of the form:

$$\text{IF } \mathbf{X} \text{ THEN } \mathbf{Y}$$

where $\mathbf{X}$ is often called the *antecedent* and $\mathbf{Y}$ the *consequent*. Thus, $\mathbf{X} \Rightarrow \mathbf{Y}$

This rule is similar and can be related to the Semantic Web Rule Language, SWRL [9] which is useful especially to provide more ontological description and enhancement to learning process model.

The SWRL rule has the form; *atom*$^\wedge$ *atom* (antecedent) ... → *atom*$^\wedge$ *atom* (consequent).

Association rule learning strongly supports the use of metrics frequently expressed in the form of *support* and *confidence*. These expressions help in measurement of the strength of the association between learning objects. *Support* determines how often a rule is applicable to a given data set which means the fraction of instances for which both antecedent and consequent hold. Thus, a rule with high support is more useful than a rule with low support. A rule that has low support may occur simply by chance and is likely to be irrelevant from a learning perspective because it may not be profitable to monitor, recommend and promote learning activities or learning patterns.

Support is used to evaluate learning process models and its execution, where:

N_x is the number of instance for which, x, learning activity holds.
N_y the number of instances for which learning activity y holds.
And N_{x_y} is the number of instances for which activity x and y holds

Consequently, *support* for the rule $X \Rightarrow Y$ is described as

Support, $s(X \Rightarrow Y) = N_{x\_y}/N$: where N is the total number of instances.

Confidence, on the other hand, measures the reliability of the inference made by a rule over a learning process. Thus, for a give rule of the form $X \Rightarrow Y$, the higher the confidence, the more likely it is for the consequent Y (learning pattern extension) to be represented within the learning process that contains X (learning patterns). Confidence measures the conditional probability that the extension Y will happen given X.

Confidence, $c(X \Rightarrow Y) = N_{x\_y}/N_x$

Overall, inferences made by an association rule learning suggest co-occurrence of relationships between items in the antecedent (X) and consequent (Y) of the rule. Therefore, for every given set of activities or item set, there exist rules having support ≥ *minSup,* and confidence ≥ *minConf,* where: *minSup* and *minConf* are respectively the corresponding *support* and *confidence* thresholds.

In Learning Process models, these metrics can be used to dramatically reduce the exploration or drilling down space when constructing the set of frequent activity logs. The simple requisite is that X and Y are non-empty and any variable appears at most once in X or Y. For instance,

IF Learner(X) **AND** hasLearning_Activities **THEN** hasPartLearning_Process(Y)

Thus,   Learner(?X), hasLearning_Activities(?X, Activity) → hasPartLearning_Process(?X, ?Y)

This approach has been used to provide process specification and expressive language formats that are logically and fundamental to knowledge representation such as the Knowledge Interchange Format (KIF) [10] which makes it possible to understand the meaning of logic expressions through Declarative Semantics. For instance, it can be expressed that "Every Learner has a Learning_Activity". Thus;

```
( forall   ( ?X )
          ( => ( Learner ?X )
               ( exists  ( ?Y )
               ( and ( someLearningActivity  ?Y)
                    ( Learning_Activity ?X ?Y ) ) ) ) )
```

Consequently, Every Learning_Activity is part of a Learning_Process and must have some kind of a Learner. Thus the expression;

```
( forall ( ?X  ?Y )
         ( => ( exist ( Learning_Process ?X )
                     ( Learner ?Y )
              (<=>   ( Learning_Activity ?X ?Y )
                          ( and   ( someLearning_Process ?X ?Y )
                               ( someLearner ?X ?Y ) ) )))
```

Such rule expressions suggest that a strong relationship exist between the Learning_Process and the Learner. This is because Learner(X) has_Activities described as a Learning_Activity, and Learning_Activity has been described as PartOfLearning_Process.

Designers of knowledge base systems can use this type of rule expressions to help identify new opportunities especially for enhancement of process models. Association rule is now recently being used in application domains such as the web mining and scientific data analysis. The association patterns reveals interesting connection among domain entities, the individual classes and object/data types to provide a better understanding of how the different elements within the Learning Process Knowledge base relate and interact with each other. In the next section, we describe and implement the concept of ontological reasoning of learning activities capable of deducing inferences based on such design rule-base semantic approach to automated learning. The focus is on implementing the learning objects/property restrictions to define the classes and relationships of the entities within the Learning Knowledge-base.

3.1 Ontology Model for Automation of Learning and Reasoning

Based on ontological vocabularies, inductive reasoning has as input data types from which a possible *someValuesFrom* or believable generalisation is computed. This technique is an existential restriction, which describes the set of individuals that have at least one specific kind of relationship to individuals that are member of a specific class. It is a relationship that exist between two individuals, i.e. concept assertion that hold between two objects. On the other hand, deductive reasoning which has been generally adopted in the Semantic Web context assumes an *all-ValuesFrom* restriction, whereby given a set of general axioms, precise and definite conclusions are drawn by the use of a formal proof. This technique is referred to as universal restriction, which describes the set of individuals that for a given property only have relationships to other individuals that are members of a specific class.

Figure 1 is an Ontology Web Language (OWL) version 2 model of our proposed Learning Ontology, implemented in Protégé 4.3 and reasoned upon using Pellet 2. Protégé 4.3 OWL editor [11] supports Description Logic (DL) Queries [12] and SWRL rules [9].

In Fig. 1, we use the protégé Editor to construct ontology that expresses the functionality of the Learning model in terms of individual learning characteristics. The *Cases* within the model were defined as sub-class of the main class *LearningProcess*. The class expression is based on the OWL syntax primarily focused on

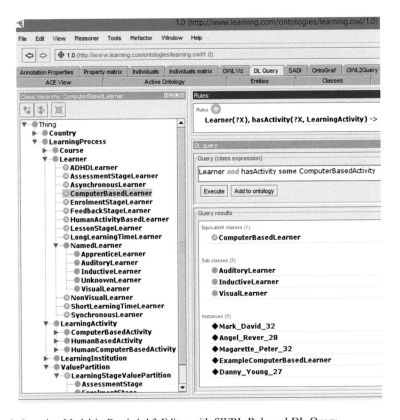

Fig. 1 Learning Model in Protégé 4.3 Editor with SWRL Rule and DL Query

collecting all information about a particular class or individual into a single construct, called a Frame. The DL Query provides the platform for searching the classified ontology to infer the learning activities of any named individual. The result of the logic expression and reasoning is what we use to show the process model and automated discovery of learning patterns. The tactics aims at discovering rules similar to the *Association Rule* [13] as previously described in Sect. 3, but then without focusing on a particular variable to discover user interaction patterns and then respond by making decisions based on adaptive rules centred on the captured user profiles. The goal is to discover and create rules of the same form;

$$X \Rightarrow Y \ (IF \ X \ THEN \ Y)$$

where **X** = *Learning pattern* (Antecedent) and **Y** = *Learning pattern extension* (Consequent)

e.g. Learner (?X) , hasActivity (?X, ?LearningActivity) -> hasLearning Process (?X, ?Y)

Learner (?X) , hasLearningProcess (?X, ?Y), hasLearningActivity (?Computer BasedActivity, ?Z) ->
hasComputerBasedLearner (?X, ?Z)

Driven by the variables as defined in the Learning Ontology in Fig. 1, and the OWL 2 XML file Fig. 2, the resulting rules expressions Fig. 3, were derived to improve the reasoning capability and semantics of the learning process model.

```xml
<?xml version="1.0"?>
<Ontology xmlns="http://www.w3.org/2002/07/owl#"
    ontologyIRI="http://www.semanticweb.org/kingsley/ontologies/2014/11/learning-ontology-25"
    <Annotation>
        <AnnotationProperty abbreviatedIRI="rdfs:comment"/>
        <Literal datatypeIRI="&rdf;PlainLiteral">Learning Ontology that classifies the various Units
            of Learning Process based on the Learning Activities</Literal>
    </Annotation>
    <EquivalentClasses>
        <Class IRI="http://www.learning.com/ontologies/learning.owl#AsynchronousLearner"/>
        <ObjectIntersectionOf>
            <Class IRI="http://www.learning.com/ontologies/learning.owl#Learner"/>
            <ObjectAllValuesFrom>
                <ObjectProperty IRI="http://www.learning.com/ontologies/learning.owl#hasActivity"/>
                <Class IRI="http://www.learning.com/ontologies/learning.owl#AsynchrounousActivity"/>
            </ObjectAllValuesFrom>
        </ObjectIntersectionOf>
    </EquivalentClasses>
    <EquivalentClasses>
        <Class IRI="http://www.learning.com/ontologies/learning.owl#ComputerBasedLearner"/>
        <ObjectIntersectionOf>
            <Class IRI="http://www.learning.com/ontologies/learning.owl#Learner"/>
            <ObjectSomeValuesFrom>
                <ObjectProperty IRI="http://www.learning.com/ontologies/learning.owl#hasActivity"/>
                <Class IRI="http://www.learning.com/ontologies/learning.owl#ComputerBasedActivity"/>
            </ObjectSomeValuesFrom>
        </ObjectIntersectionOf>
    </EquivalentClasses>
</Ontology>
```

Fig. 2 A fragment of the learning ontology OWL 2 XML file in Protégé

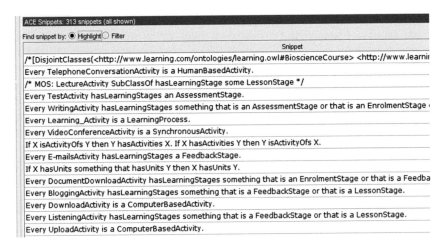

ACE Snippets: 313 snippets (all shown)

Find snippet by: ● Highlight ○ Filter

Snippet
/*[DisjointClasses(<http://www.learning.com/ontologies/learning.owl#BioscienceCourse> <http://www.learnir
Every TelephoneConversationActivity is a HumanBasedActivity.
/* MOS: LectureActivity SubClassOf hasLearningStage some LessonStage */
Every TestActivity hasLearningStages an AssessmentStage.
Every WritingActivity hasLearningStages something that is an AssessmentStage or that is an EnrolmentStage
Every Learning_Activity is a LearningProcess.
Every VideoConferenceActivity is a SynchronousActivity.
If X isActivityOfs Y then Y hasActivities X. If X hasActivities Y then Y isActivityOfs X.
Every E-mailsActivity hasLearningStages a FeedbackStage.
If X hasUnits something that hasUnits Y then X hasUnits Y.
Every DocumentDownloadActivity hasLearningStages something that is an EnrolmentStage or that is a Feedba
Every BloggingActivity hasLearningStages something that is a FeedbackStage or that is a LessonStage.
Every DownloadActivity is a ComputerBasedActivity.
Every ListeningActivity hasLearningStages something that is a FeedbackStage or that is a LessonStage.
Every UploadActivity is a ComputerBasedActivity.

Fig. 3 Fraction of rules executed in the learning ontology

4 Discussion

Automation of learning process involves the flow of activities within a learning knowledge-base technically described as Workflow. Being able to use the Reasoner to automatically compute the class hierarchy of the activities within the knowledge-base is one of the major benefits of building ontology using OWL, SWRL Rules and DL Query. Annotation properties are used to add information (Metadata—data about data) to the classes, individuals and object/data properties in the ontology. The proposed Learning Model Ontology allows the meaning of properties to be enhanced through the use of *property characteristics* and *classification of discoverable entities*. We utilize the main function offered by the Reasoner to help in checking for consistency in the model; to test whether or not a class is a subclass of another class, or checking whether or not it is possible for a class to have any instances. This means a class is said to be inconsistent if it does not have any instances.

By performing such test (i.e. Classification) it becomes possible to compute the inferred activity hierarchy. The functional property assertion allows this condition, where: for a given class, there can be at least one individual that is related to the class by means of the *restriction property*. In OWL, a Restriction describes class of individuals based on the relationship the members of the class participate in. In the proposed model, we describe the class Learner to be a subclass of the LearningProcess. The *necessary condition* is: if something is a Learner, it is *necessary* for it to be a participant of the class LearningProcess and *necessary* for it to have a kind of sufficiently defined condition and relationship with other classes e.g. LearningActivity, LearningInstitution, Course, LearningStageValuePartition etc. From the example in Fig. 1, we show that a ComputerBasedLearner is a subclass of, amongst other *NamedLearners*, a Learner and also a subclass of the

LearningActivity class that have at least one Activity that is *ComputerBased*. This assertion is achieved through the Restriction Property. The property is used to infer anonymous classes (Unnamed classes) that contains all of the individuals that satisfies the restriction, in essence, all of the individuals that have the relationship required to be a member of the class. In effect, the *necessary and sufficient Condition* makes it possible to implement and check for consistency in the model which means that it is necessary to fulfil the condition of Object/Data Property Restriction—for any individual to become a member of a class.

5 Related Works

Learning patterns or behaviours can be discovered as a consequence or condition of a Rule. d'Amato et al. [14] notes that various methods have been proposed in literature which are directed towards obtaining a more expressive model from knowledge bases [15]. The authors [14] argue that classification is a fundamental task for a lot of intelligent applications, and that classifying through logic reasoning may be both too demanding and frail because of inherent incompleteness and complexity in the knowledge bases. However, they observe that these methods adopt the availability of an initial drawing of ontology that can be automatically enhanced by adding or refining concepts, and have been shown to effectively solve learning modelling problems using Description Logics particularly those based on classification, clustering and ranking of individuals.

Learning Process modelling has been tackled over the years by customising Machine Learning methods such as Instance Based Learning [16] and Support Vector Machine (SVM) [17] to Description Logics (DLs) [12] queries which is the standard theoretical foundation upon which semantic web languages such as OWL and SWRL are based.

According to [18, 19] Bayesian models have paved way for new machine-learning algorithms with more powerful and more human-like capabilities. Semantic web ontology and its application cannot be explained without mentioning the Bayesian theory of probability [20, 21]. The Bayesian probabilistic theory have been proven to be one of the few mathematical interpretation of predictive concepts for representing a state of knowledge, thus, an extension of logic proposals that enables reasoning with hypothesis whose true or false values is uncertain. Bayesian model is based on 3 vital probes: What are the content of probabilistic theories? How can they be used to support reasoning? And how can they themselves be reasoned upon? The hypotheses are measured by computing the Bayes' rule, where: Probability, $P(x\backslash h, T)$ measures how well each argument predicts the data and the initial marking or likelihood. $P(h\backslash T)$ expresses the plausibility of the hypotheses given the users background knowledge. The posterior probability, $P(h\backslash x, T)$, is proportional to the result of the two expressions representing the level of certainty in each of the hypothesis given both the constraints of the background theory T, and observed data x.

According to Tenenbaum et al. [22], the challenge comes in specifying hypotheses and probability distributions that support Bayesian inference for a given task/domain. The authors argue that both structured knowledge and statistical inference are necessary to explain the nature, use and acquisition of such human knowledge and further introduced a theory-based Bayesian framework for modelling inductive learning and reasoning. Explicitly, the problems of modelling learning processes can be solved by transforming ontology population problem to a classification problem where, for each entity within the ontology, the concepts (classes) to which the entities belongs to have to be determined (i.e., classified) [1, 14]. Generally, these approaches assume that there already exist a probabilistic and/or fuzzy knowledge-base upon which this methods are able to predict the patterns/behaviour (hence, the classification) of new but not previously observed object/data types within the process.

Inductive and deductive reasoning methods can be used as building block towards the development of probabilistic learning knowledge-bases, by learning the probability that an inclusion axiom or concept assertion holds between two objects. The authors in [14] argue that in presence of noisy and inconsistent knowledge-bases that could be highly probable in a distributed environment such as the world wide web, that deductive reasoning is no more applicable since it requires correct/true premises; whereby if all premises are true and the rules of the deductive logic are followed, then the conclusion reached is necessarily true. On the other hand, inductive reasoning which is grounded on the generalisation of specific instances/assertions rather than correct premises, allows the formulation of conclusions even when inconsistent or noisy knowledge bases are being considered. Unlike deductive reasoning, inductive reasoning allows for the possibility that the conclusion is false, even if all of the premises are true and does not rely on universal restrictions over a closed axiom to draw conclusions. Currently, inductive reasoning is the main practice for logical reasoning, obtaining conclusions that are believed by the scientific community to be the most probable explanation of observed phenomena.

Reasoning on ontological knowledge plays an important role in the semantic representation of processes such as learning process. This is possible because semantic reasoning allows the extraction and conversion of explicit information into some implicit information, for instance, the intersection or union of classes, description of relationships and concepts/role assertions. Thom et al. [23] describes Workflow Activity Patterns (WAPS) as common structures involving the interaction between individual entities and the control-flow constructs used to model the semantics of activities as they are being performed. Workflow systems assume that a process can be divided into small, unitary actions, called Activities [3]. To perform a given process, one must perform the set (or perhaps a subset) of the activities that comprise it. Hence, an Activity is an Action that is a semantic unit at some level, which can be thought of as a function that modifies the state of the process in terms of the semantics of the patterns and can be discovered automatically by means of semantic reasoning [1].

6 Conclusion and Future Work

The work in this paper uses ontological schema/vocabularies to describe and propose a semantic rule-based approach that supports automated computing of different patterns within a learning process knowledge-Base through Semantic Reasoning. The method is proposed in order to address the problem of determining the presence of different learning patterns in process models. Any pattern or learning behaviour can be discovered as a consequence or condition of a Rule. Ontology provides us with benefits in discovery, flexible access and information integration due to the inherent connectedness (inference), concept matching and reasoning capability. This characteristic is the ability to match same idea as well as use the coherence and structure itself to inform and answer questions about relationships the learning objects (process instances) share amongst themselves within the Learning model.

Future work will focus on applying process mining techniques to provide better support for automated learning systems by means of semantic reasoning. The goal is to cover the whole spectrum of the approach presented in this paper; to help improve engagements within learning execution environments by adopting the main tools offered through conventional process mining for analysing process event data logs.

References

1. Okoye, K., Tawil, A.R.H., Naeem, U., Bashroush, R., Lamine, E.: A semantic rule-based approach supported by process mining for personalised adaptive learning. Procedia Comput. Sci. **37**(c), 203–210 (2014)
2. Van der Aalst, W.M.P.: Process Mining: Discovery, Conformance and Enhancement of Business Processes. Springer (2011)
3. Ferreira, D.L., Thom, L.H.: A semantic approach to the discovery of workflow activity patterns in event logs. J. Bus. Process Integr. Manage. Inderscience Publisher **6**(1) (2012)
4. Nguyen, T.H., Vo, B.Q., Lumpe, M., Grundy, J.: KBRE: a framework for knowledge-based requirements engineering. J. Softw. Qual. **22**(1), 87–119 (2014)
5. Lumpe, A.: Forms of Models and Induction. Seattle Pacific Uni. Lecture notes, Seattle, WA (2007)
6. Jepbarova, N.: Survey of instructional strategies. Lecture notes EDU 6526, 7 March 2007
7. Barbieri, D., Braga, D., Ceri, S., Valle, E.D., Yi Huang, Tresp, V., Rettinger, A., Wermser, H.: Deductive and inductive stream reasoning for semantic social media analytics. Intell. Syst. IEEE **25**(6), 32,41 (2010)
8. Yarandi, M., Jahankhani, H., Tawil, A.R.H.: Towards adaptive E-learning using decision support systems. Int. J. Emerg. Technol. Learn. **8**, 44–51 (2013)
9. SWRL: A Semantic Web Rule Lang "http://www.w3.org/Submission/2004/SUBM-SWRL-20040521/"
10. Knowledge Interchange Format http://cl.tamu.edu/discuss/kif-100101.pdf Accessed Dec 2014
11. Protégé Editor. http://protege.cim3.net Accessed 4 June 2014
12. Baader, F., Calvanese, D., McGuinness, D.L., Nardi, D., Patel-Schneider, P.F. (eds.): Description Logic Handbook. Cambridge University Press (2003)

13. Han, J., Kamber, M., Pei, J.: Data Mining: Concepts and Techniques. The Morgan Kaufmann Series in Data Management Systems (3rd edn). Morgan Kaufmann Publishers (2011)
14. d'Amato, C., Fanizzi, N., Esposito, F.: Query answering and ontology population: an inductive approach. In: Bechhofer, S., Hauswirth, M., Hoffmann, J., Koubarakis, M. (eds.): Proceedings of the 5th European Semantic Web Conference ESWC2008. Vol. 5021 of LNCS, pp. 288–302. Springer (2008)
15. Lehmann, J., Hitzler, P.: Concept learning in description logics using refinement operators. Mach. Learn. **78**, 203–250 (2010)
16. Mitchell, T. (ed.): Machine Learning. McGraw Hill (1997)
17. Shawe-Taylor, J., Cristianini, N. (eds.): Support Vector Machines & Other Kernel-Based Learning Methods. Cambridge University Press (2000)
18. Kemp, C. et al.: Semi-supervised learning with trees. In: Advances in Neural Information Processing System, vol. 16, pp. 257–264. MIT Press (2004)
19. Griffiths, T.L., Chater, N., Kemp, C., Perfors, A., Tenenbaum, J.B.: Probabilistic models of cognition: exploring representations & inductive biases. Trends Cogn. Sci. **14**(8), 357–364 (2010)
20. McGrayne, S.B.: The Theory That Would Not Die, p. 10. Google Books (2011)
21. Fienberg, S.E.: When did Bayesian Inference become "Bayesian"? Bayesian Anal. **1**(1), 1–40 (2006)
22. Tenenbaum, J.B., Griffiths, T.L., Kemp, C.: Theory-based Bayesian models of inductive learning and reasoning. Trends Cogn. Sci. **10**(7) (2006)
23. Thom, L.H., Reichert, M., Iochpe, C.: Activity patterns in process-aware information systems: basic concepts and empirical evidence. J. Bus. Process Integr. Mangt. **4**(2), 93–110 (2009)

Cooperation Evolution in Structured Populations by Using Discrete PSO Algorithm

Xiaoyang Wang, Lei Zhang, Xiaorong Du and Yunlin Sun

Abstract Social dilemma is a challenge to many scientists. The Prisoner's Dilemma and Snowdrift game were the most used social dilemma models in the cooperation evolution. A particularly effect to the evolutionary process comes from population structure. By comparing population structures that amplify selection with other population structures, both analytically and numerically, we show that evolution also affected by the cost to benefit ratio and neighbor number.

Keywords Evolution · Iterated prisoner's dilemma (IPD) · Iterated snowdrift game (ISD) · Discrete particle swarm optimization (PSO) · N-player

1 Introduction

The cooperation of people in social dilemmas is a hot and challenging topic among natural and social sciences [1–4]. Cooperation is a conundrum in a competitive world. The most popular theoretical frameworks that have been widely used in the past to shed some light on these issues are the Prisoner's Dilemma (PD) and Snowdrift (SD) model. In the conventional two-player games, when both players cooperate, both of them will obtain the maximum joint benefit; however, defector (free-riding) [5–9] when meeting a cooperator leads to the highest individual payoff, thus leading to a dilemma as mutual defection is worse than mutual cooperation. Cooperation can rarely be evolved by natural selection unless specific mechanisms are at work. Five such mechanisms have been proposed: direct reciprocity, indirect reciprocity, spatial selection, multilevel selection, and kin selection.

X. Wang (✉)
University of Electronic Science and Technology of China, Zhongshan Insitute,
Guangdong, China
e-mail: wxy_lele@163.com

L. Zhang · X. Du · Y. Sun
School of Physics and Engineering, Sun Yat-Sen University, Guangdong, China

© Springer International Publishing Switzerland 2016
N. Pillay et al. (eds.), *Advances in Nature and Biologically Inspired Computing*,
Advances in Intelligent Systems and Computing 419,
DOI 10.1007/978-3-319-27400-3_6

61

Spatial reciprocity has perhaps been the most widely researched in recent years [10, 11]. Several models have simulated the effects of a network structure on the promotion of cooperation, mostly in the framework of the PD, but the results of these models largely depend on details such as the type of spatial structure or the evolutionary dynamics (see, e.g., Refs. [12, 13] for reviews). On the experimental side, to our knowledge, there are only a few experiments of this kind based on different game models and different evolutionary algorithms. To contribute to this goal, in this paper we present a comparative analysis of PD and SD models in spatial structures by using discrete PSO algorithm in order to evolve the cooperation. By extracting the properties underlying all of them such as the number of neighborhood, the cost-to-benefit ratio, the spatial of population which can be considered as the details of the setups and parameters in order to give as much generality and support as possible to our conclusions.

2 Background

2.1 Conventional Social Dilemma Games

In conventional social dilemma games, each player has two choices: cooperation and defection. A player would receive payoff as the payoff matrix set when his opponent makes his choice in games [2]. A reward (R) is given when both players choose to cooperate, whereas punishment (P) will be given if both of them choose to defect. In the situation where one player defects and the other player cooperates, the one who defects is awarded a tempting reward (T) but the one who cooperates will be given the sucker's punishment (S). Accordingly, a PD game will always exist given the rules $T > R > P > S$ and $2R > T + S$. The PD game reflects a social dilemma in the strictest sense; however the SD game relaxes some of these constraints by (1) allowing players to obtain some immediate benefits from their cooperative acts and (2) sharing the cost of cooperation between cooperators. Therefore, the SD game has $T > R > S > P$.

Table 1 illustrates the PD game in terms of costs and benefits to the players. A cooperative act results in a benefit b to the opposing player and a cost c to the cooperator, where $b > c > 0$. Under this situation, if the opponent cooperates, a player gets the reward $R = (b - c)$ if he/she also cooperates, but can get $T = b$ by defecting. If the opponent defects, a player gets the lowest payoff $S = -c$ for being cooperative and $P = 0$ for being defect. As the definition of PD game, the defection is the better choice regardless of what the opponent plays. To normalize the range of cost and benefit, we define $r_{PD} = c/b$ (b, c are illustrated in 2.1) as cost-to-benefit ratio.

Table 2 illustrates the SD game in terms of costs and benefits to the players. A cooperative act results in a benefit b to the opposing player and a cost c to the cooperator, where $b > c > 0$. Under this situation, if the opponent cooperates, a

Table 1 Payoff matrix of prisoner's dilemma game

A	B	
	Cooperate	Defect
Cooperate	$b - c$	$-c$
Defect	b	0

Table 2 Payoff matrix of snowdrift dilemma game

A	B	
	Cooperate	Defect
Cooperate	$b - c/2$	$b - c$
Defect	b	0

player gets the reward $R = (b - c/2)$ if he/she also cooperates, but can get $T = b$ by defecting. If the opponent defects, a player gets the payoff $S = b - c$ for being cooperative and $P = 0$ for being defect. As the definition of SD game, the cooperation is a mixed evolutionarily stable equilibrium behaviour. To normalize the range of cost and benefit, we define $r_{SD} = c/(2b - c)$ (b, c are illustrated in 2.1) as cost-to-benefit ratio.

2.2 Multiplayer Social Dilemma Games

The n-IPD game was first proposed by Boyd and Richerson [14], the N-IPD consists of N players ($N > 2$) making decision independently based on the other players' past actions. The more players are cooperating, the more social benefit b they are receiving. And for cooperators, they always should pay cost c, and $b > c$ as in the two-player game. For example, if a player who cooperates is pitted against k other players and i of those are cooperators, then the payoff is $bi - ck$. On the other hand, a player who defects does not bear any cost but receives benefit from players who cooperate. Hence, the payoff is bi. The utility values $\prod n$, where $n \in [1, \ldots, N]$, for this scenario can be formally defined as Eq. (1).

$$\prod n = \begin{cases} b \times i - c \times (N - 1), & for\ cooperators \\ b \times i, & for\ defectors \end{cases} \quad (1)$$

In the N-player SD game, the payoff of a cooperator is dependent examined by Zheng et al. [15]. If there is only one cooperator in the group, the payoff is $b - c$. If two cooperators exist, then the payoff is $b - c/2$. With three cooperators, the payoff becomes $b - c/3$, and so on. On the other hand, if exists at least one cooperator, the free riding defector(s) will receive a payoff b without doing anything. However, if there is not any cooperator, the payoff is 0. Accordingly, the utility can be summarized as Eq. (2).

$$\prod n = \begin{cases} b - \frac{c}{i}, & \textit{for cooperators} \\ b, & \textit{for defectors when } i > 0 \\ 0, & \textit{for defectors when } i = 0. \end{cases} \tag{2}$$

2.3 Co-evolutionary Search as a Particle Swarm Process

The PSO technique was introduced by Kennedy and Eberhart [16]. Inspired by the flocking behaviour of birds, PSO has been applied successfully to function the optimization, game learning, data clustering, and image analysis and neural networks training [17, 18].

In the PSO algorithm, each particle adjusts its position in a direction toward its own personal best position in a direction toward its own personal best position and the global best position. The velocity of the particle is calculated using Eq. (3)

$$V_{id}^{k+1} = \omega V_{id}^k + c_1 r \ \text{ and } \ (0,1)\left(P_{id}^k - X_{id}^k\right) + c_2 r \ \text{ and } \ (0,1)\left(P_{gd}^k - X_{id}^k\right) \tag{3}$$

For particle i, its position x_d is changed according to:

$$X_{id}^{k+1} = \begin{cases} 1, & \textit{if } sig\left(V_{id}^{(k+1)}(j)\right) > r_{ij} \\ 0, & \textit{otherwise} \end{cases} \tag{4}$$

where,

$$sig\left(V_i^{(t+1)}(j)\right) = \frac{1}{1 + \exp\left(-V_i^{(t+1)}(j)\right)} \tag{5}$$

For particle i, the position vector can be represented by $X_i = (X_{i1}, X_{i2}, \ldots, X_{id})$, V_{id} presents the velocity of the ith particle on the specific d-dimension. ω is the inertia weight, c_1 and c_2 are acceleration coefficients. r_{ij} in Eq. (4) is a random number in range [0, 1]. P_{id}^k is the best position of particle i on dimension d at iteration k, and P_{gd}^k is the dth dimensional best position in the whole particle swarm at iteration k.

Combining with the PD and SD framework, each particle in the swarm is associated a behaviour (or "strategy"), and the purpose of the behaviours is to maximize the payoff of individual or the group. In this paper, P_{id}^k is the strategy which can make particle i have the maximum utility on dimension d during k iterations, P_{gd}^k is the strategy chosen by one particle on dimension d which have the maximum utility in the neighbourhood structure at iterated k. The value of P_{id}^k and P_{gd}^k can choose 0 or 1, which respectively means defection or cooperation.

The strategy is used to compute its position in the next step. An interpretation of multiple levels of choice between cooperation and defection is employed to the PSO approach.

2.4 Spatial Population for Evolution

In spatial evolutionary environment, the players (or agents) of a population are distributed on a regular grid and interact with other players in its neighbourhood. In our model, the players are mapped to network nodes (vertices) and the edges (or links) which dictate the interaction topology. Each agent participates in an inter-active game with $l-1$ other agent drawn from its local neighborhood. Figure 1 depicts the four examples of neighborhood structure for interaction, and single agent is located in the node (vertex) of the grid world. In Fig. 1a, b, the two examples show that agent can only interact with the agents in dotted box. The regular-network (Fig. 1d) is two-dimensional and agents are connected by the edges. As for the small-world network, we use a version similar to the one intro-duced by Watts and Strogatz [18]. From the two-dimensional regular-network substrate we rewire each link with probability ρ (Fig. 1c). In this paper, we allow neither self or repeated links nor disconnected graphs to ensure that the individuals in the population are highly clustered and have relatively short path length. When $\rho = 0$, it's essentially a regular-network. As $\rho = 1$, it is defined as full-connected network.

In general, the neighborhood best position is calculated as

$$P_{gd}^{k+1} \in \left\{ NE_i | f\left(P_{gd}^k\right) = \max\{f(x), \forall x \in NE_i\} \right\} \quad (6)$$

with the neighbourhood defined as

$$NE_i = \left\{ P_{gd}^k\left(i - \frac{l}{2}\right), \ldots, P_{gd}^k(i-1), P_{gd}^k(i), P_{gd}^k(i+1), \ldots, P_{gd}^k\left(i + \frac{l}{2}\right) \right\} \quad (7)$$

for neighbourhood size of l.

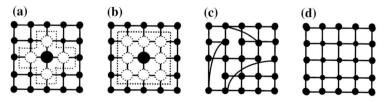

(a)   (b)   (c)   (d)

Fig. 1 Examples of neighborhood structures. **a** Von-Neumann. **b** Moore. **c** Small-World. **d** Regular-Network

3 Simulation Experiments and Results

Extensive computational simulations have been carried out to investigate the population dynamics of the games played. The experiments in the first part are conducted to analyze the general performance of strategy co-evolution. The second part of the experiments is conducted to compare the frequency of cooperation by varying the cost-to-benefit ratio and the detail analysis in all the spatial structures.

3.1 Evolution of Cooperation in Spatial Structure Environment

The following parameters are used in the game: the cost-to-benefit ratio for 2-player PD and SD game is separately fixed as $r = 0.2$ in this section. In order to keep the uniform distribution of choices during the initialization process, each choice is set to players with the same probability.

We can find that the trends of average cooperation ratios in two game models are similar from Fig. 2. The average cooperation ratio in IPD game is higher than in ISD game, perhaps because the defectors would still get enough payoffs. The spatial structure can evolve the cooperation in IPD game and persist the decreasing of cooperation in ISD game.

From Figs. 3, 4, 5, 6, 7, 8, 9 and 10, the actions of players reflect the previous conclusions. In PD game, more and more players choose cooperate as iterations increase, while in SD game, few player can persist cooperation to the end.

3.2 The Affection of Cost-to-Benefit Ratio for Multi-Player Spatial Social Dilemma Game

In the first previous sets of experiments, we compare the average cooperation ratio in spatial game by using discrete PSO algorithm. In this part of experiment, the

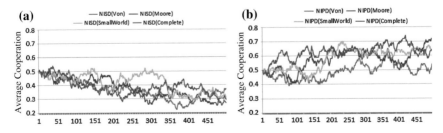

Fig. 2 The average cooperation comparison in spatial structure of **a** SD game and **b** PD game

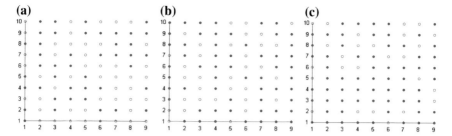

Fig. 3 The choice of each player in PD game on Moore structure. The situation is **a** on iteration 10, **b** on iteration 250, **c** on iteration 495. The *filled dots* represent the cooperators while the *hollow dots* represent the defectors

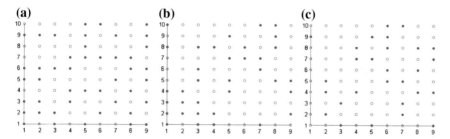

Fig. 4 The choice of each player in SD game on Moore structure. The situation is **a** on iteration 10, **b** on iteration 250, **c** on iteration 495. The *filled dots* represent the cooperators while the *hollow dots* represent the defectors

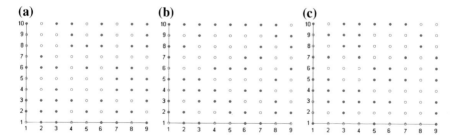

Fig. 5 The choice of each player in PD game on Von Neumann structure. The situation is **a** on iteration 10, **b** on iteration 250, **c** on iteration 495. The *filled dots* represent the cooperators while the *hollow dots* represent the defectors

cost-to-benefit ratios, r is observed to find the influence to the evolution of cooperation in multi-player spatial game. For each cost-to-benefit ratio r, the players will compete for 500 iterations, the total iteration is 5000 for r varies from 0 to 1.

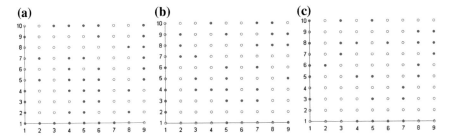

Fig. 6 The choice of each player in SD game on Von Neumann structure. The situation is **a** on iteration 10, **b** on iteration 250, **c** on iteration 495. The *filled dots* represent the cooperators while the *hollow dots* represent the defectors

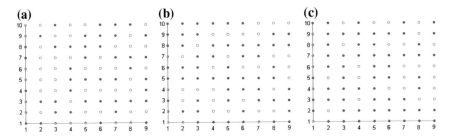

Fig. 7 The choice of each player in PD game on Small World structure. The situation is **a** on iteration 10, **b** on iteration 250, **c** on iteration 495. The *filled dots* represent the cooperators while the *hollow dots* represent the defectors

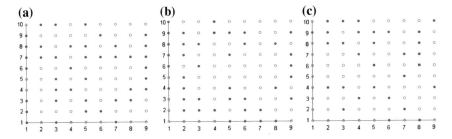

Fig. 8 The choice of each player in SD game on Small World structure. The situation is a on iteration 10, **b** on iteration 250, **c** on iteration 495. The *filled dots* represent the cooperators while the *hollow dots* represent the defectors

From Figs. 11, 12, 13, 14 and 15, we find as r increases, the average cooperation ratio drops regardless of the spatial structures. When the value of r is sufficiently high, defectors would dominate the agent population; on the other hand, when the value of r is small it would be worth to cooperate. For 3-player IPD game, the

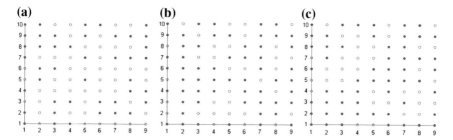

Fig. 9 The choice of each player in PD game on Complete connected structure. The situation is **a** on iteration 10, **b** on iteration 250, **c** on iteration 495. The *filled dots* represent the cooperators while the *hollow dots* represent the defectors

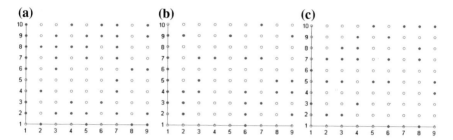

Fig. 10 The choice of each player in SD game on Complete connected structure. The situation is **a** on iteration 10, **b** on iteration 250, **c** on iteration 495. The *filled dots* represent the cooperators while the *hollow dots* represent the defectors

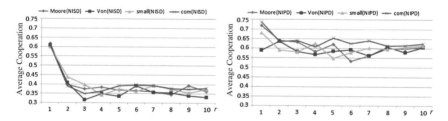

Fig. 11 The average cooperation comparison with r varies in 3-player spatial structure of **a** SD game and **b** PD game

average cooperation ratios are above 0.5; while for 3-player ISD game, the average cooperation ratios are below 0.5. In the 3-IPD game, players can keep higher cooperation ratio within smaller group sizes persisted for the change of r.

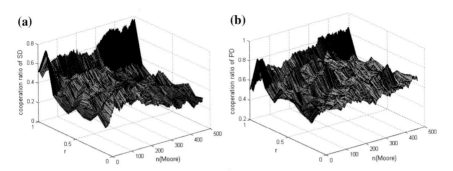

Fig. 12 The detail cooperation ratio in 3-player game on Moore structure. The real iteration in experiments is 5000, because of the limitation of the figure, we compute the average ratio every 10 iterations

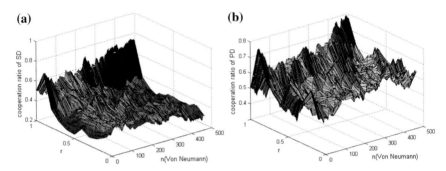

Fig. 13 The detail cooperation ratio in 3-player game on Von Neumann structure. The real iteration in experiments is 5000, because of the limitation of the figure, we compute the average ratio every 10 iterations

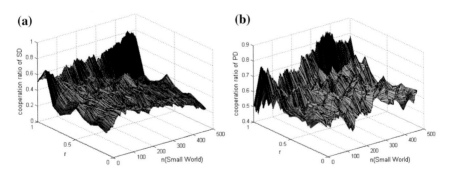

Fig. 14 The detail cooperation ratio in 3-player game on Small World structure. The real iteration in experiments is 5000, because of the limitation of the figure, we compute the average ratio every 10 iterations

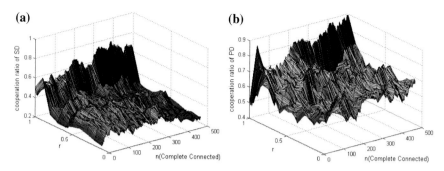

Fig. 15 The detail cooperation ratio in 3-player game on Complete Connected structure. The real iteration in experiments is 5000, because of the limitation of the figure, we compute the average ratio every 10 iterations

4 Conclusion and Discussion

This paper applied discrete PSO-based co-evolutionary training techniques to the nonzero-sum game of the PD and SD game in spatial structure environments. Two main issues were presented. Firstly, discrete PSO algorithm is used to co-evolve the cooperation as evaluators, chose cooperate or defect according to the fitness from the previous iteration in the IPD and ISD game. Secondly, with the increasing cost-to-benefit ratio r and group sizes n, the cooperative strategies become difference.

Apart from using discrete PSO algorithm to our understanding the effects of iterated interactions in the multiplayer version of the well-known IPD game, another major contribution of this paper was the new results of the iterative version of the multiplayer ISD game.

Future work will include an investigation of the effects of using different payoff matrices to influence cooperative behaviour, competing against more neighbours of the spatial environment. As the theoretical study of the PSO techniques is extremely necessary and important, more detailed studies of the combination of evolved strategies and better PSO algorithms in the co-evolutionary training technique should be carried out.

Acknowledgment This work is supported by two research program of China. One is the Zhongshan Science and Technology Development Funds under Grant no. 2014A2FC385, the Dr Startup project under Grant no. 414YKQ04, and also supported by Production, learning and research of Zhuhai under Grant no. 2013D0501990003.

References

1. Chong, S.Y., Yao, X.: Behavioral diversity, choices and noise in the iterated prisoner's dilemma. IEEE Trans. Evol. Comput. **9**(6), 540–551 (2005)
2. Chong, S.Y., Yao, X.: Multiple choices and reputation in multiagent transactions. IEEE Trans. Evol. Comput. **11**(6), 689–711 (2007)

3. Chong, S.Y., Tiño, P., Yao, X.: Measuring generalization performance in coevolutionary learning. IEEE Trans. Evol. Comput. **12**(4), 479–505 (2008)
4. Chong, S.Y., Tiño, P., Yao, X.: Relationship between generalization and diversity in coevolutionary learning. IEEE Trans. Comput. Intell. AI Games **1**(3), 214–232 (2009)
5. Wang, X.Y., Chang, H.Y., Yi, Y., Lin, Y.B.: Co-evolutionary learning in the n-choice iterated prisoner's dilemma with PSO algorithm in a spatial environment. In: 2013 IEEE Symposium Series on Computational Intelligence, pp. 47–53. IEEE press, Singapore, 2013
6. Darwen, P.J., Yao, X.: Co-evolution in iterated prisoner's dilemma with intermediate levels of cooperative: Application to missile defense. Int. J. Comput. Intell. Appl. **2**(1), 83–107 (2002)
7. Ishibuchi, H., Namikawa, N.: Evolution of iterated prisoner's dilemma game strategies in structured demes under random pairing in game playing. IEEE Trans. Evol. Comput. **9**(6), 552–561 (2005)
8. Zheng, Y., Ma, L., Qian, I.: On the convergence analysis and parameter selection in particle swarm optimization. In: Processing of International Conference of Machine Learning Cybern., pp. 1802–1807 (2003)
9. Franken, N., Engelbrecht, A.P.: Comparing PSO structures to learn the game of checkers from zero knowledge. In: The 2003 Congress on Evolutionary Computation, pp. 234–241 (2003)
10. Ishibuchi, H., Takahashi, K., Hoshino, K., Maeda, J., Nojima, Y.: Effects of configuration of agents with different strategy representations on the evolution of cooperative behaviour in a spatial IPD game. In: IEEE Conference on Computational Intelligence and Games, 2011
11. Axelrod, R.: The Evolution of Cooperation. Basic Books, New York (1984)
12. Moriyama, K.: Utility based Q-learning to facilitate cooperation in Prisoner's Dilemma games, Web Intelligence and Agent Systems: An International Journal, vol. 7, pp. 233–242. IOS Press, 2009
13. Chen, B., Zhang, B., Zhu, W.D.: Combined trust model based on evidence theory in iterated prisoner's dilemma game. Int. J. Syst. Sci. **42**(1), 63–80 (2011)
14. Chong, S.Y., Tiño, P., Ku, D.C., Yao, X.: Improving generalization performance in co-evolutionary learning. IEEE Trans. Evol. Comput. **16**(1), 70–85 (2012)
15. Zheng, D.F., Yin, H.P., Chan, C.H., Hui, P.M.: Cooperative behavior in a model of evolutionary snowdrift games with N-person interactions. Europhys. Lett. **80**(1), 18002 (2007)
16. Chiong, R., Kirley, M.: Iterated N-player games on small-world networks. In: GECCO'11, Jul. 2011
17. Nowark, M.A.: Five rules of the evolution of cooperation. Science **314**, 1560–1563 (2006)
18. Watts, D., Stogatz, S.H.: Collective dynamics of small-world networks. Natrue **393**, 440–442 (1998)

Memetic and Opposition-Based Learning Genetic Algorithms for Sorting Unsigned Genomes by Translocations

Lucas A. da Silveira, José L. Soncco-Álvarez, Thaynara A. de Lima
and Mauricio Ayala-Rincón

Abstract A standard genetic algorithm ($\mathcal{GA}_S$) for sorting unsigned genomes by translocations is improved in two different manners: firstly, a memetic algorithm ($\mathcal{GA}_M$) is provided, which embeds a new stage of local search, based on the concept of mutation applied in only one gene; secondly, an opposition-based learning ($\mathcal{GA}_{OBL}$) mechanism is provided that explores the concept of internal opposition applied to a chromosome. Both approaches include a convergence control mechanism of the population using the Shannon entropy. For the experiments, both biological and synthetic genomes were used. The results showed that $\mathcal{GA}_M$ outperforms both $\mathcal{GA}_S$ and $\mathcal{GA}_{OBL}$ as confirmed through statistical tests.

Keywords Sorting permutations · Sorting unsigned genomes · Genetic algorithms · Memetic algorithms · Opposition-based learning algorithms

1 Introduction

In order to estimate the evolutionary relationships between species, molecular biologists compare gene and genome sequences of different species to reveal the degree of similarity between them. While algorithms for gene comparison are based on local mutations such as insertions and deletions of biological data, algorithms for genome comparison are based on global mutations such as reversals, transpositions and translocations. In this paper we adopt the model of genetic evolution in genomes by translocations. The translocation operation involves the interchange of

L.A. da Silveira · J.L. Soncco-Álvarez (✉) · M. Ayala-Rincón
Departaments of Computer Science, Universidade de Brasília, Brasília, Brazil
e-mail: jose.soncco.alvarez@gmail.com

M. Ayala-Rincón
Departaments of Computer Science and Mathematics, Universidade de Brasília,
Brasília, Brazil

T.A. de Lima
Department of Mathematics, Universidade de Goiás—Campus II, Goiás, Brazil

© Springer International Publishing Switzerland 2016
N. Pillay et al. (eds.), *Advances in Nature and Biologically Inspired Computing*,
Advances in Intelligent Systems and Computing 419,
DOI 10.1007/978-3-319-27400-3_7

73

blocks of genes between the chromosomes of a genome. So the genome comparison
can be modelled as an optimisation problem which consists in finding the minimum
number of translocations necessary to transform one genome into another one. This
problem is known as the translocation distance problem, which has two versions:
one using *signed* genomes and the other using *unsigned* genomes, where the difference remains resp. in considering or not the orientation of the genes between the
chromosomes.

The first polynomial algorithm ($O(n^3)$) for the signed translocation distance problem (STD) was proposed by Hannenhalli [9]. Almost a decade later, Wang et al. [17]
proposed an $O(n^2)$ algorithm taking advantage of the techniques proposed in [9].
More recently, Bergeron et al. [2] proposed an $O(n)$ approach.

While STD belongs to the complexity class $\mathcal{P}$, the unsigned translocation distance
problem (UTD) is a problem of high complexity. Indeed, Zhu and Wang proved that
this is an $\mathcal{NP}$-hard problem in [18]. Thus, developments of approximate solutions
for this problem are necessary. In [4] Cui et al. presented a 1.75-approximation algorithm for solving UTD, and further improved the ratio to $1.5 + \varepsilon$ in [5]. The best
known approximation algorithm in the literature has ratio $1.408 + \varepsilon$ and was proposed in [10]. More recently, we proposed a genetic algorithm $\mathcal{GA}_S$, introduced in
[6], in which the fitness function is given as the translocation distance for signed
versions of the unsigned genomes, that is linearly computed as in [2]. To verify the
quality of the solutions computed by $\mathcal{GA}_S$, an implementation of the $1.5 + \varepsilon$ approximation algorithm was used as control mechanism. Results of the experiments in [6]
showed that on average $\mathcal{GA}_S$ computes better results than the $1.5 + \varepsilon$-approximation
algorithm.

Two hybrid evolutionary algorithms for UTD are proposed: $\mathcal{GA}_M$ and $\mathcal{GA}_{OBL}$. The
former is a memetic algorithm and the latter an opposition-based learning (OBL)
genetic algorithm. They embed the local search and OBL in the following stages
of $\mathcal{GA}_S$: improvement of the initial population, restart of the population, and after
the breeding cycle. The quality of the results obtained by the proposed algorithms
were compared with the results computed with the algorithms proposed in [6] and
[5] for groups of hundred unsigned genomes of different length, single unsigned
genomes, and genomes based on biological data (built from [3]). Results from these
tests showed that $\mathcal{GA}_M$ has the best performance, and that there is no significant dif-
ference in the performance of $\mathcal{GA}_{OBL}$ and $\mathcal{GA}_S$. The C source code of $\mathcal{GA}_S$, $\mathcal{GA}_M$ and
$\mathcal{GA}_{OBL}$ as well as an extended report are available at http://www.mat.unb.br/~ayala/
publications.html.

2 Background

Definitions and Terminology

In a simplified model, a *signed* chromosome can be represented by a non empty
sequence of integer numbers of the form $X = (x_1, \ldots, x_l)$, where $x_i \in \{\pm 1, \ldots \pm n\}$
for all $i \in \{1, \ldots, l\}$, each integer denotes a gene and $|x_i| \neq |x_j|$, whenever $i \neq j$,

$i, j \in \{1, \ldots, l\}$. With the restriction that for all $i \in \{1, \ldots, l\}$, $x_i \in \{1, \ldots, n\}$, the chromosome is called *unsigned*. Chromosomes do not have orientation, thus if $X = (x_1, \ldots, x_l)$ is a signed chromosome then X and $X' = (-x_l, \ldots, -x_1)$ are equal, whereas if X is an unsigned chromosome, X and $X'' = (x_l, \ldots, x_1)$ are equal. A genome G with N chromosomes and n genes is a list of chromosomes of the form $(x_{11}, \ldots, x_{1r_1}) \ldots (x_{N1}, \ldots, x_{Nr_N})$, where for all its genes $|x_{ij}| \neq |x_{km}|$ whenever $i \neq k$ or $j \neq m$, and $\sum_{i=1}^{N} r_i = n$. Unsigned and signed genomes consist resp. of unsigned and signed chromosomes.

Let $X = (x_1, \ldots, x_i, \ldots, x_l)$ and $Y = (y_1, \ldots, y_j, \ldots, y_m)$ be two chromosomes of a signed genome. On the one hand, a *translocation by prefix-prefix*, denoted as $\rho(X, Y, x_i, y_j)$, transforms X and Y into the chromosomes $X^\rho = (x_1, \ldots, x_i, y_{j+1}, \ldots, y_m)$ and $Y^\rho = (y_1, \ldots, y_j, x_{i+1}, \ldots, x_l)$. On the other hand, a *translocation by prefix-suffix*, $\theta(X, Y, x_i, y_j)$, produces the new chromosomes $X^\theta = (x_1, \ldots, x_i, -y_j, \ldots, -y_1)$ and $Y^\theta = (-y_m, \ldots, -y_{j+1}, x_{i+1}, \ldots, x_l)$. Notice that a translocation by prefix-suffix $\theta(X, Y, x_i, y_j)$ can be mimicked by a translocation $\rho(X, Y', x_i, -y_{j+1})$ by prefix-prefix and vice-versa, where $Y' = (-y_m, \ldots, -y_1)$. For $X = (x_1, \ldots, x_i, \ldots, x_l)$ and $Y = (y_1, \ldots y_j, \ldots, y_m)$ of an unsigned genome, a *translocation by prefix-prefix* $\rho(X, Y, x_i, y_j)$ transforms X and Y into $X^\rho = (x_1, \ldots, x_i, y_{j+1}, \ldots, y_m)$ and $Y^\rho = (y_1, \ldots, y_j, x_{i+1}, \ldots, x_l)$ and, a *translocation by prefix-suffix* $\theta(X, Y, x_i, y_j)$ creates the chromosomes $X^\theta = (x_1, \ldots, x_i, y_j, \ldots, y_1)$ and $Y^\theta = (y_m, \ldots, y_{j+1}, x_{i+1}, \ldots, x_l)$.

The translocation distance problems are defined as follows.

Signed translocation distance problem (STD): consider two signed genomes A and B, with the same number of genes and chromosomes, where the genes of B are positive integers in increasing order. The STD consists in finding the minimum number of translocations needed to transform A into B.

Unsigned translocation distance problem (UTD): the UTD is defined as the STD but restricted to unsigned genomes.

Genomes as B above, are called *identity* genomes. Given a signed chromosome $X = (x_1, \ldots, x_l)$, the elements x_1 and $-x_l$ are called *tails* of X. Two signed genomes A and B are said to be *co-tails* if the sets T_A and T_B composed by the tails of the chromosomes of A and B, resp., are equal. For example, $A = (+1, +6, +7)(-4, -3, -2, -5)$ and $B = (+1, +2, +3, +4)(+5, +6, +7)$ are co-tails. When $X = (x_1, \ldots, x_l)$ is an unsigned chromosome, x_1 and x_l are its *tails* and *co-tails* are correspondingly defined.

Notice that translocations by prefix-prefix or prefix-suffix do not modify the tails of a genome. Thus, in order to be able to transform a genome A into B by translocations, A and B must satisfy the property below.

Property 1 *The genomes A and B have the same number of genes (and chromosomes) and A and B are co-tails.*

Notice that when A and B are co-tails by renaming the genes, B can be rewritten as an identity. Thus, without loss of generality, whenever A and B are co-tails one can assume that B is an identity.

STD belongs to the complexity class $\mathcal{P}$ [9], while UTD is $\mathcal{NP}$-hard [18]. The construction of polynomial algorithms for solving STD as well as the proof that UTD

(a)   **(b)**

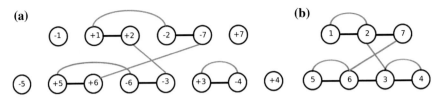

Fig. 1 Breakpoint graphs for **a** $A = (+1, -2, +7)(+5, -6, +3, +4)$ and $B = (+1, +2, +3, +4)$ $(+5, +6, +7)$; and **b** $A = (1, 2, 7)(5, 6, 3, 4)$ and $B = (1, 2, 3, 4)(5, 6, 7)$

is $\mathcal{N}P$-hard explore the relation between these problems and the problem of maximum cycle decomposition of the *breakpoint graph* of a genome. This data structure is defined below and illustrated in Fig. 1.

Given a (signed or unsigned) chromosome $X = (x_1, \ldots, x_l)$ of a genome, the elements x_i and x_{i+1}, $1 \leq i \leq l - 1$ are said to be *adjacent*; otherwise, the elements are *not adjacent*. Elements in different chromosomes are not adjacent.

Consider two signed genomes A and B satisfying the Property 1. The breakpoint graph $G_s(A, B)$ is built as follows: for each chromosome $X = (x_1, \ldots, x_l)$ of A, we associate to each x_i an ordered pair of vertices $(l(x_i), r(x_i)) = (-x_i, +x_i)$; there is a black edge between $r(x_i)$ and $l(x_{i+1})$, i.e., between $+x_i$ and $-x_{i+1}$, for every $1 \leq i \leq l - 1$; if $+i$ and $+(i + 1)$ are adjacent in B, there is a gray edge between $+i$ and $-(i + 1)$.

Now, consider two unsigned genomes A and B satisfying the Property 1. The breakpoint graph $G_u(A, B)$ is built as follows: the vertices are given by the genes of A; furthermore, for each chromosome $X = (x_1, \ldots, x_l)$ of A, there is a black edge between x_i and x_{i+1}, $1 \leq i \leq l - 1$, and there is a gray edge between i and $i + 1$ (since i and $i + 1$ are adjacent in B).

The decomposition of the breakpoint graph of signed genomes into cycles with edges of alternating colors (alternating cycles) is unique, because each vertex has at most one black and one gray incident edge. However, the same is not true for unsigned genomes: roughly, the latter observation is the key to understand why STD is a polynomial problem, whereas UTD is $\mathcal{N}P$-hard; indeed, the translocation distance between two genomes is closely related with a maximum decomposition into alternating cycles of their breakpoint graph (see [18] and [9]).

A Standard Genetic Algorithm $\mathcal{GA}_S$ for UTD

A standard genetic algorithm for UTD called $\mathcal{GA}_S$ was proposed in [6] which works as follows for an input genome A with n genes. Initially, a random population of $n \log n$ individuals, that are signed genomes with n genes, is generated based on the unsigned input genome, where each individual is obtained by randomly assigning either a positive or negative sign to each gene of A. Solutions for one signed genome are also consistent solutions for the input, so $\mathcal{GA}_S$ searches for the minimum solution of this population. In each generation: the fitness function calculates the translocation distance for each individual of the generated offspring. This is done using the linear algorithm proposed in [2] for solving the STD. Then, the population is classified according to the fitness value of each individual, maintaining the best individuals at the top. The individuals selected for reproduction are part of the current best solutions for which crossover and mutation are applied producing new individuals (offspring). Finally, these new individuals are incorporated into the current population. $\mathcal{GA}_S$ finishes after n generations have been completed. See pseudo-code of $\mathcal{GA}_S$ in Algorithm 1. In [6] it was proved that $\mathcal{GA}_S$ computes solutions that have better quality than those computed by the approximate algorithm in [5].

3 Memetic and OBL Algorithms for UTD

$\mathcal{GA}_M$: Memetic algorithms (MA) combine population-based search as in the genetic algorithms with one or more phases of local search, and even heuristics or approximate methods [11, 13]. MAs maintain a population of individuals which perform independent explorations (local optimisation), cooperating by means of reproduction operators, and continuously competing by means of selection and substitution operators [12].

The proposed $\mathcal{GA}_M$ for UTD is based on $\mathcal{GA}_S$, maintaining, for inputs of length n, a population of $n \log n$ signed genomes, that are signed versions of the input and the fitness is the same as the one used in the standard $\mathcal{GA}_S$.

Algorithm 1: Genetic algorithm for computing UTD—$\mathcal{GA}_S$

Input: Unsigned genomes A and B, where B is an identity satisfying Prop. 1
Output: Number of translocations to sort genome A
1 Generate the initial population of signed genomes;
2 Compute fitness of the initial population;
3 **for** $i = 1$ *to numberGenerations* **do**
4   Perform the selection and save the best solution found;
5   Apply the crossover operator;
6   Apply the mutation operator;
7   Compute the fitness of the offspring;
8   Perform replacement of the worst individuals;

$\mathcal{GA}_M$ includes phases of local search into the stage of generation of the initial population and the stage of restarting population, and also, it applies local search at the end of the breeding cycle. The stage of restarting the population is executed whenever the population converges to a degenerate state, which is measured using the Shannon entropy [15] and is defined as follows: $H(S) = - \sum_i p_i \log_2 p_i$, where S is the set of different elements of the current population, and p_i is the probability of occurrence of the element i in the current population. When the entropy has a high value it is said that the population has a good diversity, but when the entropy tends to lower values it is said that the population is converging to a degenerate state. The pseudo-code for $\mathcal{GA}_M$ for UTD is shown in Algorithm 2 excluding lines 3, 10 and 12.

The local search consists in a very simple step of modifying the sign of an element of one signed genome at a random position, and verify whether this change improves

---

Algorithm 2: Memetic and OBL Algorithms for UTD - $\mathcal{GA}_M$ and $\mathcal{GA}_{OBL}$

Input: Unsigned genomes A and B, where B is an identity satisfying Prop. 1
Output: Number of translocations to sort genome A
1 Generate the initial population of signed genomes;
2 Compute fitness of the initial population;
3 Improve initial population by applying **Local Search**; /*   For $\mathcal{GA}_M$ */
3 Improve initial population by applying **OBL Heuristic**; /* For $\mathcal{GA}_{OBL}$ */
4 **for** $i = 1$ *to numberGenerations* **do**
5 Perform the selection and save the best solution found;
6 Apply the crossover operator;
7 Apply the mutation operator;
8 Compute the fitness of the offspring;
9 Perform replacement of the worst individuals;
10 Apply **Local Search** to the current population; /* For $\mathcal{GA}_M$ */
10 Apply **OBL Heuristic** to the current population; /* For $\mathcal{GA}_{OBL}$ */
11 **if** *entropyThreshold is reached* **then**
12   Restart population improved by **Local Search**; /* For $\mathcal{GA}_M$ */
12   Restart population improved by **OBL Heuristic**; /* For $\mathcal{GA}_{OBL}$ */

---

the fitness. In the case that the fitness is improved it is updated, otherwise the last state of the signed genome is recovered. For each genome, the number of possible modifications was restricted to two. The pseudo-code of the local search for one signed genome is shown in Algorithm 3.

Algorithm 3: Local Search

Input: A signed genome A
Output: An improved signed genome A
1 *bestFitness* = Compute fitness of A;
2 **for** i to *numberIterations* **do**
3 generate a random position k for A;
4 swap the sign of the element at position k;
5 *fitness* = calculate fitness of A;
6 **if** *fitness* < *bestFitness* **then**
7   update new fitness for A;
8   break;
9 **else**
10   recover last state of A;

Algorithm 4: OBL Heuristic

Input: A signed genome A
Output: An improved signed genome A
1 *bestFitness* = Compute fitness of A;
2 $\tilde{A}$ = generate a type-I opposite for genome A;
3 *fitness* = calculate fitness of $\tilde{A}$;
4 **if** *fitness* < *bestFitness* **then**
5 keep the genome $\tilde{A}$;
6 discard genome A;
7 **else**
8 discard genome $\tilde{A}$;

$\mathcal{GA}_{OBL}$: OBL is a searching technique proposed by Tizhoosh [16]. The main idea about OBL is that when one is searching for a solution in one direction, it would be a good idea to search in an opposite way. This increases the chance of improving the solution specially when one is in a worst case scenario.

The proposed opposition-based genetic algorithm ($\mathcal{GA}_{OBL}$) for UTD is also based on $\mathcal{GA}_S$, and applies OBL exactly in the stages where $\mathcal{GA}_M$ applies local search: generation of the initial population, restarting of the population and after the breeding cycle. Also, convergence of the population is controlled using the Shannon Entropy [15]. The fitness function used in $\mathcal{GA}_{OBL}$ is the same as the used by $\mathcal{GA}_S$ and $\mathcal{GA}_M$. The pseudo-code of $\mathcal{GA}_{OBL}$ is shown also in the Algorithm 2 excluding lines 3, 10 and 12.

The OBL heuristic consists in applying the type-I opposition for one signed genome. This new opposite genome is kept in the population whenever its fitness value is better than the original. Otherwise, it is discarded. The pseudo-code of the OBL heuristic is shown in Algorithm 4.

The number of generations for $\mathcal{GA}_M$ and $\mathcal{GA}_{OBL}$ as for $\mathcal{GA}_S$ is fixed as n. The stages completed by $\mathcal{GA}_M$ and $\mathcal{GA}_{OBL}$ in the lines 1, 2, 6, 7, 8 and 9 have time complexity $O(n^2 \log n)$ and in line 5 $O(n \log n)$ as for $\mathcal{GA}_S$ (see [6]). Also, restarting the population (lines 12 and 12) has complexity $O(n^2 \log n)$. Applications of Algorithms 3 and 4 (lines 3, 3, 10 and 10) over (resp. 60 and 70 %) of the current population have complexity $O(n^2 \log n)$. For calculating the entropy value (line 11), a hash table was implemented indexing each different fitness value found in the current population obtaining time complexity $O(n^2)$. Thus, $\mathcal{GA}_M$ and $\mathcal{GA}_{OBL}$ have both time complexity in $O(n^3 \log n)$.

4 Experiments

Setting the Parameters
To obtain a configuration of parameters that provides the best results (number of translocations) an exhaustive sensitivity analysis was performed.

The parameters obtained for $\mathcal{GA}_M$ are: crossover with single point and probability 90 %, mutation probability 2 %, percentages of selection, replacement, local search and preservation are resp. 80, 70, 60 and 60 % and the threshold entropy 0.3. The parameters for $\mathcal{GA}_{OBL}$ are the same, except that the percentage of OBL and preservation are resp. 70 and 50 %, and the threshold entropy 0.1.

Experiments with Synthetic Genomes
Experiments with $\mathcal{GA}_{OBL}$, $\mathcal{GA}_M$, $\mathcal{GA}_S$ and the $1.5 + \varepsilon$-approximation algorithm, as implemented in [6], were conducted: initially, hundred unsigned genomes with n genes and N chromosomes, $n \in \{20, 30, \ldots, 150\}$ and $N \in \{2, 3, 4, 5\}$ were randomly generated. All GAs were executed ten times and the $1.5 + \varepsilon$-approximation algorithm once for each genome. Then, the average of the results (translocations) for each genome was calculated for the three GAs. Finally, for each family of hundred genomes of size (n, N) the average was computed (Table 1).

Table 1 Avg. translocations for synthetic genomes with 2, 3, 4 and 5 chromosomes

n	2 chrom.				3 chrom.			
	1.5App	$\mathcal{GA}_S$	$\mathcal{GA}_{OBL}$	$\mathcal{GA}_M$	1.5App	$\mathcal{GA}_S$	$\mathcal{GA}_{OBL}$	$\mathcal{GA}_M$
20	11.240	10.107	10.106	10.090	10.210	9.186	9.185	9.180
30	19.130	16.886	16.861	16.792	17.710	15.602	15.586	15.552
40	26.910	23.538	23.536	23.367	25.530	22.327	22.289	22.200
50	35.540	30.620	30.602	30.358	34.080	29.650	29.627	29.424
60	44.150	37.931	37.908	37.577	42.100	36.422	36.371	36.093
70	52.220	45.250	45.180	44.745	50.040	43.410	43.312	42.944
80	59.580	51.602	51.548	51.065	58.400	50.659	50.592	50.117
90	68.250	59.098	59.072	58.493	66.220	57.331	57.341	56.769
100	76.650	66.601	66.520	65.852	74.740	64.935	64.905	64.303
110	85.860	74.905	74.846	74.165	83.230	72.372	72.337	71.648
120	93.880	81.859	81.749	81.039	91.720	79.921	79.842	79.105
130	102.010	89.192	89.065	88.292	99.240	86.971	86.880	86.102
140	109.930	96.153	96.175	95.324	108.100	94.625	94.571	93.740
150	118.560	103.940	103.801	103.092	116.420	101.792	101.769	100.943

n	4 chrom.				5 chrom.			
	1.5App	$\mathcal{GA}_S$	$\mathcal{GA}_{OBL}$	$\mathcal{GA}_M$	1.5App	$\mathcal{GA}_S$	$\mathcal{GA}_{OBL}$	$\mathcal{GA}_M$
20	9.240	8.461	8.460	8.460	7.830	7.320	7.320	7.320
30	16.330	14.619	14.619	14.612	15.730	14.070	14.064	14.061
40	23.870	20.925	20.896	20.862	22.240	19.792	19.784	19.753
50	31.680	27.554	27.527	27.396	29.710	26.298	26.275	26.197
60	39.370	34.156	34.131	33.913	36.930	32.275	32.249	32.112
70	47.940	41.361	41.320	41.036	44.870	39.321	39.300	39.069
80	55.350	48.024	48.011	47.614	51.600	44.912	44.900	44.619
90	62.240	54.404	54.391	53.930	59.670	52.036	52.036	51.659
100	71.110	62.033	61.956	61.407	66.790	58.399	58.343	57.911
110	78.660	68.544	68.477	67.846	74.940	65.646	65.639	65.084
120	87.330	76.177	76.101	75.420	81.820	71.828	71.747	71.143
130	95.010	83.198	83.195	82.432	90.190	79.062	79.086	78.380
140	101.970	89.313	89.328	88.515	97.280	85.325	85.273	84.515
150	111.210	97.480	97.511	96.651	104.150	91.729	91.646	90.858

Experiments with Single Genomes

Nine benchmark unsigned genomes were proposed. The genomes have the following nomenclature *GenomeLxCy*, where L stands for length, C stands for number of chromosomes. For each genome, $\mathcal{GA}_S$, $\mathcal{GA}_M$ and $\mathcal{GA}_{OBL}$ were executed fifty times, and then the following measures calculated: mean, median, minimum and maximum. See Table 2, where the best values are highlighted.

Table 2 Results for GA_S, GA_M, and GA_{OBL} using benchmark genomes from [6]

	Mean			Median			Minimum			Maximum		
	GA_S	GA_M	GA_{OBL}	GA_S	GA_M	GA_{OBL}	GA_S	GA_M	GA_{OBL}	GA_S	GA_M	GA_{OBL}
GL150C2	102.52	**101.58**	102.48	102.50	**102.00**	**102.00**	101.00	**100.00**	**100.00**	106.00	**104.00**	106.00
GL150C3	109.54	**108.82**	109.40	**109.00**	**109.00**	109.50	108.00	**107.00**	**107.00**	**111.00**	**111.00**	112.00
GL150C4	100.94	**100.04**	101.08	101.00	**100.00**	101.00	99.00	**98.00**	99.00	104.00	**102.00**	104.00
GL150C5	96.92	**96.32**	96.82	97.00	**96.00**	97.00	**95.00**	**95.00**	**95.00**	99.00	**98.00**	99.00
GL150C6	84.8	**84.06**	84.58	85.00	**84.00**	**84.00**	**84.00**	**84.00**	**84.00**	87.00	**85.00**	86.00
GL150C7	91.08	**90.38**	91.14	91.00	**90.00**	91.00	**90.00**	**90.00**	**90.00**	93.00	**92.00**	94.00
GL150C8	77.52	**76.88**	77.46	77.50	**77.00**	78.00	**76.00**	**76.00**	**76.00**	79.00	**78.00**	**78.00**
GL150C9	78.80	**78.22**	78.74	79.00	**78.00**	79.00	**78.00**	**78.00**	**78.00**	81.00	**79.00**	81.00
GL150C10	77.52	**77.10**	77.48	**77.00**	**77.00**	**77.00**	**77.00**	**77.00**	**77.00**	81.00	**78.00**	79.00

Table 3 Results using biological data

Genome	1.5-App	$\mathcal{GA}_S$	$\mathcal{GA}_M$	$\mathcal{GA}_{OBL}$
Human–Cat	43	43	43	43
Human–Mouse	53	51	51	51
Cat–Mouse	53	50	50	50

Experiments with Genomes based on Biological Data

Three different mammals species were chosen: cat, mouse, and human. Their genomes were taken from [3], considering only the genetic material in common. Modifications in these genomes were done to fulfill the Property 1: genes 82, 83 and 88 were removed in the original mapping proposed in [3], then each genome remains with 18 chromosomes; also, auxiliary genes were added at the extremes of each chromosome in each genome, in order to obtain genomes co-tails.

Initially, the human genome was fixed as an identity, and the corresponding mapping of genes over the cat and mouse genomes were performed to generate the human-cat and human-mouse data. Then, the cat genome was fixed as the identity and the corresponding mapping of genes was applied over the mouse genome to generate the cat-mouse data. $\mathcal{GA}_S$, $\mathcal{GA}_M$ and $\mathcal{GA}_{OBL}$ were executed ten times for these biological data (the $1.5 + \varepsilon$-algorithm only once), and then the average was calculated. See Table 3.

5 Discussion

From the experiments using families of hundred randomly generated genomes (Table 1), one can observe that $\mathcal{GA}_M$ outperforms the results of the other algorithms. $\mathcal{GA}_{OBL}$ presents just a small improvement regarding $\mathcal{GA}_S$.

From experiments using biological data (Table 3), it can be observed that $\mathcal{GA}_M$, $\mathcal{GA}_{OBL}$ and $\mathcal{GA}_S$ give the same results. This is explained because these genomes are instances that are very easy to solve since cat, human and mouse, have very similar sequences regarding randomly generated inputs. Also, the worst results are those computed by the $1.5 + \varepsilon$-approximation algorithm.

From experiments using specific genomes (Table 2), $\mathcal{GA}_M$ computes the best results compared with those computed by $\mathcal{GA}_S$ and $\mathcal{GA}_{OBL}$. For the same measures $\mathcal{GA}_{OBL}$ computes slightly better results than $\mathcal{GA}_S$, and for certain instances the results are the same.

Regarding running time, observing the time necessary to compute the most difficult instances (Table 2), the experiments showed that $\mathcal{GA}_M$ and $\mathcal{GA}_{OBL}$ take approximately 323 and 160 % of the running time of $\mathcal{GA}_S$, resp.. Despite this, both $\mathcal{GA}_M$ and $\mathcal{GA}_{OBL}$ are of practical interest since the first algorithm takes just 1 min to sort genomes with 150 genes and 10 chromosomes in a modest OSX Intel Core I5 processor platform.

Statistical Analysis The analysis was performed following the methodology discussed in [7, 8, 14]; that is, initially, apply the *Kolgomorov-Smirnov* test to determine non-normal distribution of the samples (50 runs of $\mathcal{GA}_S$, $\mathcal{GA}_M$ and $\mathcal{GA}_{OBL}$); then, apply the *Wilcoxon Rank Sum* to compare the medians of pairs of algorithms. The results of the *Wilcoxon Rank Sum* test for $\mathcal{GA}_M$ and $\mathcal{GA}_{OBL}$ with the other algorithms are shown resp. in Tables 4 and 5. Two columns are included: one with the median and the other with the result of the statistical test: the symbols "$s+$" and "$s-$" indicate resp. that two algorithms have different performance and that there is not statistical significance difference ($p\text{-}value > 0.05$) between them. Also, the lowest median values indicate better performance.

Table 4 $\mathcal{GA}_M$ versus other Algorithms using the Wilcoxon Rank Sum Test

	$\mathcal{GA}_M$	$\mathcal{GA}_S$		$\mathcal{GA}_{OBL}$	
GL150C2	**102.00**	102.50	$s+$	**102.00**	$s+$
GL150C3	**109.00**	**109.00**	$s+$	109.50	$s+$
GL150C4	**100.00**	101.00	$s+$	101.00	$s+$
GL150C5	**96.00**	97.00	$s+$	97.00	$s+$
GL150C6	**84.00**	85.00	$s+$	**84.00**	$s+$
GL150C7	**90.00**	91.00	$s+$	91.00	$s+$
GL150C8	**77.00**	77.50	$s+$	78.00	$s+$
GL150C9	**78.00**	79.00	$s+$	79.00	$s+$
GL150C10	**77.00**	**77.00**	$s+$	77.00	$s+$

Table 5 $\mathcal{GA}_{OBL}$ versus other Algorithms using the Wilcoxon Rank Sum Test

	$\mathcal{GA}_{OBL}$	$\mathcal{GA}_S$		$\mathcal{GA}_M$	
GL150C2	**102.00**	102.50	$s-$	**102.00**	$s+$
GL150C3	109.50	**109.00**	$s-$	**109.00**	$s+$
GL150C4	101.00	101.00	$s-$	**100.00**	$s+$
GL150C5	97.00	97.00	$s-$	**96.00**	$s+$
GL150C6	**84.00**	85.00	$s-$	**84.00**	$s+$
GL150C7	91.00	91.00	$s-$	**90.00**	$s+$
GL150C8	78.00	77.50	$s-$	**77.00**	$s+$
GL150C9	79.00	79.00	$s-$	**78.00**	$s+$
GL150C10	**77.00**	**77.00**	$s-$	**77.00**	$s+$

6 Conclusion

Two hybrid evolutionary algorithms for UTD were proposed, that are based on a standard GA, in which the search space consists of signed versions of the initial unsigned genome being sorted, exploring the property that a sorting translocation sequences for any signed genome is also a feasible solution for the initial unsigned genome. Thus, as fitness function the translocation distance of signed genomes is applied. The main feature of the algorithms, $\mathcal{GA}_M$ and $\mathcal{GA}_{OBL}$, is the application of local search and the OBL heuristic, resp.. These heuristics are embedded for the improvement of the initial population, for restarting the population whenever the entropy limit is reached, and as a new stage after the breeding cycle. In both algorithms the convergence of the population is controlled using the Shannon entropy. Experiments were performed to verify the quality of results computed by both algorithms regarding $\mathcal{GA}_S$ [6] and a $1.5 + \varepsilon$-approximation algorithm [5]. Experiments showed that $\mathcal{GA}_M$ outperforms the other algorithms, while $\mathcal{GA}_{OBL}$ showed just a small improvement regarding $\mathcal{GA}_S$. Additionally, three genomes based on biological data (cat, mouse, and human) were processed showing that $\mathcal{GA}_M$, $\mathcal{GA}_{OBL}$ and $\mathcal{GA}_S$ compute the same results. A statistical analysis was performed using the Wilcoxon Rank Sum test concluding that $\mathcal{GA}_M$ has the best performance and that $\mathcal{GA}_{OBL}$ and $\mathcal{GA}_S$ did not show different performance.

As future work, is it of great interest adding biological data to the experiments, generated from sequences taken from the GeneBank database. Also, it is of interest implementing a sharing population parallel version of $\mathcal{GA}_M$ in order to improve the quality of the solutions reducing the running time. Despite its practical limitations, since the $1.408 + \varepsilon$ approximation algorithm, proposed in [10], has a better approximation ratio than the $1.5 + \varepsilon$ one, it is relevant implementing this algorithm and comparing its results with those obtained by $\mathcal{GA}_M$.

References

1. Al-Qunaieer, F.S., Tizhoosh, H.R., Rahnamayan, S.: Opposition based computing - a survey. In: Neural Networks (IJCNN), pp. 1–7. IEEE (2010)
2. Bergeron, A., Mixtacki, J., Stoye, J.: On sorting by translocations. J. Comput. Biol. **13**(2), 567–578 (2006)
3. Bourque, G., Pevzner, P.A.: Genome-scale evolution: reconstructing gene orders in the ancestral species. Genome Res. **12**(1), 26–36 (2002)
4. Cui, Y., Wang, L., Zhu, D.: A 1.75-approximation algorithm for unsigned translocation distance. J. Comput. Sys. Sci. **73**(7):1045–1059 (2007)
5. Cui, Y., Wang, L., Zhu, D., Liu, X.: A $(1.5+\varepsilon)$-approximation algorithm for unsigned translocation distance. IEEE/ACM Trans. Comput. Biol. Bioinf. **5**(1):56–66 (2008)
6. da Silveira, L.A., Soncco-Álvarez, J.L., de Lima, T.A., Ayala-Rincón, M.: Computing translocation distance by a genetic algorithm. In: Acc. Proc, CLEI (2015)
7. Demšar, J.: Statistical comparisons of classifiers over multiple data sets. J. Mach. Learn. Res. **7**, 1–30 (2006)

8. Durillo, J.J., García-Nieto, J., Nebro, A.J., Coello, C.A., Luna, F., Alba, E.: Multi-objective particle swarm optimizers: an experimental comparison. In: Evolutionary Multi-Criterion Optimization, pp. 495–509. Springer (2009)
9. Hannenhalli, S.: Polynomial-time algorithm for computing translocation distance between genomes. Discrete App. Math. **71**(1), 137–151 (1996)
10. Jiang, H., Wang, L., Zhu, B., Zhu, D.: A $(1.408+ \varepsilon)$-approximation algorithm for sorting unsigned genomes by reciprocal translocations. In: Frontiers in Algorithmics, pp. 128–140. Springer (2014)
11. Krasnogor, N., Smith, J.: A tutorial for competent memetic algorithms: model, taxonomy, and design issues. IEEE Trans. Evol. Comput. **9**(5), 474–488 (2005)
12. Moscato, P., Cotta, C.: An introduction to memetic algorithms. Inteligencia Artificial, Revista iberoamericana de Inteligencia Artificial **19**, 131–148 (2003)
13. Moscato, P., et al.: On evolution, search, optimization, genetic algorithms and martial arts: towards memetic algorithms. Caltech concurrent computation program, C3P. Report 826, 1989 (1989)
14. Muñoz, D.M., Llanos, C.H., Coelho, L.S., Ayala-Rincón, M.: Opposition-based shuffled PSO with passive congregation applied to FM matching synthesis. In: IEEE Congress on Evolutionary Computation (CEC), pp. 2775–2781 (2011)
15. Shannon, C.E.: A mathematical theory of communication. Bell Syst. Tech. J. **27**(3), 379–423 (1948)
16. Tizhoosh, H.R.: Opposition-based learning: a new scheme for machine intelligence. In: International Conference on Computational Intelligence for Modelling, Control and Automation, and Intelligent Agents, Web Technologies and Internet Commerce, vol.1, pp. 695–701. IEEE (2005)
17. Wang, L., Zhu, D., Liu, X., Ma, S.: An $O(n^2)$ algorithm for signed translocation. J. Comput. Syst. Sci. **70**(3), 284–299 (2005)
18. Zhu, D., Wang, L.: On the complexity of unsigned translocation distance. Theor. Comput. Sci. **352**(1), 322–328 (2006)

Aesthetic Differential Evolution Algorithm for Solving Computationally Expensive Optimization Problems

Ajeet Singh Poonia, Tarun Kumar Sharma, Shweta Sharma
and Jitendra Rajpurohit

Abstract The applications of Differential Evolution (DE) and the attraction of researchers towards it, shows that it is a simple, powerful, efficient as well as reliable evolutionary algorithm to solve optimization problems. In this study an improved DE called aesthetic DE algorithm (ADEA) is introduced to solve Computationally Expensive Optimization (CEO) problems discussed in competition of congress of evolutionary computation (CEC) 2014. ADEA uses the concept of mirror images to produce new decorative positions. The mirror is placed near the most beautiful (global best) individual to accentuate its attractiveness (significance). Simulated statistical results demonstrate the efficiency and ability of the proposal to obtain good results.

Keywords Differential evolution · DE · Optimization · Computationally expensive optimization problems

1 Introduction

This study is concentrated on Differential Evolution Algorithm (DEA), a simple, reliable and efficient meta-heuristic algorithm especially for continuous functions [1]. DE introduced in 1995 by Storn and Price [1], have managed to seek the attention of many researchers to solve optimization problems in versatile domains.

A.S. Poonia
Govt. College of Engineering and Technology, Bikaner, India
e-mail: pooniaji@gmail.com

T.K. Sharma (✉) · S. Sharma · J. Rajpurohit
Amity University Rajasthan, Noida, India
e-mail: taruniitr1@gmail.com

S. Sharma
e-mail: shweta_sharma0287@yahoo.com

J. Rajpurohit
e-mail: jiten_rajpurohit@yahoo.com

© Springer International Publishing Switzerland 2016
N. Pillay et al. (eds.), *Advances in Nature and Biologically Inspired Computing*,
Advances in Intelligent Systems and Computing 419,
DOI 10.1007/978-3-319-27400-3_8

87

In general synchronous evolution strategy (based on generation) is employed in DE [2]. The generation is updated with vectors having better fitness than their counterparts after evaluation of all individuals of population in DE. But, the choice of trial vector and associated parameter values used decides the performance of the DE algorithm. There are only two control factors apart from population size in DE namely the scaling factor (F) and Crossover rate (Cr). Sometimes inappropriate parameter tuning or selection of strategy may limit to convergence speed or may trigger premature convergence escaping local minima. The details are reasonably demonstrated with the values of F and Cr in [3, 4].

Similar to other metaheuristics, DE also suffers weakness of slow convergence or losing efficiency while solving noisy problems. This study presents aesthetic differential evolution algorithm (ADEA) to enhance the efficiency and performance of the conventional DE to solve optimization problems.

ADEA uses the concept of mirror images to produce new decorative positions. The mirror is placed near the most beautiful (global best) individual to accentuate their attractiveness. This idea facilitates in providing the displacement of optimal solutions (based on fitness value). The individual solution in the population is updated only when a better fitness value is achieved.

This dynamic behavior assists in dealing with complex optimization problems. This mechanism helps in following optimal solutions.

ADEA is tested and validated on ten Computationally Expensive Optimization (CEO) problems discussed in competition of congress of evolutionary computation (CEC) 2014 [5].

The paper is structured as follows: Sect. 2 presents a brief introduction of DE. The proposed algorithm, ADEA is detailed in Sect. 3. Section 4 demonstrates benchmark problems and parameter settings. Results are discussed in Sect. 5. The conclusions drawn from the study are summarized in Sect. 6.

2 DE: An Overview

DE is arguably a most frequently used evolutionary algorithm that has been successfully applied to solve wide range of optimization problems. However, like other meta-heuristics, DE also suffers with some intrinsic drawbacks that lead to slow convergence. To enhance its ability and efficacy, various variants of DE have been proposed. The next paragraph focuses on the modifications in basic DE and applications in the recent past.

In 2011, Ghosh et al. [6] introduced a self adaptive variant of DE tunes both F and Cr on run time. Zou et al. [7] also proposed a modified DE where the values of F and CR are dynamically adjusted with successive iterations and later applied o solve assignment problem.

In 2012, Wang et al. [8] proposed modified DE variant to solve binary coded optimizations problems that balances exploration and exploitation capabilities while maintain diversity of the population. Later it is applied to solve multidimensional knapsack problems. Mohamed and Sabry [9] introduced a new directed mutation rule to solve constrained optimization problems.

In 2013, Zou et al. [10] proposed modified DE that employs Gaussian and uniform distribution to adjust the values of F and Cr. Wu et al. [11] introduced self adaptive DE to improve global searching capability where the population is dynamically divided and control parameters are also dynamically adjusted on run time.

In 2014, Ali et al. [12] applied basic DE in search of optimal parameters to enhance the quality and robustness of watermarked image. Coelho et al. [13] proposed self adaptive DE that applied Gaussian probability, gamma distribution and chaotic sequence.

In 2015, Mohamed [14], presented an improved DE that employs a new mutation rule, called triangular mutation, based on convex combination vector of the triplet. Xiang et al. [15] introduced a variant inspired from Shuffled frog leap algorithm in order to increase the convergence of basic DE. Mallipeddi and Lee [16] presented a surrogate model based DE to tune the control parameters. Interested readers may consult [17] for recent enhancements in DE.

DE is executed in three different phases' mutation, crossover and selection. Like other evolutionary algorithms DE also starts with randomly generated population NP D dimensional search variable vectors. The basic structure of DE is given below.

Basic dE

Initialization

Population (S) (solutions) of NP candidate solutions is initialized:

$$X_{i,G} = \{x_{1,i,G}, x_{2,i,G}, x_{3,i,G}, \ldots, x_{D,i,G}\}Z \tag{1}$$

where the index i represents ith individual in the population and $i = 1, 2,..., NP$; G is generation and D is dimension of the problem.

Uniformly randomizing individuals within the search space constrained by the prescribed lower and upper bounds: $l = \{l_1, l_2,..., l_D\}$ and $u = \{u1, u2,..., u_D\}$ are initialized.

jth component of the ith vector may be initialized as follows:

$$x_{j,i} = l_j + rand_{i,j}[0, 1] \times (u_j - l_j)Z \tag{2}$$

where $rand_{i,j}$ [0, 1] is uniformly distributed random number between 0 and 1.

Mutation

In the current generation (G), corresponding to each target vector $(X_{i,j})$ a donor (perturbed) vector $(V_{i,G})$ is created. There are several schemes defined in DE for this. Here the scheme called DE/rand/bin is used throughout the study and defined as:

$$V_{i,G} = X_{r1,G} + F \times (X_{r2,G} - X_{r3,G})Z \qquad (3)$$

where r_1, r_2, r_3 are random numbers taken randomly from [1, NP] in such a way that $r_1 \neq r_2 \neq r_3 \neq i$. F denotes scaling (amplification) factor and is a positive controlling parameter which lies between 0 & 1.

Crossover
This phase helps in increasing diversity of the population and gets activated after the mutation phase where the donor or perturbed vector $V_{i,G} = \{v_{1,i,G}, v_{2,i,G}, v_{3,i,G}, \ldots, v_{D,i,G}\}$ exchanges its components with that of target vector $X_{i,G} = \{x_{1,i,G}, x_{2,i,G}, x_{3,i,G}, \ldots, x_{D,i,G}\}$ to introduce trial vectors $U_{i,G} = \{u_{1,i,G}, u_{2,i,G}, u_{3,i,G}, \ldots, u_{D,i,G}\}$. The binomial crossover is performed to produce trail vectors.

$$u_{j,i,G} = \begin{cases} v_{j,i,G} & if\ rand_{i,j}[0,1] \leq Cr \vee j = j_{rand} \\ x_{j,i,G} & otherwise \end{cases} \qquad (4)$$

where $j = 1,\ldots, D$ and $j_{rand} \in \{1,\ldots, D\}$. Cr is second positive control parameter and known as Crossover rate and is $Cr \in [0, 1]$. Comparing Cr to $rand_{i,j}$ [0, 1] determines the source for each remaining trial parameter. If $rand_{i,j}[0,1] \leq Cr$, then the parameter comes from the mutant vector; otherwise, the target vector is the source.

Selection
Selection is the final phase of DE, where the population for the next generation is selected from the individual in the current population and its corresponding trail vector based on greedy mechanism and is given as:

$$X_{i,G+1} = \begin{cases} U_{i,G} & if\ f(U_{i,G}) \leq f(x_{i,G}) \\ X_{i,G} & Otherwise \end{cases} Z \qquad (5)$$

3   Proposed ADEA

ADEA is proposed to enhance the conventional DE performance to overcome the weakness of slow convergence or losing efficiency while solving noisy problems.

ADEA uses the concept of mirror images to produce new decorative positions and is inherited from Interior Search Algorithm [18]. The mirror is placed near the most beautiful (global best) individual to accentuate their attractiveness. This idea facilitates in providing the displacement of optimal solutions (based on fitness value). The individual solution in the population is updated only when a better fitness value is achieved. This mechanism helps in following optimal solutions.

In ADEA, similar to basic DE, initial set of individuals called population is generated randomly between upper (u) and lower (l) bounds. Then fitness of each is estimated using fitness (objective) function. The fittest (x_{gbest}) individual (having

Fig. 1 Randomly generated population with global best (*Red mark*), optimal (*green mark*) and *blue arrows* presents possible random walk

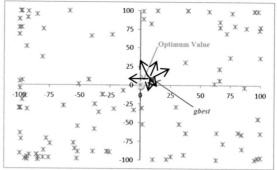

Random solutions in the search space (**x**); *Random Walk* (→);
gBest (■); *Optimum Value* (⬤)

the minimum objective value) is identified. Then a random walk is performed to slightly change the position of x_{gbest} in order to perform local search in feasible search space (Figs. 1 and 2a) using Eq. (6). This helps in enhancing exploitation.

$$x^k_{gbest} = x^{k-1}_{gbest} + r_{NP} + \lambda \qquad (6)$$

where r_{NP} is random number (uniformly distributed) and the scale factor λ, which depends on the size of search region and set as:

$$\lambda = 0.01 * (u - l) \qquad (7)$$

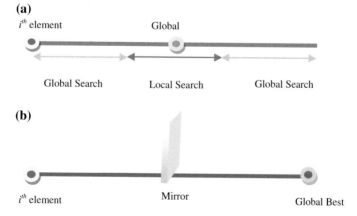

Fig. 2 a Random Walk and **b** Mirror Searching process in ADEA (Adapted from [18])

Secondly, to give equal consideration to exploration and exploitation a parameter α is introduced in the mutation phase. The value of α is in the ranges of 0 and 1. In mutation process of ADEA a mirror is placed randomly between each individual and *gbest*. The idea is demonstrated in Fig. 2b. The position of the mirror for *ith* individual in *kth* iteration is given as:

$$x^k_{m,i,G} = r_5 x^{k-1}_{i,G} + (1 - r_5) x^k_{gbest,G} Z \tag{8}$$

where r_5 is a random number between 0 and 1.

The reflected image of the individual by mirror is given as:

$$v^k_{i,G} = 2x^k_{m,i,G} - x^{k-1}_{i,G} Z \tag{9}$$

If the reflected image (Fig. 2b) is near the *gbest* than it performs local search else global search is performed, which helps in exploitation and exploration. Sometimes it is quite possible that reflected images may be fall outside the limited bounds.

If the value of α is less than 0.5 (which is randomly generated r_6) then mutant vectors generated using mirror images are taken forward for crossover otherwise the mutants are generated using original mutation process given in Eq. (3).

The pseudo code of ADEA is detailed in Fig. 3.

Aesthetic Differential Evolution Algorithm

Initialize the population of solutions using equation (2).
while stopping criterion not satisfied
Evaluate the fitness function and find global best (x^k_{gbest}), where k is iteration.

for $i = 1$ to NP
    if x_{gbest}
        $x^k_{gbest,G} = x^{k-1}_{gbest,G} + r_{NP} + \lambda$
    if $r_5 \leq \alpha$
        $x^k_{m,i,G} = r_6 x^{k-1}_{i,G} + (1 - r_5) x^k_{gbest,G}$
        $v^k_{i,G} = 2x^k_{m,i,G} - x^{k-1}_{i,G}$
    else
        $v^k_{i,G} = x_{r1,G} + F \times (x_{r2,G} - x_{r3,G})$
    end if
    Check the boundaries.
    Perform Crossover
    Selection
    end for
end while

Fig. 3 Pseudo-code of ADEA

4 CEO Benchmark Problems and Experimental Settings

The efficiency of the proposed ADEA is tested and validated on 10 computationally expensive benchmark optimization problems consulted from [5] with the following settings:

- PC Configuration

    System: Microsoft Windows XP Professional (2002)
    CPU: Celeron (R) Dual Core CPU T3100@1.90 GHz
    RAM: 1.86 GB; Simulated in Deb C++.

- DE Parameters
    The population size is fixed to 100. F & Cr are fixed as 0.5 and 0.9 respectively. Maximum numbers of function evaluations (NFE) are fixed to 10^5. 30 independent runs are performed for all the experiments. Random numbers are generated using inbuilt function called $rand$ () in C++.
    Boundary constraints are handled using the given formulation [19]

$$f(x_i) = \begin{cases} r_7 \times l_i + (1 - r_7)x_{gbest,i} & if \ x_i < l_i \\ r_8 \times u_i + (1 - r_8)x_{gbest,i} & if \ x_i > u_i \end{cases} Z \qquad (10)$$

where r_7 and r_8 are random numbers between 0 and 1. x_{gbest} is global best solution.

- Simulation Strategy
    In order to evaluate the performance of the ADEA, the statistical results in terms of best, worst function value, median, mean and standard deviation (Std. Dev.) are evaluated.
- Algorithm taken for performance comparisons

    - Fireworks Algorithm with Differential Mutation (FWA–DM) [20] and
    - Surrogate-assisted Differential Evolution Algorithm with Dynamic Parameters Selection (Sa–DE–DPS) [21]

5 Result Discussion

The performance of the proposed ADEA is tested on the ten computationally expensive single objective benchmark problems that were presented in CEC 2015. Three problems are unimodal and seven are simple multimodal. All the problems are solved for only 10 dimensions (10D) because of brevity of space. The simulated statistical results in term of best, worst, median, mean and Std. Dev. are illustrated

Table 1 Simulated results of CEO benchmark problems (Prob.) for 10D

Prob.	Best	Worst	Median	Mean	Std. dev.
P_1	1.89E–02	9.86E+03	2.98E+02	4.06E+03	8.92E+02
P_2	0.00E+00	8.62E-03	7.14E-22	1.10E-04	6.70E-04
P_3	0.00E+00	3.87E-08	3.19E-19	1.07E-09	4.09E-09
P_4	0.00E+00	5.17E+00	9.02E-02	9.79E-01	1.06E+00
P_5	1.29E+01	1.90E+01	1.44E+01	1.61E+01	3.47E-02
P_6	1.32E+00	5.88E+00	3.71E+00	4.01E+00	5.61E-01
P_7	6.32E-03	1.96E-01	6.99E-02	8.18E-02	1.74E-02
P_8	0.00E+00	1.89E+00	7.65E-04	8.75E-02	1.23E-01
P_9	1.78E+00	1.48E+01	2.99E_00	8.79E+00	1.22E+00
P_{10}	8.71E-12	5.90E+00	8.19E-01	1.27E+00	1.19E+00

in Table 1. The proposed ADEA has shown the efficiency in solving all the ten CEO problems. ADEA performed extremely well on problem P_3. For rest of the problems the proposal performed good and very close to the optimal solutions. Further to compare the performance of ADEA with the latest state-of art algorithms, two algorithms namely FWA–DM and Sa–DE–DPA are consulted from literature. The parameter settings of both of the algorithms remain same for unbiased comparison. The simulated results in terms of best function value and Std. Dev are presented in Table 2. From the table it can be analyzed that the ADEA performed equally good for almost all the 10 problems, where as for P_1 and P_7 it performed extremely well in comparison. Similarly, while comparing Std. Dev., ADEA proves it performance especially for the Problems P_5, P_6, P_8, P_9 and P_{10}.

Table 2 Comparative Simulated results of CEO benchmark problems (Prob.) for 10D

Prob.	FWA–DM		Sa–DE–DPS		ADEA	
	Best	Std. dev.	Best	Std. dev.	Best	Std. dev.
P_1	2.48E-02	1.67E+04	0.00E+00	0.00E+00	1.89E-02	8.92E+02
P_2	0.00E+00	9.49E-04	0.00E+00	0.00E+00	0.00E+00	6.70E-04
P_3	0.00E+00	9.02E-09	0.00E+00	0.00E+00	0.00E+00	4.09E-09
P_4	0.00E+00	1.60E+00	0.00E+00	0.00E+00	0.00E+00	1.06E+00
P_5	2.00E+01	4.17E-02	1.35E-03	6.47E-01	1.29E+01	3.47E-02
P_6	3.40E-03	6.40E-01	4.82E+00	4.30E+00	1.32E+00	5.61E-01
P_7	1.72E-02	4.92E-02	0.00E+00	0.00E+00	6.32E-03	1.74E-02
P_8	0.00E+00	8.09E-01	1.18E-03	1.49E-01	0.00E+00	1.23E-01
P_9	1.99E+00	2.45E+00	1.46E+01	1.03E+01	1.78E+00	1.22E+00
P_{10}	9.09E-13	2.08E+00	0.00E+00	1.21E+00	8.71E-12	1.19E+00

6 Conclusions and Future Work

This study presents a modification in mutation phase of original DE. The modification is intended to improve its performance in terms of convergence rate. The proposal uses the concept of mirror images to produce new decorative positions inspired by Interior search algorithm. The mirror is placed near the most beautiful (global best) individual to accentuate their attractiveness. The proposal is named as aesthetic DE algorithm (ADEA). ADEA is implemented on ten CEO benchmark problems discussed in CEC 2014. The proposal is consistently able to reach the optimal solutions. The computed statistical results indicate equally good performance of the proposal with respect to state-of-art algorithms.

In future ADEA will be implemented on real world problems.

References

1. Storn, R., Price, K.: Different Evolution—A Simple and Efficient Heuristic for global Optimization over Continuous spaces, pp. 341–359, Dec 1997
2. Storn, R.., Price, K.: Differential evolution-a simple and efficient adaptive scheme for global optimization over continuous spaces, TR-95-012 (1995). http.icsi.berkeley.edu/~storn/litera.html
3. Lampinen, J., Zelinka, I.: On stagnation of the differential evolution algorithm. In: Oˇsmera, P. (ed.) Proceedings of the 6th International Mendel Conference on Soft Computing, pp. 76–83 (2002)
4. Gämperle, R., Müller, S., Koumoutsakos, P.: A parameter study for differential evolution. In: Grmela, A., Mastorakis, N.E. (eds.) Advances in Intelligent Systems, Fuzzy Systems, Evolutionary Computation, pp. 293–298. WSEAS Press, Interlaken, Switzerland (2002)
5. Liu, Q.C.a.Q.Z.B., Liang, J.J., Suganthan, P.N., Qu, B.Y.: Problem Definitions and Evaluation Criteria for Computationally Expensive Single Objective Numerical Optimization, Computational Intelligence Laboratory and Nanyang Technological University, Zhengzhou and Singapore, Technical Report 2013
6. Arnob, G., Swagatam, D., Aritra, C., Ritwik, G.: An improved differential evolution algorithm with fitness-based adaptation of the control parameters. Inf. Sci. **181**(18), 3749–3765 (2011)
7. Dexuan, Z., Haikuan, L., Liqun, G., Steven, L.: An improved differential evolution algorithm for the task assignment problem. Eng. Appl. Artif. Intell. **24**(4), 616–624 (2011)
8. Ling, W., Xiping, F., Yunfei, M., Ilyas, M.M., Minrui, F.: A novel modified binary differential evolution algorithm and its applications. Neurocomputing **98**, 55–75 (2012)
9. Mohamed, A.W., Sabry, H.Z.: Constrained optimization based on modified differential evolution algorithm. Inf. Sci. **194**(1), 171–208 (2012)
10. Dexuan, Z., Jianhua, W., Liqun. G., Steven, L.: A modified differential evolution algorithm for unconstrained optimization problems. Neurocomputing **120**, 469–481 (2013)
11. Deng, W., Xinhua, Y., Li, Z., Meng, W., Yaqing, L., Yuanyuan, Li: An improved self-adaptive differential evolution algorithm and its application. Chemometr. Intell. Lab. Syst. **128**(15), 66–76 (2013)
12. Musrrat, A., Wook, A.C., Patrick, S.: Differential evolution algorithm for the selection of optimal scaling factors in image watermarking. Eng. Appl. Artif. Intell. **31**, 15–26 (2014)
13. dos Santos Coelho L., Ayala H.V.H., Mariani V.C.: A self-adaptive chaotic differential evolution algorithm using gamma distribution for unconstrained global optimization. Appl. Math. Comput. **234**, 452–459 (2014)

14. Mohamed Ali Wagdy: An improved differential evolution algorithm with triangular mutation for global numerical optimization. Comput. Ind. Eng. **85**, 359–375 (2015)
15. Wan-li, X., Ning, Z., Shou-feng, M., Xue-lei, M., Mei-qing, A.: A dynamic shuffled differential evolution algorithm for data clustering. Neurocomputing **158**(22), 144–154 (2015)
16. Rammohan, M., Minho, L.: An evolving surrogate model-based differential evolution algorithm. Appl. Soft Comput. **34**, 770–787 (2015)
17. Mashwani Wali Khan, Enhanced versions of differential evolution: state-of-the-art survey. Int. J. Comput. Sci. Math. **5**(2), 107–126
18. Gandomi, A.H.: Interior search algorithm (ISA): a novel approach for global optimization. ISA Trans. Elsevier **53**(4), 1168–1183 (2014)
19. Gandomi, A.H., Yang, X.S.: Evolutionary boundary constraint handling scheme. Neural Comput. Appl. **21**(6), 1449–1462 (2012)
20. Yu, C., Kelley, L., Zheng, S., Tan, Y.: Fireworks algorithm with differential mutation for solving the CEC 2014 competition problems, In: 2014 IEEE Congress on Evolutionary Computation (CEC), pp. 3238–3245. Beijing, China, 6–11 July 2014
21. Elsayed, S.M., Ray, T., Sarker, R.A.: A surrogate-assisted differential evolution algorithm with dynamic parameters selection for solving expensive optimization problems. In: 2014 IEEE Congress on Evolutionary Computation (CEC), pp. 1062–1068. Beijing, China, 6–11 July 2014

Automatic Discovery and Recommendation for Telecommunication Package Using Particle Swarm Optimization

Shanshan Liu, Bo Yang, Lin Wang and Ajith Abraham

Abstract Telecommunication package is a product produced by telecom operator to satisfy different consumer groups. Telecom operator should not only consider the users' acceptability, but also enable to maximize profits. However, how to balance the relationship of both and designing a package are very important tasks and complicated problems. At present, design of telecom package is affected greatly by designer's subjective experience which is blind. In this paper, a new idea of automatic discovery and recommendation for telecom package is proposed. This idea is combined user's acceptance with operator's profit. The package model and customer model are set up based on consumption of customers. Particle swarm optimization is used for discovering an inverse package. Meanwhile, the potential customers of the targets are selected by calculating proportion of package attribute usage. Experimental results show that the proposed method has favorable performance.

Keywords Telecom packages · Particle swarm optimization algorithm · Automatic discovery · Recommendation

1 Introduction

In recent years, the telecom market is gradually maturing, the technology is developing rapidly, and competition among the telecom operators is becoming fiercer and fiercer. With customer's high demand, the design of package is much more diverse, and the type of packages is also more complete, then telecom

S. Liu · B. Yang (✉) · L. Wang
Shandong Provincial Key Laboratory of Network Based Intelligent Computing,
University of Jinan, Jinan 250022, China
e-mail: yangbo@ujn.edu.cn

A. Abraham
Machine Intelligence Research Labs (MIR Labs),
Scientific Network for Innovation and Research Excellence, Auburn, USA

© Springer International Publishing Switzerland 2016   97
N. Pillay et al. (eds.), *Advances in Nature and Biologically Inspired Computing*,
Advances in Intelligent Systems and Computing 419,
DOI 10.1007/978-3-319-27400-3_9

packages have already dazzled in mobile client. However, these packages do not
bring better advantages to telecom market, in contrast, they increase the difficulties
of the operator's management, while consumers do not decide which they need.
The main reason lies in that the operator ignores principle of design and lacks
normative steps in designing process. Secondly, operator does not know which
package really satisfies users, and which brings revenue to the operators themselves. This behavior makes the design of packages blinder. Beside, customer's
psychology has become more and more mature with continuous selection. They
will consider their actual consumption, product quality and many other factors
when they choose a package. Therefore, the package which resembles the actual
consumption of customers interests them [1, 2].

However, the traditional telecom packages were influenced by the policymakers'
subjective experience deeply which is blind. At present, both domestic and foreign
research on the telecom package is relatively little. Most of them are about telecom
product introduction, market survey, the qualitative analysis, profit calculation,
guiding strategy, customer churn prediction and so on [3, 4]. A system of automatic
discovery and recommendation is needed.

Particle Swarm Optimization (PSO) is introduced by Kennedy and Eberhart [5]
as an optimization method for continuous nonlinear functions. Zulal Sevkli and
Erdoğan Sevilgen [6] use a hybrid particle swarm optimization algorithm for
function optimization and show rationality of the improved optimization by
experimental results. Dub and Stefek [7] use PSO method for system identification.
In addition, particle swarm optimization algorithm is also used in many fields of
data mining, such as neural network training, classifier design, clustering analysis,
and network community discovery, etc.

PSO is simple and implements easily. It is used in engineering applications
widely. It has less adjustable parameters and the choice of parameters has been a
mature theoretical results [8, 9]. In this research, a novel approach which optimizes
telecom packages automatically and inversely by particle swarm optimization is
proposed. Profits taken from customers are defined as the fitness value in the
optimization process. The profits are combined with the possibility of customers
choosing an inverse package. It maximizes operator's profits. Then select potential
customers according to proportion of package attribute usage.

Rest of the paper is arranged as follows. Section 2 gives detailed methods of
package discovery and recommendation. Section 3 outlines and discusses experiments followed by conclusions in Sect. 4.

2 Methodology

The main goal of this method is to discover an inverse package, and find out the
potential customers in the target customers. Firstly, select reference group and
target group of customers. Secondly, set up package model and customer model to
store their information. Then optimize an inverse package with PSO. Finally, select

Fig. 1 The process of package discovery and recommendation

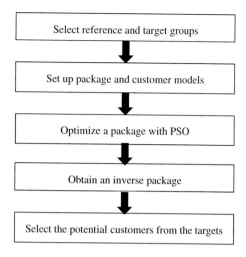

potential customers of the inverse package from target group. The process is shown in the following Fig. 1.

2.1 Models

Through investigation and research, three indexes are extracted to describe a package: calls, data traffic and monthly subscription fees. The package model and customer model are shown in the following Tables 1 and 2 separately.

The package model is used for storing information of inverse package. It includes monthly subscription fee (yuan), free calls of a package (min), free data traffic of a package (MB), extra unit price of calls (yuan/min) and extra unit price of data traffic (yuan/KB).

The customer model is used for storing customer's consumption data. It includes customer's number, information of existing package and information of average

Table 1 Package model

monthly subscription fee	free quantity of a package		extra unit price of a package	
	calls	data traffic	fen/min	fen/KB

Table 2 Customer model

customer number	monthly subscription fee	free quantity of existing package		extra unit price of existing package		average consumption		
		calls	data traffic	fen/min	fen/KB	calls time	data traffic	monthly cost

consumption. Customer's number is used for distinguishing different customer. Existing package refers to customer's original package. The average consumption contains average calls time, average data traffic and average cost of a customer.

2.2 Particle Swarm Optimization

A particle represents a package. Each particle has a position and a velocity vector. The process is shown as following.

(a) *Initialization.* The particle is initialized randomly according to the interval of customer's consumption.
(b) *Calculating fitness value of a particle.*
(c) *Comparing the fitness value with particle's local best value (pbest).* If the fitness value is greater than pbest, we update particle's position and the pbest.
(d) *Comparing pbest with particle's global best value (gbest).* If pbest is greater than gbest, we update particle's position and the gbest.

The position and velocity vector are updated as following Eqs. (1) and (2).

$$V_{i+1} = w * V_i + c1 * rand1 * (pbest - X_i) + c2 * rand2 * (gbest - X_i) \qquad (1)$$

$$X_{i+1} = X_i + V_{i+1} \qquad (2)$$

V_{i+1}, X_{i+1} : velocity and position at iteration $i + 1$;
w : the inertia weight which is the balancing factor between exploration and exploitation;
$c1, c2$: coefficients of learning factors, which are the weights of contributions of personal experience and social interaction, generally at 2;
$rand1, rand2$: random numbers, generally in $(0, 1)$

If the velocity and position exceed their maximum or minimum, their boundaries are taken.

The population size is 10, and iteration is 100. The package is optimized iteratively until the stopping criterion. Usually, the stopping criterion is the maximum number of times or the desired fitness value. The fitness value is defined by the expression as follows Eq. (3).

$$Z = \sum_{j=1}^{N} Pj * Fj \qquad (3)$$

Z : income that operator get from the inverse package
Pj : probability that customer j use the inverse package

Fj : cost that customer j spend on the inverse package. It consists of three parts: monthly subscription fee of the new package, extra cost of free calls and data traffic

N   : number of target customers

For a particle, a customer j from targets, select people from reference customers whose average consumption is similar to the customer j and calculate the number of people as Mj. Then select people from Mj customers whose existing package is similar to the particle and calculate the number as Nj. Regard Nj/Mj as the probability Pj that customer j chooses the inverse package. Multiply Pj and Fj as an operator's income from the customer. Use the same method to all target customers and then get the total sum Z as the income on the inverse package.

2.3 Recommendation

The inverse package should be extended to target customers. The main job is to make sure that if a customer is going to use the inverse package. Then, calculate the proportion of a customer using the existing package and the inverse package. The expression is shown as follows Eq. (4) and Eq. (5).

$$R_0 = 1 - \sum_i [(t_i - T_i)/T_i] \qquad (4)$$

$$R_1 = 1 - \sum_i [(t_i - T_i')/T_i'] \qquad (5)$$

R_0 : proportion of a customer using the existing package.
R_1 : proportion of a customer using the inverse package.
i : an attribute of the package, for example, calls.
t_i : the value of attribute i of customer's average consumption, for example, calls time.
T_i : the value of attribute i of customer's existing package.
T_i' : the value of attribute i of the inverse package.

If $R_0 < R_1$, it means that a customer make more full use of the inverse package than the existing one. So, the inverse package is better to a customer. Then the customer is considered as a potential people of the inverse package. Select all potential people from the targets in order to extend the inverse package.

3   Experiment

The experiments run on Visual Studio2010, and the test data is provided by a telecom operator in China. Customer's consumption information is taken for initialization of PSO. Then an inverse package is obtained and extended to targets.

In the experiments, the percentage of potential customers in the targets is calculated by Eq. (6). Operator's income from the inverse package is counted, and the proportion of the changes is also calculated. The expression is shown as follows Eq. (7).

$$r1 = n/N * 100\% \tag{6}$$

r1   : the percentage of potential customers in the targets
n   : the number of potential customers
N   : the number of target customers

$$r2 = (y2 - y1)/y1 * 100\% \tag{7}$$

r2   : the proportion of the changes of the operator's income. If $r2 > 0$, the income rise by r2 percent. If $r2 < 0$, the income fall by r2 percent
y1   : operator's income from existing package
y2   : operator's income from inverse package

The inverse package is shown in an array. Attributes in the array represent monthly subscription fee, free calls of a package, free data traffic of a package, extra unit price of calls, and extra unit price of data traffic respectively. For different targets, the inverse package is different.

In 100 experiments, the mean convergence process is shown in Fig. 2. The inverse package distribution is shown in Fig. 3.

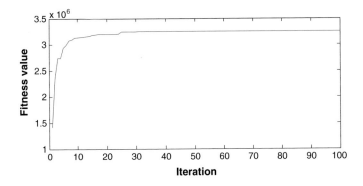

Fig. 2 Mean convergence process

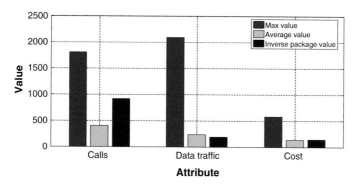

Fig. 3 Inverse package distribution

From Fig. 2, the fitness value is increasing. After iteration 25, it is stable. The convergence in this method is fast which indicates that the setting of parameters is reasonable.

From Fig. 3, the attributes' comparison among inverse package, maximum value, and average value can be seen clearly. Taking into account the user's acceptance, attribute value of inverse package locates in the range of customer actual monthly consumption. A part of customers need more calls and a small part of them need much traffic data. From average value of attribute, customers depend on calls more than traffic data, and their average cost are about 136 yuan. Accordingly, the calls percentage of inverse package is greater. It indicates that the inverse package applies to groups whose calls are more while data traffic is less. The monthly subscription fee of inverse package is about 141yuan. There is not a big difference to customer's average cost. It can not only ensure the operator's income, but also improve the customer's acceptance.

There are 65 % good packages in 100 experiments that show this method is valuable. Inverse package's income is more than the existing one. The mean value of r2 is 42 % and the standard deviation is 0.18. It implies that operator's income is increasing steadily with an inverse package. The mean r1 is 21 %, and r1's standard deviation is 0.02. It implies that this method can be able to select potential customers more stably. Above all, this method is auxiliary for package design and extension.

4 Conclusions

In this paper, a new idea of telecom package's automatic discovery and recommendation is proposed. Operator's income is defined as the fitness value in optimization which maximizes operator's income. Potential customers are selected by calculating the proportion of a customer's use of the existing package and the inverse package. The experiments imply that this method is valuable for automatic discovery and recommendation of telecom package. The intelligent method avoids blindness of artificial experience and brings reference significance to operators when they design and extend a package to a market.

Research on the discovery and recommendation technology based intelligent calculation is an important way to promote the service intelligence. It helps to digging customer's needs and expending a package to markets effectively and accurately. It also helps to improve the efficiency of the customer's resources, service level and management level of telecom operators.

Acknowledgments This work was supported by National Natural Science Foundation of China under Grant No. 61573166, No. 61572230, No. 61373054, No. 61472164, No. 81301298, No.61302128. Shandong Provincial Natural Science Foundation, China, under Grant ZR2015JL025. National Key Technology Research and Development Program of the Ministry of Science and Technology under Grant 2012BAF12B07-3. Jinan Youth Science & Technology Star Project under Grant No. 2013012.

References

1. Ancheng, H.: Research on the evaluation system of A Telecom Operator, University of Electronic Science and Technology of China (2011) (in Chinese)
2. Jiang, Xin Kuang, Chen, Xu: Research on prediction model of the impact of new telecom services tariff based on the customer choice behavior. J Adv. Mater. Res. **765–767**, 3249–3252 (2013)
3. Amin, A., Shehzad, S., Khan, C., Ali, I., Anwar, S.: Customer churn prediction in telecommunication industry: with and without counter-example. Nature-Inspired Computation and Machine Learning. LNCS, vol. 8857, pp. 206–218. Springer, International Publishing (2014)
4. Amin, A., Shehzad, S., Khan, C., Ali, I., Anwar, S.: Churn prediction in telecommunication industry using rough set approach. In: New Trends in Computational Collective Intelligence Studies in Computational Intelligence, vol. 572, pp. 83–95 (2015)
5. Kennedy, J., Eberhart, R.C.: Particle swarm optimization. In: International Conference on Neural Networks (Perth, Australia), Piscataway, pp. 1942–1948. IEEE Service Center, Los Alamitos (1995)
6. Sevkli, Z., Sevilgen, F.E.: A hybrid particle swarm optimization algorithm for function optimization. Applications of Evolutionary Computing. LNCS, vol. 4947, pp. 585–595. Springer, Berlin (2008)
7. Dub, M., Stefek, A.: Using PSO method for system identification. Mechatronics **2013**, 143–150 (2014)
8. Liang, Xiaolei, Li, Wenfeng, Zhang, Yu., Zhou, MengChu: An adaptive particle swarm optimization method based on clustering. J Soft Comput. **19**, 431–448 (2015)
9. Jordehi, A.R.: Particle swarm optimization for dynamic optimization problems: a review. J. Neural Comput. Appl. **25**, 1507–1516 (2014)

EEG Signals of Motor Imagery Classification Using Adaptive Neuro-Fuzzy Inference System

Shereen A. El-aal, Rabie A. Ramadan and Neveen I. Ghali

Abstract Brain Computer Interface (BCI) techniques are used to help disabled people to translate brain signals to control commands imitating specific human thinking based on Electroencephalography (EEG) signal processing. This paper tries to accurately classify motor imagery imagination tasks: e.g. left and right hand movement using three different methods which are: (1) Adaptive Neuro Fuzzy Inference System (ANFIS), (2) Linear Discriminant Analysis (LDA) and (3) k-nearest neighbor (KNN) classifiers. With ANFIS, different clustering methods are utilized which are Subtractive, Fuzzy C-Mean (FCM) and K-means. These clustering methods are examined and compared in terms of their accuracy. Three features are studied in this paper including AR coefficients, Band Power Frequency, and Common Spatial pattern (CSP). The classification accuracies with two optimal channels C3 and C4 are investigated.

Keywords Brain Computer Interface · Classification · Adaptive neuro fuzzy inference system

1 Introduction

Brain Computer Interface (BCI) provides new communication and control technology to people who suffer from disabilities and neuromuscular disorders. It enables disabled persons to operate, for example, word processing programs and control devices in some special circumstances. Signal processing algorithms are used to translate brain signals to control commands without using muscles. Non-invasive EEG reads electrical signals using wired or wireless BCI [1] from cortical surface

S.A. El-aal (✉) · N.I. Ghali
Faculty of Science, Alazhar University, Cairo, Egypt
e-mail: shereen_mhmd86@yahoo.com

R.A. Ramadan
Computer Engineering Department, Cairo University, Cairo, Egypt

Hail University, Hail, Kingdom of Saudi Arabia

© Springer International Publishing Switzerland 2016
N. Pillay et al. (eds.), *Advances in Nature and Biologically Inspired Computing*,
Advances in Intelligent Systems and Computing 419,
DOI 10.1007/978-3-319-27400-3_10

through the scalp. Then, it describes band powers in terms of rhythmic brain waves: 8–13 Hz for Mu rhythm, 14–30 Hz for Beta rhythm.

Different classifiers are utilized for EEG classification including LDA [2], Support Vector Machine (SVM) [3], Fuzzy SVM (FSVM) [4], KNN [5] and ANFIS [6].

This paper is organized as follows: Sect. 2 introduces BCI categories, some of researches done in BCI signal classification and feature extraction, EEG features, different classifiers and different clustering methods used. Section 3 defines the problem. Section 4 surveys BCI processing phases and the methods used in every phase. Section 5 assess and compare classification accuracy using different classifiers and different features and finally Sect. 6 gives the conclusion.

2 Related Work

BCI is a communication system that allows disabled users to control computer application without a need to peripheral muscle and nerves, just by thoughts.

In all cases, brain waves are categorized to five categories: Delta, Theta, Alpha, Beta and Gamma [7]. They are described by their frequency where Delta waves falls in the range 0.5–4 Hz, Theta is in the range 4–7 Hz, Alpha is within the range 8–13 Hz, Beta is in the range 14–30 Hz and Gamma is supposed to be in the above 30 Hz range. Alpha and Beta are the two most prominent frequency bands used in motor Imagery task. Alpha peaks around 10 Hz and it appears in mediation and relaxation state. Beta rhythm is associated with conscious and focuses states as well as it occurs in motor task movements.

Due to the importance of the extracted data accuracy, different steps have introduced for this reason. In the signal preprocessing step filter techniques are used to convert recorded signals to a special form in order to be easier in categorization. Some of the used techniques are Laplacian Spatial Filtering (LSF) [8] and Kalman filter [9].

In feature extraction step, suitable features with dimension reducing are used in order to gets good classification accuracy. The most prominent feature for imagery task is band power frequency. One of the issues of EEG signals is its nature is non-stationary signals; that mean its frequency changes over a time. Classical Fourier transform is not appropriate for such EEG signals because it produces only frequency component variable. In addition, Discrete Wavelet Transform (DWT) [4] was able to decompose the EEG signal into different level of resolution (frequency bands) using wavelet functions.

In classification step, several methods are utilized for BCI classification in which they are employing to predict class label for EEG signals patterns. In addition, supervised linear classifiers are commonly used especially, LDA [2] and Support Vector Machine (SVM) [3]. It depends on linear functions to discriminate classes. SVM is good when using features with high dimension, however it is sensitive to noise or outliers. So Fuzzy SVM (FSVM) [4] is used which promotes the effect of noise. In addition, it employs fuzzy membership for each training sample which is used for

Fig. 1 Brain signal
processing

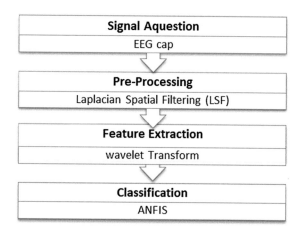

demonstrating decision. ANFIS is a hybridization method in which the system is trained using back propagation learning method.

3 Problem Definition

BCI becomes a new challenge in many of the recent applications. One of these applications is helping disabled users to move just by thinking of movement. The process requires brain signals related to motor imagery movement to be extracted and classified especially, hand movement. The purpose is to sense if disabled person able to move his/her hands through imagination. Consequently, the subject can use the same system for playing video games such as games depends on joystick movements. Certainly, this will help people who suffer from Amyotrophic lateral sclerosis (ALS) or spinal injuries. The purpose of this paper is to introduce a complete framework for the brain signal processing including different modules for accurate classification to EEG signals. The framework utilizes the Laplacian Spatial Filtering (LSF), wavelet transform, and the ANFIS for accurate classification. In addition, we tend to report the best EEG signal feature for imagery tasks based on our experiments.

4 EEG Proposed Classification and Clustering Algorithms

Before going deeply into the utilized techniques and algorithms, Fig. 1 elaborates on our approach in extracting accurate EEG signals. As can be seen in the figure, our proposed approach consists of four major phases which are signal acquisition, signal preprocessing, feature extraction, and classification.

Fig. 2 Alpha in C3 and C4 channels for *left* and *right* hand

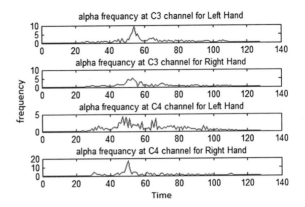

In the signal acquisition phase, the EEG signals are collected using very sensitive sensors on the scalp. The second phase is the preprocessing signals in which the data noise is filtered using specific filters such as: Kalman [9] and LSF [8] using high pass filter. In addition, channel selection is used in preprocessing.

In feature extraction phase, suitable features with dimension reducing are extracted to show differences between various classes. Band power feature, especially Alpha and Beta, is one of the most important features for motor imagery classification [8]. Alpha also known as mu rhythm and it is increased in cortex comparable to the arm used for imagination. In addition, Beta power linked to motor movement and appears in anxious thinking and active concentration. Wavelet Transform (WT) [4] is the most prominent method for extracting band power features because it considers the frequency of the signal in different band power. Four WT coefficients can be extracted which are D1:D4 and final approximation A4. WT coefficients are extracted using 'db8' wavelet function.

After analysis, as we will see later, we discovered that, if the user imagines moving left hand, the signal frequency in channel C3 becomes quite stronger than the signal frequency in channel C4 as shown in Fig. 2. In addition, if the user imagine moves his/her right hand, the signal frequency in channel C3 becomes quite weaker than signal frequency in channel C4 while channel C3 located in left side and C4 in right side.

The last phase is the classification in which the classifier is used to specify the brain signals for each class. In this research, ANFIS is the essential classifier used for data categorization. Moreover, LDA and KNN are used for comparing results. Performance of classifiers is measured by classification accuracy in which it is presented by the percentage of the number of correctly classified test patterns.

4.1 Autoregressive (AR) Model

AR model method [10] is used to extract linear feature for EEG signals which it describes the signal characteristics. AR model of order p is given by:

$$x[n] = -\sum_{k=1}^{p}(a_k \cdot x[n-k] + e[n]) \tag{1}$$

where p is the model order, $e[n]$ represents white noise error and $x[n]$ is signal data at point n. If EEG data is recorded using multiple channels, AR coefficients are produced for every channel and then combined to be a feature vector. Consequently, if the brain signals are recorded using C3 and C4 channels, AR coefficients for these signals are defined by $a_{c3}[n]$, $a_{c4}[n]$ respectively.

4.2 Common Spatial Pattern(CSP)

CSP [11] is a spatial filter method for preprocessing EEG signals. The main idea is to decompose brain signals into spatial patterns. Signal decomposition aims to minimize similarity between both classes and maximize differences by assigning the signal into patterns. This spatial pattern can be determined using covariance matrices conjoint diagonal of EEG signals for both classes. If the brain signals have length N and recorded through n electrodes, the dimension for covariance matrix will be $n \times N$.

CSP only suits two classes of discrimination because it depends on Fisher Discriminant Analysis (FDA) criteria. So CSP is modified to Bisearch CSP (BCSP) to discriminate four classes [12]. Details of CSP Equations are illustrated in [11].

4.3 Adaptive Neuro-Fuzzy Inference System (ANFIS) Classifier

Fuzzy inference system is a model that maps input characteristic to output decision. The shape of input/output membership function changes according to the input/ output parameters (fuzzy sets and rules); so it is easy to model fuzzy inference system. However, occasionally it is hard to extract parameters to tailor input/output membership functions when looking at data. Therefore, it will be appropriate if learning algorithm such as NN is used along with the fuzzy logic.

ANFIS [13] is a Sugeno model implementation for fuzzy inference in which it has the same view of NN with multiple hidden layers in between. ANFIS has five layers [14] as shown in Fig. 3. Details for each layer are explained in [13].

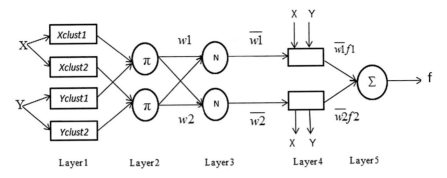

Fig. 3 ANFIS architecture for Sugeno method

Clustering process partitions the data into number of groups to facilitate data analysis where data in the same group have similar proprieties.

In the following subsection, different clustering techniques are explored including Subtractive, FCM and K-means.

4.3.1 Subtractive Clustering

Subtractive clustering [6, 15] fits under ANFIS technique. It defines cluster center based on the density around the data points. Candidates' data points are considered as cluster center then density of data point x_i, for instance, is computed by:

$$D_i = \sum_{j=1}^{n} \exp(-\frac{\|x_j - x_i\|^2}{(r_a/2)^2}) \tag{2}$$

where r_a is a positive constant specifying neighborhood for every cluster. After measuring all data points density, the largest density is selected as first cluster center. Assume the density for first data points x_{c1} is D_{c1}, density measure for data point x_i is revised by:

$$D_i = D_i - D_{c1} \exp(-\frac{\|x_i - x_{c1}\|^2}{(r_b/2)^2}) \tag{3}$$

where r_b is a positive constant determine neighborhood that has measurable reductions in density measure, generally $r_b = 1.5r_a$.

The subtractive clustering selects the highest density as the next cluster center after revising. Then, it continues until reaching sufficient number of clusters.

4.3.2 K-Means Clustering

One of data clustering algorithm is the K-means. K-means clustering [16] partitions data collection of n vectors $x_j, j = 1, 2,...n$ to a certain number of groups G_i, $i = 1, 2,...c$ and it defines k centroids one for each cluster. k centroids change their location step by step until they reach suitable locations. It is based on minimizing the cost (objection) function of dissimilarity using Euclidean distance to measure distance between data point x_k and cluster center c_i. Cost function is given by:

$$J = \sum_{i=1}^{c} J_i = \sum_{i=1}^{c} (\sum_{k,x_k \in G_i} \|x_k - c_i\|^2) \tag{4}$$

where J_i is the cost function for cluster i. Details of K-means clustering algorithm equations and steps are illustrated in [16].

4.3.3 Fuzzy C-Mean Clustering

(FCM) [16] is also classified under clustering algorithm. FCM is usually used in pattern recognition applications. In FCM, each data point belongs to more than one cluster with membership grades between 0 and 1. FCM objective function is represented by:

$$J_m = \sum_{i=1}^{n} \sum_{j=1}^{c} u_{ij}^m \|x_i - c_j\|^2 \quad , \quad 1 \leq m \leq \infty \tag{5}$$

where m is the weighting exponent and U_{ij} is the membership grade for data point x_i in the cluster j and c_j is the centroid for the cluster j. Details of FCM clustering algorithm equations and steps are illustrated in [16].

4.4 K-Nearest Neighbor (KNN) Classifier

KNN [3] is simple machine learning method based on the close neighbors with the same properties may belong to the same class. Each data point is assigned to a class according to the majority voting of its neighbors where KNN specifies k neighbors training points for every test data point using metrics as Euclidean, Manhattan, Mahalanobis distances. KNN calculates the distance for all training data vector to each test data vector then it selects the closest k neighbor. Therefore, the output value is the average of k nearest neighbors' values.

KNN is a simple classifier but it is known as lazy learner because it uses training data for classification. It takes the most common label in k training neighbors to be predicted label for new test data point. In addition, changes in the k closest nearest neighbor will change the output class.

4.5 Linear Discriminant Analysis (LDA) classifier

LDA [2] is a supervised linear classifier algorithm is used for dimensionality reduction. It is based on finding linear discrimination function to maximize multiple class separation by representing axes component. LDA Specifies data point to a class according to decision hyperplane function $g(x)$ given by:

$$g(x) = w^T x + w_0 \qquad (6)$$

where w is a weight vector for combining features, x is the feature vector and w_0 called Bias that determines the position of decision surface. Consider we have two classes $C1$ and $C2$, then $x \in C1$ if $g(x) \geq 0$ or $x \in C2$ otherwise.

5 Experimental Results

In this section, data set used for testing our approach is described and achieved results are analyzed according to classifiers performance and features used.

5.1 Graz Dataset

The used EEG data set in this paper is provided by Medical Informatics department, University of Technology at Graz. Female subjects with 25 years old were recorded imagery left and right hand movement dataset during sessions. Each session have 7 runs with 40 trials of 9 seconds length i.e. (280 trials). The subjects sat in a relaxing chair putting their arms on armrests. First two seconds was quite; at t = 2 an acoustic stimulus refers to the starting of the trial and cross '+' displayed on the screen for only one second. At t = 3 an arrow (left or right) was displayed for 1s, then subject asked to imagine left or right hand. The recording feedback was based on C3, Cz and C4 channels.

5.2 Classification with Varying Signal Intervals

In this subsection, classification is performed for hand imagination movement using C3 and C4 channels. Subsequently LSF is used to filter the EEG signals. Thereafter, WT is used to produce band power feature. In the present research, data set is digitized at 128 Hz; so fourth level of decomposition is chosen which is based on D1:D4 and A4. Moreover, the results are obtained by applying ANFIS, KNN and LDA using wavelet based features, AR coefficients and CSP.

In this experiment, the subject was idle in the first 3 seconds; so there is no enough information for classification. Therefore, intervals used for classification are from t = 4 to t = 8. Furthermore, these intervals are used in order to reduce the feature

Table 1 Classification accuracy using different signal intervals

ANFIS clustering methods

Signal length (%)	Subtractive (%)	FCM (%)	K-means (%)	KNN (%)	LDA (%)
600	95	90	90	87.5	90
500	82.5	80	80	70	80
300	72.5	70	70	75	75
200	92.5	92.5	92.5	82.5	85
Average	85.6	83.1	83.1	78.8	82.5

space to utilize signal information for classification. Every second is represented by 128 samples (signal length); so intervals used in classification have 640 samples. In this experiment, band power feature is extracted using WT and the classification was done with varying feature space. With hypothesis, subjects may lose their focus for moments so that affects acquired signals accuracy. Therefore, different intervals are selected randomly to add more classifiable information; then, the average will be taken after all.

Table 1 demonstrates ANFIS, LDA and KNN classifiers performance using band power feature with different signal intervals. It shows that, LDA accuracy is within the range from 75 % to 90 % with average 82.5 %. Likewise, ANFIS accuracy along different clustering algorithms varies from 70 % to 95 % and average about 85.6 %. As shown in Table 1, ANFIS and LDA classifiers achieved satisfactory results where, ANFIS produced the best result for subtractive method.

5.3 Classification for Every Second Interval

In this experiment, classification was done using band power feature while signal intervals are represented at every second from t = 4 to t = 8. Furthermore, 4 seconds was used as one interval. Assuming that, signals may involve coincide with more thinking process so the next second interval may have beneficial information for classification. Therefore, the classification process is tested at every second interval; then the average is compared with interval t = 4 to t = 8.

Table 2 demonstrates the accuracy for ANFIS, KNN and LDA classifiers. It shows that, the results at every second interval as well as the average. It seems that the average do not provide satisfactory result as shown in highlighted row. Consequently, all classifiers accuracy varies from 67 to 76 %. On the other hand, signal interval form t = 4 to t = 8 provides much better results for all classifiers. In particular, ANFIS using subtractive clustering algorithm is giving the best accuracy in which it achieves accuracy of 95 % compared to other researchers using the same data set [5].

Table 2 Classification accuracy for every second

ANFIS clustering methods

Time sample (%)	Subtractive (%)	FCM (%)	K-means (%)	KNN (%)	LDA (%)
T = 4	75	75	75	60	62.5
T = 5	80	82.5	82.5	77.5	77.5
T = 6	85	85	85	87.5	82.5
T = 7	72.5	75	75	57.5	67.5
T = 8	67.5	62.5	62.5	55	72.5
Average	76	76	76	67.5	72.5
T = 4:8	95	92.5	92.5	92.5	92.5

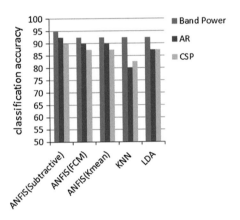

Fig. 4 Classification accuracy using band power AR and CSP features

5.4 Classification Accuracy Using Different Features

In this set of experiments, classifiers performance is evaluated using band power, AR coefficients and CSP features.

Band Power: Alpha and Beta are the most bands frequency that modulate the EEG signals for imagination activity. Classification accuracy using band power feature is demonstrated in Table 2 and Fig. 4 using feature space form t = 4 to t = 8 as one interval. As can be noticed, LDA, KNN and ANFIS classifiers provide the same results in which the produced accuracy is equal to 92.5 % while the subtractive method achieves the peak accuracy equal to 95 %.

AR Coefficients: As can be seen in the Fig. 4, KNN classifier does not perform well using AR in which it achieves only 80 %. LDA, FCM and K-means produce good results where LDA has 87.5 % accuracy while FCM and K-means gives accuracy of 90 %. However, the best accuracy is provided by ANFIS using subtractive clustering in which its average accuracy is 92.5 %.

CSP: CSP is the most popular algorithm in BCI field for learning spatial filters for oscillatory processes using frequency band and time window. CSP classification results for different classifiers are shown in the Fig. 4. Here, almost all of the algorithms are giving the same results including ANFIS using FCM, K-means and LDA. Their results approach 87.5 % of classification accuracy. However, it seems that KNN did not reach the same accuracy; therefore, it is not recommended to be used with CSP.

6 Conclusion

This paper has presented various EEG features and BCI classifiers using motor imagery dataset. This dataset was proposed to discriminate left and right hand imagination movement. WT based features was used in which it decomposes the signals to different levels. The number of decomposition level is based on EEG signal frequency. In addition, AR coefficients and CSP features were used.

FCM and K-means clustering algorithms provide stable classification result when the number of clusters is only two. However they produce different results for more than two clusters. This is because k-means chooses centroid for each cluster randomly and FCM depends on membership grade initialization for every cluster. Subtractive clustering algorithm generates the numbers of clusters according to provided cluster radius. Subsequently, it calculates cluster centroid density based on surrounding data points. Subtractive clustering produces stable result whatever the numbers of clusters.

The paper investigated the differences of three classifiers (ANFIS, LDA, and KNN) and achieved classification accuracies were reported. ANFIS classifier implementation depends on using clustering algorithms (subtractive, FCM and K-means). Clustering algorithms are used to tailor input data to output rules. The maximum classification accuracy is obtained using ANFIS classifier. In particular, using subtractive clustering algorithm in which it is provided 95 % for band power and 92.5 % for AR coefficients features. Furthermore, FCM and K-means are almost provided the same results.

References

1. Lin, C.T., Ko, L.W., et al.: Wearable and wireless brain-computer interface and its applications. Found. Augment. Cognit. **5638**, 741–748 (2009)
2. Lotte, F., Congedo, M., et al.: A review of classification algorithms for EEG-based brain-computer interfaces. Neural Eng. **4**, R1–R13 (2007)
3. Dokare, I., Kant, N.: Performance analysis of SVM, KNN and BPNN classifiers for motor imagery. Eng. Trends Technol. **10**, 19–23 (2014)
4. Xu, Q., Zhou, H., et al.: Fuzzy support vector machine for classification of EEG signals using wavelet-based features. Med. Eng. Phys. **31**, 858–865 (2009)

5. Yang, R., Gray, D. *et al*: Comparative analysis of signal processing in brain computer interface. IEEE Ind. Electron. Appl. 580–585 (2009)
6. Tawafan, A., Sulaiman, M., et al.: Adaptive neural subtractive clustering fuzzy inference system for the detection of high impedance fault on distribution power system. AI **1**, 63–72 (2012)
7. Larsen, E.A.: Classification of EEG Signals in a Brain Computer Interface System. University of Science and Technology, Norwegians (2011)
8. Qin, L., Ding, L., et al.: Motor imagery classification by means of source analysis for brain computer interface applications. Neural Eng. **1**, 144–153 (2004)
9. Aznan, NKN., Yang, YM.: Applying kalman filter in eeg-based brain computer interface for motor imagery classification. ICT Converg. 688–690 (2013)
10. Yang, R.: Signal Processing for a Brain Computer Interface. School of Electrical and Electronic Engineering. University of Adelaide, Adelaide (2009)
11. Gutierrez, D., Salazar, R.: EEG signal classification using time-varying autoregressive models and common spatial patterns. IEEE EMBS, 6585–6588 (2011)
12. Fang, Y., Chen, M., et al.: Extending CSP to detect motor imagery in a four-class BCI. Inf. Comput. Sci. **9**, 143–151 (2012)
13. Guler, I., Ubeyli, E.: Adaptive neuro-fuzzy inference system for classification of EEG signals using wavelet coefficients. Neurosci. Methods **148**, 113–121 (2005)
14. bin Othman, MF., Shan Yau, TM.: Neuro fuzzy classification and detection technique for bioinformatics problems. Model. Simul. 375–380 (2007)
15. Priyono, A., Ridwan, M., et al.: Generation Of fuzzy rules with subtractive clustering. Teknologi **43**, 143–153 (2005)
16. Ghosh, S., Dubey, S.K.: Comparative analysis of K-means and fuzzy C-means algorithms. J. Adv. Comput. Sci. Appl. **4**, 34–39 (2013)

Modeling Insurance Fraud Detection Using Imbalanced Data Classification

Amira Kamil Ibrahim Hassan and Ajith Abraham

Abstract This paper proposes an innovative insurance fraud detection method to deal with the imbalanced data distribution. The idea is based on building insurance fraud detection models using Decision tree (DT), Support vector machine (SVM) and Artificial Neural Network (ANN), on data partitions derived from under-sampling (with-replacement and without-replacement) of the majority class and merging it with the minority class. Throughout the paper, ten-fold cross validation method of testing is used. Its originality lies in the use of several partitioning under-sampling approaches and choosing the best. Results from a publicly available automobile insurance fraud detection data set demonstrate that DT performs slightly better than other algorithms, so DT model was used to compare between different partitioning-under-sampling approaches. Empirical results illustrate that the proposed model gave better results.

Keywords Insurance fraud detection · Imbalanced data · Decision tree · Support vector machine and artificial neural network

A.K.I. Hassan (✉) · A. Abraham
Department of computer science, Sudan University of Science and Technology,
Khartoum, Sudan
e-mail: amirakamil2@yahoo.com

A. Abraham
e-mail: ajith.abraham@ieee.org

A.K.I. Hassan
Machine Intelligence Research Labs (MIR Labs), Auburn, WA, USA

A. Abraham
IT4Innovations, VSB - Technical University of Ostrava, Ostrava, Czech Republic

© Springer International Publishing Switzerland 2016
N. Pillay et al. (eds.), *Advances in Nature and Biologically Inspired Computing*,
Advances in Intelligent Systems and Computing 419,
DOI 10.1007/978-3-319-27400-3_11

117

1 Introduction

Fraud is a major problem causing a lot of losses for many insurance companies. Data mining can minimize some of these losses by making use of the massive collections of customer data. Besides scalability and effectiveness, the fraud-detection task is faced with technical problems that include imbalanced dataset, which have not been widely studied in the insurance fraud detection community. The insurance fraud detection or generally fraud detection data is imbalanced. The fraudulent cases are minority class while the legitimate cases are big majority class. Using the data as it is results in high success rate for predicting legitimate cases but without predicting any fraudulent cases. There are two methods that are used to solve this problem [1]. The first method is to apply different algorithms (meta-learning). Each algorithm has its unique strengths, so that it may perform better on particular data instances than the rest [2]. The second method is to manipulate the class distribution (sampling). The minority class training examples can be equal or nearly equal in proportion to the majority class in order to raise the chances of correct predictions by the model(s) [3]. This paper introduces the new fraud detection method by using partitioning-under-sampling technique. The innovative use of decision tree, support vector machine and artificial neural network classifiers to process the partitioned data has the possibility of getting better results. One related problem caused by imbalanced data includes measuring the performance of the classifiers. Recent work on imbalanced data sets was evaluated using better performance metrics such as recall, precision [3–5] and area under curve [6]. In this paper, recall, precision and *PRC Curves (the Precision-Recall Characteristic curve)* were used to evaluate the performance of models. Section 2 contains related works, mentioning all recent researches in the area of insurance fraud detection and also includes a brief description of data mining algorithms used. Section 3 contains the proposed models and the data set and the experimental plan. Section 4 contains data pre-processing and experimental setup and Section 5 illustrates the results and analysis followed by Conclusions.

2 Related Works

The main problems facing insurance fraud detection are imbalanced data and the choice of data mining classifiers that give the best results [7]. Sternberg and Reynolds (1997) solved the problem by searching manually for the features that cause fraud [8]. Brockett et al. (1998) and Tennyson et al. improved this method a little by sorting the data into categories by an expert [9, 10]. Artis et al. and Caudill et al. (2005) used oversampling of the fraud claims class [11–13]. Pérez et al. also used oversampling of the fraud claims but testing with different fraud percentage [3]. Belhadji et al. used simple random sampling [14]. Pinquet et al. divided the dataset randomly into holdout sample (Random auditing) and working sample (usual auditing strategy) [15]. Viaene et al. and Farquad et al. randomly resample the dataset using a stratified sampling with blocked ten-fold cross-validation [4, 16].

Viaene et al. used partitioning-sample [17] and improved by randomly partitioning the data into k disjoint sets of approximately equal size [18]. Xu et al. generated multiple training subsets by rough set reduction technique [19]. Vasu and Ravi proposed a hybrid under-sampling approach, that employs kRNN method to detect the outliers then using K-means clustering [20]. Sundarkumar and Ravi further improved the proposed method by including one-class support vector machine to reduce the majority class even more [5]. Sternberg and Reynolds designed fraud detection expert system [8]. Brockett et al. applied Kolionen's method used Self-organizing feature map to classify automobile bodily injury claims by degree of fraud suspicion; neural network and a back propagation were used to investigate the validity [9]. Several other research works were also reported [10–13, 16, 21]. Neural network was used in [16, 18]. Xu et al. used Neural network classifier then improved the performance of the classifier by designing an ensemble neural network [19]. Several works used support vector machine[4, 5, 16, 20] and other used decision tree [3–5, 16, 20, 22]. Belhadji et al. and Pinquet et al. used Probit model to design expert system [14, 15].

Viaene et al. compared several data mining classifiers, Neural networks, support vector machine (SVM), K-nearest neighbor, Naïve Bayes (NB), Bayesian belief network, decision trees and Logistic model (LM), the results show that there is no great difference between the different classifiers, except SVM, LM and NB result were a little better [16]. Viaene et al. used (smoothed) naive Bayes (NB), Ada-Boosted naive Bayes (AB), and AdaBoosted weights of evidence (ABWOE) [17]. Pérez et al. compared two decision tree induction algorithms C4.5 and CTC [3]. Bhowmik used Naïve Bayesian Classification and Decision Tree-Based (C4.5 algorithms and Consolidated Trees) [22]. Farquad et al. used SVM-RFE for feature selection and for rule generation using decision tree (DT) and Naive Bayes tree (NB Tree) [4]. Vasu and Ravi compared several classifiers, support vector machine (SVM), logistic regression (LR), multi layer perceptron (MLP), radial basis function network (RBF), group method of data handling (GMDH), genetic programming (GP) and decision tree (J48) [20]. Sundarkumar and Ravi improved the previous proposed new under-sampling method tested its efficiency by comparing the new results with previous result, again using several classifier algorithms support vector machine (SVM), logistic regression (LR), multi layer perceptron (MLP), radial basis function network (RBF), group method of data handling (GMDH), genetic programming (GP) and decision tree (J48), it was shown that the proposed method gave better results. DT and SVM produced the best results [5].

3   Proposed Models

The methodology adopted in this paper is divided into three stages. In the first stage, a novel partitioning undersampling technique is proposed to solve the problem of imbalanced data. Six different groups of ten sub-samples are formed. In the second stage the undersampling techniques is evaluated using decision tree (DT) models and insurance fraud detection decision tree (IFDDT) model is designed. In the third stage other insurance fraud detection (IFD) classifier models are contracted using support vector machine (SVM) and artificial neural network (AAN). The diagram of the proposed approach is shown in Fig. 1.

3.1   Data Partitioning

According to Chan et al. [23], the desired distribution of the data partitions belonging to a particular fraud detection data set must be determined experimentally. In a related study, it is recommended by Chan and Stolfo [24] that data partitions should neither be too large for the time complexity of the learning algorithms nor too small to produce poor classifiers [1]. Given this information, the approach adopted in this paper is to randomly select a fixed number of legal examples and merge them with the entire fraud examples. The data partitions are formed by merging all the available X fraud instance (923) with a different set of Y legal instances to form ten *X:Y* partitions, (923:923) with a fraud: legal distribution of 50:50. The two other distributions used are 40:60 (923:1385) and 30:70

Fig. 1 Experiment Phases

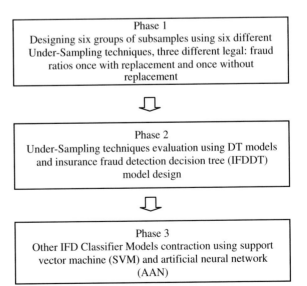

Phase 1
Designing six groups of subsamples using six different
Under-Sampling techniques, three different legal: fraud
ratios once with replacement and once without
replacement

Phase 2
Under-Sampling techniques evaluation using DT models
and insurance fraud detection decision tree (IFDDT)
model design

Phase 3
Other IFD Classifier Models contraction using support
vector machine (SVM) and artificial neural network
(AAN)

(923:2154). Therefore, there are ten data partitions with 1846 examples, another set of ten data partitions with 2308 examples, and the last ten data partitions with 3077 examples. Imbalanced data is transformed into partitions with more of the rarer examples and fewer of the common examples using the procedure known as majority under-sampling. The two random sampling techniques used were sampling with-replacement and sampling without-replacement

Sampling-with-replacement the majority class where the instances are selected randomly, the same instance could be chosen more than once. Sampling-without-replacement the majority class where the instances are selected randomly, the same instance could be chosen only once. In the case where the total size of the ten subsamples is larger than the dataset, the examples are chosen randomly until the whole dataset is chosen then the process start all over again (Fig. 2).

3.2   Algorithms Used in the Proposed Models

3.2.1   Decision Tree (DT)

Building a decision for any problem doesn't need any type of domain knowledge. Decision Tree is a classifier that uses tree-like graphs. The most common use of Decision Tree is in operations research analysis for calculating conditional probabilities [25]. Decision makers can choose best alternative and traversal using

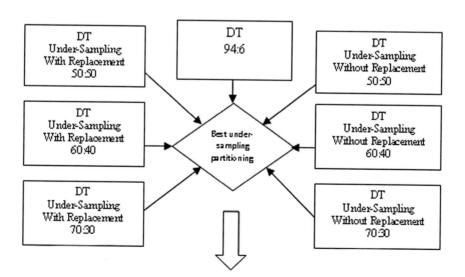

Fig. 2 Choosing the best Under-Sampling Technique

Decision Tree where root to leaf indicates exclusive class division based on maximum information gain [26].

3.2.2 Support Vector Machine (SVM)

Vapnik et al. explained SVMs technique, which is based on statistical learning theory SVMs were originally designed for binary classification but they could be well extended for multiclass problems [27, 28]. SVM perform by designing a hyper plane in input space to separate the data points. Various kernel functions such as polynomial, Gaussian, sigmoid etc., are used for this purpose. The data points are classified using a hyper plane, which is constructed with the help of support vectors [25].

3.2.3 Neural Network (NN)

It is an algorithm for classification that is designed based on biological nervous system having multiple elements that are interrelated, known as neurons, functioning in unity to solve specific problems. NN uses gradient descent method [29]. An NN adapts its structure and adjusts its weights in order to minimize the error [25].

4 Data Pre-processing and Experimental Setup

4.1 Data description

In this research the insurance dataset used was originally used by Phua et al. (2004) [30]. This dataset mainly contains the information regarding the various automobile insurance claims during the period 1994–1996, comprises 32 variables, with 31 predictor variables and one class variable. It consists of 15,420 samples of which 14,497 are legitimate cases and 923 are fraudulent cases, which mean there are 94 % legitimate cases and 6 % fraudulent cases.

4.2 Data Cleaning and Preparation

It is observed that the age attribute in the dataset appeared twice in numerical and categorical form as well; hence the categorical attribute was chosen since it is simpler.

Further, the date of the accident was represented by the four attributes year, month, week of the month and day and date of the insurance claim was represented by three attribute. Thus, a new attribute gap was derived from all these attributes; it represents the time difference between the accident occurrence and insurance claim. Hence, 15,420 examples with 24 predictor variables and one class variable formed the final dataset.

4.3 Experimental setup

Phase 1: Designing subsamples using partitioning under-sampling techniques

The dataset was divided into majority class and minority class. The minority class was untouched. The majority class was divided into ten subsamples (partitions), to choose the cases of each subsample (partition), two sampling methods were used, sampling with replacement and sampling without replacement. With each sampling methods three different (legal: fraud) ratios, were used, i.e. 50:50, 60:40 and 70:30. This means that the majority class set is sampled using six techniques; sampling with replacement and 50:50 ratio, with replacement 60:40, with replacement 70:30, without replacement 50:50, without replacement 60:40 and without replacement 70:30. Ten subsamples were selected using each one of the six sampling methods. As a result there were 60 subsamples plus the original dataset.

Phase 2: Evaluating the partitioning under-sampling technique by designing IFD Decision tree model

The DT algorithm is extremely time-efficient so it is used to compare the different partitioning under-sampling techniques and the original dataset. Insurance Fraud Detection Decision Tree (IFDDT) models were designed. All 60 subsamples and the original dataset were used to design IFDDT models. The models designed using the 60 subsamples were compared with the model of the original dataset, to show the benefit of using imbalance sampling techniques. The partitioning-under-sampling technique that produces the IFDDT model with the highest Recall is used through the rest of the experiment.

Phase 3: Other IFD models constriction using support vector machine (SVM) and artificial neural network (ANN)

The support vector machine classifier was repeated ten times each time with different Kernel; REF kernel, Polynomial Kernel, Person Universal Kernel and Normalized Polynomial Kernel. The SVM models are evaluated and the model which yields the highest recall is chosen. Then an artificial neural network (ANN) model is trained by a back propagation algorithm. Several parameters of the ANN model were tested by trial and error to get the best results that were recorded. The Three IFD models are compared with previous similar studies to choose the best model. To obtain statistically reliable results, 10-fold cross validation method is used throughout this study. The training dataset is divided into 10 subsets of equal size using sampling technique explained above in phase 1. The 9 subsamples used as training data, and the remaining subsample as testing. Then the results on test fold are averaged.

5 Results and Analysis

The proposed six methods of partitioning-under-sampling techniques resample the dataset into six groups of subsamples each containing ten partitions. For each group a decision tree model is designed and evaluated. Also a decision tree model is designed using the original data. Ten cross validation is use throughout the experiment. The average of the ten partitions is recorded and shown in Table 1 (Tables 2 and 3).

Then a IFDSVM (Insurance Fraud Detection Support Vector Machine) model is designed. The different parameters of the classifiers are adjusted using trial and error. IFDSVM models are designed using four kernel types (REF Kernel, Polynomial Kernel, Person Universal Kernel and Normalized Polynomial Kernel) to choose the best model. IFDSVM model using Normalized Polynomial Kernel was chosen because it yields a recall of 94.1 %. Although the model using Person Universal Kernel also yield a recall of 94.1 but its precision was less 68.7 % compared to other precision 68.9 %. Then at the last phase of the experiment IFDANN (Insurance Fraud Detection Artificial Neural Network) model was

Table 1 IFD model using decision trees

Under-sampling type	Recall (%)	Precision (%)	PRC area (%)
With-replacement 50:50	95.8	69.4	73.0
Without-replacement 50:50	94.6	69.0	73.9
With-replacement 60:40	90.3	60.3	74.2
Without-replacement 60:40	89.4	60.5	74.6
With-replacement 70:30	47.4	59.3	80.0
Without-replacement 70:30	51.1	57.7	78.2
Original dataset	3.6	86.8	93.6

Table 2 IFD model using SVM

Kernel type	Recall (%)	Precision (%)	PRC area (%)
REF kernel	92.2	68.5	69.4
Polynomial kernel	93.0	68.9	70.2
Person universal kernel	94.1	68.7	70.4
Normalized polynomial kernel	94.1	68.9	70.7

Table 3 IFD model using ANN

No of hidden layer	No. of units in each layer	learning rate	Momentum	Recall (%)	Precision (%)	PRC area (%)
1	2	0.3	0.2	89.7	69.0	76.7
1	3	0.3	0.2	85.6	69.3	76.8
2	2; 2	0.3	0.2	88.6	69.1	75.8
2	2; 3	0.3	0.2	89.6	68.7	76.2
1	2	0.25	0.2	86.8	70.3	76.8
1	2	0.35	0.2	89.3	69.1	76.8
1	2	0.3	0.15	89.6	68.8	76.7
1	2	0.3	0.25	88.9	69.1	76.7

Table 4 Results using German Dataset

Classifier type	Recall	Precision	PRC area
Artificial neural network	76.8	75.8	80.5
Support vector machine	77.0	76.1	68.1
Decision tree	73.5	72.3	68.2

Table 5 Results comparison with other works in the literature [31]

Classifier type	Previous model using NN	Proposed model using NN
Artificial neural network	68.0	**76.8**

designed. The parameters of the network (number of hidden layer, no of units in each layer, learning rate and momentum) were adjusted using trial and error method. The best result was recorded. For testing the proposed mode it was applied on another imbalanced dataset (German dataset), the results reported in Table 4.

Those results were compared with the result published in previous paper [31]. The highest recall (Default% as referred to in the previous paper) is 68.0, which is much less than 76.8, which is the recall of the proposed model using NN (Table 5).

6 Conclusions

In this paper, we proposed a novel partitioning-under-sampling-technique of the majority class in imbalanced datasets in order to improve the performance of classifier models. This paper demonstrates the significance of majority class partitioning in the case of imbalanced datasets. In the proposed approach, random sampling with-replacement was applied on majority class using three different subsample sizes (923, 1385, and 2154). For each subsample size 10 subsamples are selected. Then, another three groups of 10 subsamples were randomly selected using random sampling without replacement this time. Total of 60 subsamples were selected. Finally, each one of the subsamples of the majority class samples was merged with the minority class and the experiments were carried out with the modified dataset. The effectiveness of the proposed approach was tested with three classifiers DT, ANN and SVM using tenfold validation throughout the whole experiment. The best model was IFDDT model that used DT yield 95.8 recall. The proposed IFDNN model was compared with previous NN model showing an improvement in the proposed model. The results show that using partitioning-under-sampling method, the classifiers performed better compared to when original imbalanced data was presented to them. The proposed models can be further improved by using ensemble classifiers.

References

1. Phua, C., Alahakoon, Damminda, Lee, Vincent: Minority report in fraud detection: classification of skewed data. ACM SIGKDD Explor. Newsl. **6**, 50–59 (2004)
2. Wolpert, D.H., Macready, W.G.: No free lunch theorems for optimization. IEEE Trans. Evol. Comput. **1**, 67–82 (1997)
3. Pérez, J.M., Muguerza, J., Arbelaitz, O., Gurrutxaga, I., Martín, J.I.: Consolidated tree classifier learning in a car insurance fraud detection domain with class imbalance. In: Pattern Recognition and Data Mining (ed), pp. 381–389. Springer (2005)
4. Farquad, M., Ravi, V., Raju, S.B.: Analytical CRM in banking and finance using SVM: a modified active learning–based rule extraction approach. Int. J. Electron. Customer Relat. Manag. **6**, 48–73 (2012)
5. Sundarkumar, G.G., Ravi, V.: A novel hybrid undersampling method for mining unbalanced datasets in banking and insurance. Eng. Appl. Artif. Intell. **37**, 368–377 (2015)
6. Ibarguren, I., Pérez, M., Muguerza, J., Gurrutxaga, I., Arbelaitz, O.: Coverage based resampling: building robust consolidated decision trees. Knowl.-Based Syst. (2015)
7. Hassan, A.K.I., Abraham, A.: Computational intelligence models for insurance fraud detection: a review of a decade of research. J. Netw. Innovative Comput. **1**, 341–347 (2013)
8. Sternberg, M., Reynolds, R.G.: Using cultural algorithms to support re-engineering of rule-based expert systems in dynamic performance environments: a case study in fraud detection. IEEE Trans. Evol. Comput. **1**, 225–243 (1997)
9. Brockett, P.L., Xia, X., Derrig, R.A.: Using Kohonen's self-organizing feature map to uncover automobile bodily injury claims fraud. J. Risk Insur. 245–274 (1998)
10. Tennyson, S., Salsas-Forn, P.: Claims auditing in automobile insurance: fraud detection and deterrence objectives. J. Risk. Insur. **69**, 289–308 (2002)
11. Artís, M., Ayuso, M., Guillén, M.: Modelling different types of automobile insurance fraud behaviour in the Spanish market. Insur.: Math. Econ. **24**, 67–81 (1999)
12. Artís, M., Ayuso, M., Guillén, M.: Detection of automobile insurance fraud with discrete choice models and misclassified claims. J. Risk Insur. **69**, 325–340 (2002)
13. Caudill, S.B., Ayuso, M., Guillen, M.: Fraud detection using a multinomial logit model with missing information. J. Risk Insur. **72**, 539–550 (2005)
14. Belhadji, E.B., Dionne, G., Tarkhani, F.: A model for the detection of insurance fraud. In: Geneva Papers on Risk and Insurance. Issues and Practice, pp. 517–538 (2000)
15. Pinquet, J., Ayuso, M., Guillen, M.: Selection bias and auditing policies for insurance claims. J. Risk Insur. **74**, 425–440 (2007)

16. Viaene, S., Derrig, R.A., Baesens, B., Dedene, G.: A comparison of state-of-the-art classification techniques for expert automobile insurance claim fraud detection. J. Risk Insur. **69**, 373–421 (2002)
17. Viaene, S., Derrig, R.A., Dedene, G.: A case study of applying boosting Naive Bayes to claim fraud diagnosis. IEEE Trans. Knowl. Data Eng. **16**, 612–620 (2004)
18. Viaene, S., Dedene, G., Derrig, R.A.: Auto claim fraud detection using Bayesian learning neural networks. Expert Syst. Appl. **29**, 653–666 (2005)
19. Xu, W., Wang, S., Zhang, D., Yang, B., Random rough subspace based neural network ensemble for insurance fraud detection. In: 2011 Fourth International Joint Conference on Computational Sciences and Optimization (CSO), pp. 1276–1280 (2011)
20. Vasu, M., Ravi, V.: A hybrid under-sampling approach for mining unbalanced datasets: applications to banking and insurance. Int. J. Data Min. Model. Manage. **3**, 75–105 (2011)
21. Viaene, S., Ayuso, M., Guillen, M., Van Gheel, D., Dedene, G.: Strategies for detecting fraudulent claims in the automobile insurance industry. Eur. J. Oper. Res. **176**, 565–583 (2007)
22. Bhowmik, R.: Detecting auto insurance fraud by data mining techniques. J. Emerg. Trends Comput. Inf. Sci. **2**, 156–162 (2011)
23. Chan, P.K., Fan, W., Prodromidis, A.L., Stolfo, S.J.: Distributed data mining in credit card fraud detection. Intell. Syst. Appl. IEEE **14**, 67–74 (1999)
24. Chan, P.K., Stolfo, S.J.: A comparative evaluation of voting and meta-learning on partitioned data. In ICML, pp. 90–98 (1995)
25. Tomar, D., Agarwal, S.: A survey on Data Mining approaches for Healthcare. Int. J. Bio-Sci. Bio-Technol. **5**, 241–266 (2013)
26. Apté, C., Weiss, S.: Data mining with decision trees and decision rules. Future Gener. Comput. Syst. **13**, 197–210 (1997)
27. Cristianini, N., Shawe-Taylor, J.: An Introduction to Support Vector Machines and Other Kernel-based Learning Methods. Cambridge university press (2000)
28. Cristianini, N., Shawe-Taylor, J.: An Introduction to Support Vector Machines (ed). Cambridge University Press (2000)
29. Silver, M., Sakata, T., Su, H.-C., Herman, C., Dolins, S.B., Shea, M.J.O.: Case study: how to apply data mining techniques in a healthcare data warehouse. J. Healthc. Inf. Manage. **15**, 155–164 (2001)
30. Phua, C., Alahakoon, D., Lee, V.: Minority report in fraud detection: classification of skewed data. ACM SIGKDD Explor. Newsl. **6**, 50–59 (2004)
31. Hassan, A.K.I., Abraham, A.: Modeling consumer loan default prediction using neural netware. In: 2013 International Conference on Computing, Electrical and Electronics Engineering (ICCEEE), pp. 239-243 (2013)

Using Standard Components in Evolutionary Robotics to Produce an Inexpensive Robot Arm

Michael W. Louwrens, Mathys C. du Plessis and Jean H. Greyling

Abstract Evolutionary Robotics (ER) makes use of evolutionary algorithms to evolve controllers and morphologies of robots. Despite successful demonstrations in laboratory experiments, ER has not been widely adopted by industry as means of robot design. A possible reason for this is that current ER approaches ignore issues that are important when designing robots for practical use. For example, the availability and cost of components used for robot construction should be considered. A robot designed by the ER process may require specialised custom components to be built to support the physical functioning of the design, if the components selected by an ER process are not widely available. Alternatively, the ER designed robot may be too expensive to be constructed. This paper demonstrates that standard off-the-shelf components can be used by the ER process to design a robot. A robot arm is used as a sample problem, which is successfully optimised to use components from a fixed list while minimising cost.

Keywords Evolutionary robotics · Evolutionary computing · Standard components

The financial assistance of the National Research Foundation (NRF) towards this research is hereby acknowledged. Opinions expressed and conclusions arrived at, are those of the authors and are not necessarily to be attributed to the NRF.

M.W. Louwrens (✉) · M.C. du Plessis · J.H. Greyling
Nelson Mandela Meteropolitan University, University Way,
Port Elizabeth 6031, South Africa
e-mail: michael.louwrens@nmmu.ac.za
URL: http://www.nmmu.ac.za

M.C. du Plessis
e-mail: mc.duplessis@nmmu.ac.za

J.H. Greyling
e-mail: jean.greyling@nmmu.ac.za

© Springer International Publishing Switzerland 2016
N. Pillay et al. (eds.), *Advances in Nature and Biologically Inspired Computing,*
Advances in Intelligent Systems and Computing 419,
DOI 10.1007/978-3-319-27400-3_12

129

1 Introduction

Evolutionary Robotics is a subset of Evolutionary Algorithms(EA) and deals with evolving morphologies and controllers for robots. Morphologies are the physical layout of the robot. The physical layout would depend on the robot type as well as the task to be performed. Each robot type can have multiple layouts where the layout of the robot includes the specific components, their placement and mounting to form a robot. The controller describes how the physical layout must work to perform the desired task.

ER has been successfully applied to evolve morphologies and controllers for several robots. The ER process has shown great promise by creating novel controllers and unique morphologies with minimal human input. However, ER has not been widely adopted by industry and the use of EA is mostly limited to merely optimising design parameters of physical robots. A possible reason for the limited adoption of ER is that it does not fit in well with the standard design process. In the traditional robotic design process the available components and overall budget are known to the designer and the design will be done within these limitations taken into account from the start. In contrast, evolving a morphology and controller using ER generates a design for a robot which must be constructed after the fact. This design does not necessarily take into account component availability or cost. The resultant robot may thus be infeasible to construct in the real world or simply be too expensive.

This research attempts to address the above mentioned shortfalls in the ER process by being a proof of concept study to demonstrate that standard off-the-shelf components can be incorporated into ER throughout the entire process. A simple robot morphology and controller is evolved which makes use of standard off the shelf components to achieve the design. This will enable the optimisation of cost along with the creation of the robot.

The robot evolved in this study is a robot arm of which the end effector must travel from a start point to an end point while avoiding an obstacle in the environment. The morphology of this arm will not be a fixed morphology. Rather the entire morphology will be evolved to suit the task and optimise obstacle avoidance, positional accuracy and cost. As this is a preliminary study this will be evolved only in simulation and the mechanical strength of the components will be ignored. Motors are thus assumed to be strong enough to lift and support the required weights.

The rest of this paper is structured as follows: Related work is discussed in Sect. 2 followed by the new proposed approach in Sect. 3. This approach will be studied using a sample problem described in Sect. 4. The implementation of this sample problem will be described in Sect. 5 after which the results will be discussed in Sect. 6 and conclusions drawn in Sect. 7.

2 Related Work

Simulation is typically used to evolve robots after which the final design is transferred to the real world. The algorithm uses a simulator to test the fitness of the robot produced. New generations are produced in the same manner that they are produced in an EA: reproduction operators. Generations of robots are created until a stopping condition is reached. The algorithm then outputs the controller and/or the body plan of the robot which then may be used in the creation of a physical robot [4].

ER can evolve novel controllers and morphologies. Lego morphologies are created by Funes [3] who shows that arbitrary morphologies can be created from basic building blocks to perform specific tasks while Lipson [8] co-evolves the robot's controller and morphology simultaneously, creating the robot's morphology and controller in a single process. Feasible robot evolution [2] investigations have shown results such that the robots produced through the ER process can be applied to real world applications. The robots used modular components which could be assembled and used to create robots designed to perform a described task.

There is lack of research which takes the physical construction of the robot into account during the ER process. An exception to this is using modular robots in the ER process. Modular robots are self-contained units which are assembled to create larger robots [1]. The modular robots are used as the base component for the chromosome in the evolutionary process [6]. The modular robot components are mirrored in the real world and the result is a robot which can be immediately assembled. Several researchers have investigated using modular robots in ER [1, 2, 6]. Despite the promising research on modular robots, modularity may limit the optimal performance of the robot [7].

The sample problem for this paper is the creation of a robot arm. Although no previous work has been done on evolving the entire structure of a robot arm, previous work has been done in optimising fixed robot arm morphologies [12, 13]. Optimising fixed arm morphologies is done by optimising link parameters such as link length or angle of twist. The goal in parameter optimisation is often to maximise the workspace of the robot arm [12, 13].

3 Proposed Approach

This paper proposes the use of standard components for robot evolution. Standard components refers to non-custom parts which are commercially available and typically inexpensive. Evolved robots can thus be simply constructed in the real world while various factors such as cost can be included in the optimisation process.

The described problem is a multi-objective optimisation problem. This is challenging as the various objectives of the algorithm must be balanced. However, in this particular instance the objectives can be categorised as either primary or secondary. Primary goals are prerequisite requirements while secondary goals are not vital to task completion but do affect the robot. Cost would usually be a secondary goal while avoiding collisions would be a primary goal. An expensive robot can complete the task while a robot which collides with an obstacle does not.

4 Sample Problem

To show that standard components can be used in ER a sample problem of evolving a robot arm is examined. The robot arm is required to perform a single task. This task is to move from an initial position to a final position. In order to make the task and the controller more complex, an obstacle is inserted between the start and end position. The controller thus requires a path planning component which generates a safe path for the robot to traverse without colliding with the obstacle.

In papers by Rout [11], Panda [9] and Patel [10] it can be seen that typically in robot arm optimisation the structure is fixed and the parameters are continuous. This differs from this study where the structure is dynamic and the parameters are fixed by the components. Link length is not a continuous variable, it is a discrete parameter of the component selected.

The sample problem will use an obstacle starting at $(0.3, 0.3)$ and made from the two line segments $y = 0.3$ and $y = 0.3x$. This obstacle is infinitely defined in the z-axis thus the obstacle can be seen as an infinitely high wall. The path planning algorithm was required to not create new points any closer than 0.1 m to the obstacle. The target points the robot was required to reach were $(0.35, 0.1, 0.1)$ and $(0.35, 0.5, 0.2)$. The robot will have an accuracy threshold that defines how close the end-effector of the arm must be to the target point to be deemed to have successfully reached that point. This threshold value was selected to be 0.03m. This arm has primary objectives of reaching the threshold value without the arm colliding with the obstacle and a secondary objective of minimising the cost of the arm. This task is shown in Fig. 1.

A robot arm or manipulator in this scenario is a set of motors connected serially to provide a large working area. The motors are connected by links which are simple connectors which only differ in length. A discrete set of links and motors is available for the creation of the robot arm. The physical characteristics of a catalogue of commercially available servo motors was captured [5]. Mounting the components to each other is determined by three factors: a mounting type, a side to which it is mounted and an angle at which it is mounted.

5 Implementation

The goals of the sample problem were solved by splitting the problem into two free-standing evolutionary algorithms. The purpose of the first algorithm is to generate the sequence of points which would result in a collision free path. The second algorithm takes the result of the first algorithm as an input and evolves a robot arm along with a set of motor angles which will result in the end effector reaching each point in the path. Section 5.1 describes the first algorithm while Sect. 5.2 describes the second. Various parameter choices were motivated through preliminary parameter optimisation tests. The fitness was maximised in both algorithms.

5.1 Path Finding Algorithm

5.1.1 Representation and Initialisation

The path finding EA is represented by a variable-length chromosome containing a sequence of points in three dimensional space. The chromosomes were initialised to have one to three genes upon creation. The first point in the sequence is the starting point and the final point the end point of the path. Genes were randomly initialised in a cube with one corner at the base of the robot and sides of length equal to R. R is defined as the distance between the base of the robot and the furthest goal point (either the start or end point depending on the problem).

5.1.2 Operators

A two parent crossover was used to generate a single offspring individual. Two separate random points were chosen in each parent. The offspring individual is made up of the genetic material of the first parent up to the first parent's crossover point combined with the genetic material from the second parent's crossover point up to the end of the second parent's chromosome. Tournament selection was used with a tournament size of 20 % of the population size. Two mutation operators are used, the first to introduce small amounts of local variation, while the second is used to introduce large changes. The first mutation operator adds random values from a normal distribution with a standard deviation of $\frac{R}{10}$. The second mutation operator consisted of reinitialising genes flagged for mutation. The mutation operator was applied with a probability of 0.05 with each operator having an equal chance of being used. Elitism was implemented by saving the best result of the previous generation in the new generation.

5.1.3 Fitness Function

The fitness function was created to reward individuals for creating shorter paths and punish individuals for collisions. The algorithm functions by determining if the line between two successive points intersects with the obstacle. If an intersection is found the individual is punished. The general algorithm is given in Algorithm 1

```
D = 0; P = 0;
for each successive pair of points in the chromosome: p₁, p₂ do
    D = D + Distance(p₁,p₂);
    if a collision occurs between p₁ and p₂ then
        P = P + 10000;
    end
end
Fitness = -(D + P);
```

Algorithm 1: Fitness of path finding algorithm

5.2 Robot Arm Evolution

5.2.1 Representation and Initialisation

Each gene in the chromosome represents a component. A component can either be
a servo motor or a link, with the first component in the chromosome being a motor.
The successive components in the chromosome alternate between motors and links.
For each motor a sequence of N angles is stored, where N is the number of destination points found by the Path Finding Algorithm. Components are mounted based
on a mounting type and variables which determine mounting information for each
component is stored in the chromosome. Angles are randomly initialised within the
valid travel range of the relevant servo motor, while mounting values are randomly
selected from the list of valid mounting combinations. The chromosome itself was
initialised to have two to ten components on creation.

5.2.2 Operators

The crossover, selection operators as well as elitism used in the path finding algorithm were used in the robot arm evolution. A restriction on the crossover was added
such that the resulting child conforms to a valid representation. Two mutation operators are used in this EA. The first mutation operator potentially adds random values
from a normal distribution with a standard deviation of 5° to each angle in the gene
with a probability of 0.1. The second mutation operator was applied with a probability of 0.05 and consisted of reinitialising the component in genes flagged for
mutation.

5.2.3 Fitness Function

The fitness function aims to create a robot arm that can successfully perform the
task at the cheapest cost. The end effector of the arm must reach the threshold for
all targets without colliding with the obstacle at any point while travelling through
the path. The fitness of the arm then depends on the distance from the target at each

target point, the magnitude of the penetration into the obstacle and the overall cost of the arm. These separate fitnesses are weighted and added to create a single fitness for the individual. Additionally, the fitness function for the arm was designed such that it could create robots which could achieve largest distance remaining (LDR) values even lower than the threshold value. This was achieved by lowering the influence of the position of the arm once it has reached the threshold. The general algorithm is given in Algorithm 2.

Target = target list;
N = length of target list;
Component = list of components in chromosome x;
L_x = length of chromosome x;
$P_{collision} = 0$; $P_{targetdistance} = 0$; $C = 0$;
for 1 to N **do**
    for $c = 1$ to $L_x - 1$ **do**
        $line$ = line between Component[c] and Component[c+1];
        if $line$ intersects with the obstacle **then**
            $P_{collision} = P_{collision}$ + Penetration Amount;
        end
    end
end
for $n = 1$ to N **do**
    q = the end effector position when attempting to reach Target[n];
    if Distance(q,Target[n]) > Threshold **then**
        $P_{targetdistance} = P_{targetdistance}$ + Distance(q,Target[n]);
    end
    else
        $P_{targetdistance} = P_{targetdistance} + \frac{\text{Distance(q,Target[n])}}{1000}$;
    end
end
for $c = 1$ to L_x **do**
    $C = C + $ Cost(Component[c]);
end
Fitness $= -(P_{targetdistance} \times 1000000 + P_{collision} \times 1000000 + C)$;

Algorithm 2: Fitness of chromosome x

---

6 Evaluation

The ER process used to evaluate the sample problem was run for 30 instances. The path finding and robot arm EAs had population sizes of 100 and 50 respectively and were run for 2000 and 20,000 generations respectively. The path finding EA produced a path which added a single intermediate point to produce a collision free path as shown in Fig. 1.

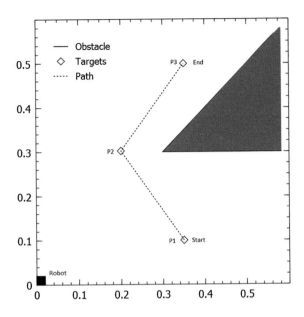

Fig. 1 Solution to the path-finding problem

Table 1 contains the output data of the 30 instances. This data is divided into multiple columns: the total magnitude of all vectors from the end effector to each target point i.e. the distance remaining summation (DRS); the largest distance remaining (LDR) which is the magnitude of the largest vector from the end effector to any target point; the sum of penetration distances(SPD) which indicates if the arm has penetrated the obstacle at any point along its path of travel; the total cost of all components in the arm; finally, the table contains the total components used to construct the arm i.e. the number of motors and links. A solution was considered successful if and only if it satisfies the primary objectives such that LDR was lower than 0.03 m and SPD was zero.

Twelve instances produced successful robots. The average cost for all thirty instances was evolved from an initial value of R1501.92 down to R601.38. The lowest cost for an arm was R332.76. This arm was the most successful by having the lowest cost while satisfying primary objectives. If minimising LDR was the goal, then the arm which had an LDR of 3.44×10^{-5}m and a cost of R482.08 would have been the most successful.

Figure 2 shows a 3D view of the best result. The LDR for this result was 0.0236 m and had a total cost of R332.76. This arm's structure had a motor at the base which rotated a link around the z-axis. The next motor was mounted at 90° to the link and would raise or lower the next link as required. The third motor was mounted in-line with the previous link allowing for further height modification with a shorter link. The last motor was mounted at 90° around a different axis, rotating its link in the xy plane instead of the yz plane the previous two motors worked in. Using this layout, three configurations are shown. Each configuration represents the end effector attempting to reach P1, P2 and P3. Starting in configuration 1 the robot

Table 1 Results

DRS [mm]	LDR [mm]	SPD	Total cost	Nr. components	Notes
0.07	0.03	0	482	10	Success—most accurate
2.68	1.88	0	635	10	Success
11	10.1	0	594	13	Success
21.3	11.7	0	591	8	Success
14.4	14.3	0	603	10	Success
31.9	17	0	605	9	Success
23.3	23.2	0	566	9	Success
23.6	23.6	0	333	8	Success—least expensive
32.4	27.4	0	644	11	Success
51	28.5	0	487	12	Success
29.3	29	0	791	14	Success
36.3	29.6	0	670	10	Success
33.4	32.5	0	796	14	
34.6	34.5	0	680	8	
78.4	69.6	0	1100	10	
77.5	77.5	0	670	8	
78.2	78.1	0.00232	495	10	
102	79.6	0	700	9	
199	91	0	513	7	
114	96.3	0	563	7	
108	108	0	630	7	
122	111	0	624	8	
142	141	0	653	13	
177	146	0.000319	552	12	
149	149	0	715	9	
182	154	0	414	6	
206	186	0	449	8	
211	189	0	453	6	
253	199	0	598	6	
279	262	0	433	4	

would rotate the base and lift up the midsection of the arm to move to configuration 2. This movement shifts the end effector of the arm from P1 to P2. Moving from configuration 2 into configuration 3 requires the rotation of the base towards the obstacle and the lowering the midsection to point at P3 with the final section rotating to point in the same direction as the rest of the arm. Configuration 2 allows the arm to safely retract and extend without colliding with the obstacle. The solution shown in Fig. 2 created a workspace that could reach all the target points within the threshold value while not colliding with the obstacle at any point.

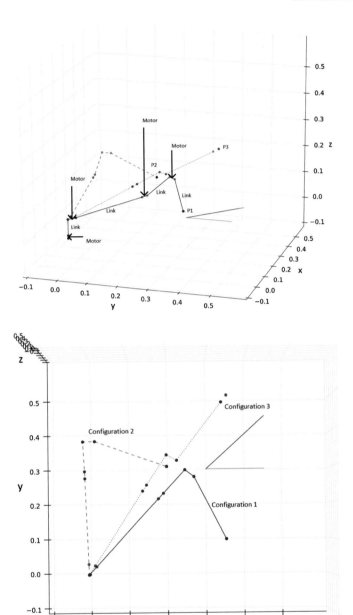

Fig. 2 Lowest cost arm *top* and orthographic views

7 Conclusion

The algorithm succeeded in producing a robot arm using the ER process and standard off-the-shelf components. 40 % of the results produced in the evaluation set were successes. The ER process succeeded in solving a multiple objective optimisation problem in that results which satisfied the primary objectives were produced while reducing cost. This is another step towards industrial application of ER.

Future work would be constructing and testing a solution in the real world. This would require the integration of torque to the fitness function. The task may or may not have a load associated with it but for the arm to be constructed in the real world, the forces acting on the arm must be calculated to ensure that the motors selected can handle the forces and possibly loads from the arm and task. This would add torque to the fitness function as a similar factor like SPD in that a robot would fail if any motor could not handle the torque acting on it.

Only rotational servos were present in the current data set, adding linear servos in future work would be an interesting addition as the results could potentially mimic a more traditional robot arm design.

References

1. Christensen, D.J., Schultz, U.P., Stoy, K.: A distributed and morphology-independent strategy for adaptive locomotion in self-reconfigurable modular robots. Robot. Auton. Syst. **61**(9), 1021–1035 (2013). Sept
2. Faiña, A., Bellas, F., Orjales, F., Souto, D., Duro, R.: An evolution friendly modular architecture to produce feasible robots. Robot. Auton. Syst. **63**, 195–205 (2015). Jan
3. Funes, P., Pollack, J.: Computer evolution of buildable objects for evolutionary design by computer, pp. 1–20 (1999)
4. Harvey, I., Husbands, P., Cliff, D., Thompson, A., Jakobi, N.: Evolutionary robotics: the Sussex approach **20**, 205–224 (1997)
5. HobbyKing. Servos and Parts: http://www.hobbyking.com/hobbyking/store/__189__189__Servos_Parts.html?idCategory=189Źpc= (2015)
6. Hornby, G., Lipson, H., Pollack, J.: Generative representations for the automated design of modular physical robots. IEEE Trans. Robot. Autom. **19**(4), 703–719 (2003). Aug
7. Lipson, H.: Evolutionary robotics and open-ended design automation. Biomimetics **17**(9), 129–155 (2005)
8. Lipson, H., Pollack, J.B.: Automatic design and manufacture of robotic lifeforms. Nature **406**(6799), 974–978 (2000)
9. Panda, S., Mishra, D., Biswal, B.: Revolute manipulator workspace optimization: a comparative study. Appl. Soft Comput. **13**(2), 899–910 (2013). Feb
10. Patel, S., Sobh, T.: Task based synthesis of serial manipulators. J. Adv. Res. (2015)
11. Rout, B., Mittal, R.: Optimal manipulator parameter tolerance selection using evolutionary optimization technique. Eng. Appl. Artif. Intell. **21**(4), 509–524 (2008). June
12. Stan, S., Balan, R., Maties, V.: Multi-objective design optimization of mini parallel robots using genetic algorithms. Ind. Electron. 2007 ISIE (1995), 2173–2178 (2007)
13. Toz, M., Kucuk, S.: Dexterous workspace optimization of an asymmetric six-degree of freedom Stewart Gough platform type manipulator. Robot. Auton. Syst. **61**(12), 1516–1528 (2013). Dec

Sinuosity Coefficients for Leaf Shape Characterisation

Jules R. Kala, Serestina Viriri and Deshendran Moodley

Abstract The design of an efficient and automated model for plant recognition and classification will give the possibility to people with little or no botanical knowledge to conduct field work. In this paper a feature for leaf shape analysis is presented: the Sinuosity coefficients. This new feature is based on the sinuosity measures, which is a value expressing the degree of meandering of a curve. The sinuosity coefficients is a vector of sinuosity measures characterising a given shape. The proposed shape feature is translation and scale invariant. This feature achieved a classification rate of 92 % with the Multilayer Perceptron (MLP) classifier, a rate of 91 % with the K-Nearest Neighbour (KNN) and a rate of 94 % with the Naive Bayes classifier, on a set of leaf images from the FLAVIA dataset.

Keywords Sinuosity measure · Leaf recognition · MLP

1 Introduction

Plants are the key elements for life on earth. They are food sources for animals and insects, help in the process of climate control and are a source of energy. Leaves are one of the visible parts of a plant that have valuable information that can help to identify the species to which the plant belongs. Botanists use information from leaves to identify a plant. They consider the teeth pattern on the leaf margin and whether it is jagged or smooth. It is claimed that there is no clear algorithm to detect and characterise teeth pattern on the leaf margin [1] as not all plant leaves have teeth.

In this paper a new approach for the characterisation of the shape of the leaf margin is presented. The method is based on the sinuosity measure. The sinuosity of a given curve represents the degree of meandering of that curve. It has been extensively used in the domain of medical science for the analysis of the meandering of

J.R. Kala (✉) · S. Viriri · D. Moodley
University of Kwazulu-Natal, Durban, South Africa
e-mail: raymondkala1@gmail.com
URL: http://www.ukzn.ac.za

© Springer International Publishing Switzerland 2016
N. Pillay et al. (eds.), *Advances in Nature and Biologically Inspired Computing*,
Advances in Intelligent Systems and Computing 419,
DOI 10.1007/978-3-319-27400-3_13

a spinal column [3] and in hydrography to evaluate the degree of meandering of a river [2]. A leaf margin is a closed curve and in order to use the sinuosity measure for the analysis of the leaf shape, it has to be divided into four different sections. The sinuosity measure of each section is used to create a vector of sinuosity coefficients of the considered shape. The rest of this paper is organized as follows: Sect. 2 discusses some related works and works about the use of the sinuosity measure. Section 3 present the sinuosity measure and the sinuosity coefficients. Experimental results and discussions are examines in Sect. 4. Conclusion and future work in Sect. 5.

2 Related Works

The following subsection will discuss related work about plants classification using leaves and the use of the sinuosity measure.

2.1 Plant Recognition Using Leaf Shape

Panagiotis et al. [14] proposed a morphological approach to plant classification based on leaf analysis. The goal of this approach was the design of a system that is able to extract specific morphological and geometrical features from a plant leaf. They further use fuzzy surface selection to select the relevant features. This approach reduces the dimensionality of the feature space leading to a very simplified model that is more adapted to real time classification application. The model obtained is scale and orientation invariant and yields to a system that achieved a classification rate of 99 % with a neural network classifier, even with deformed leaves on a dataset of four species of plant leaves. The main drawback is the fact that this approach is not translation invariant.

João et al. [13] developed an approach of leaf shape analysis based on Elliptic Fourier (EF analysis) for plant classification. They used the principal component analysis to select the Fourier coefficients that will be used for the classification. The classification process was performed using linear discriminant analysis with an average accuracy rate of 89.2 % based on a leaves dataset of four species of plant. The advantage of this method is that it is fast and accurate during the classification process. The main drawbacks of the method are that the EF is computationally expensive and not invariant in orientation.

Cerutti et al. [4], inspired by the criteria used in botany, developed a method based on the computation of explicit leaf shape descriptors using the Curvature-Scale Space (CSS) representation. The CSS is a multiscale organization of shape inflexion points. This method achieved a classification rate of 84 % on a dataset of the limitation of this method is the fact that it cannot be used to characterise deformed leaves.

2.2 Plant Recognition Using Leaf Texture

Esma et al. [12] proposed a system based on Dendritic Cell algorithms derived from the Danger Theory [10] as a classifier. The wavelet transform is used to extract leaf features for the classification algorithm. This approach achieved a classification accuracy of 94 %.

Ahmed et al. [9] developed an approach that combines texture features based on Discrete wavelet transformation with an entropy measurement to construct an efficient leaf identifier. An accuracy of 92 % was achieved. The main advantage of this approach is noise removal from the image background which improved the accuracy of the extracted features. The drawback of the method is the size of the data set as it is based on less than 10 species.

Andre et al. [11] proposed a method based on the analysis of the complexity of the surface generated from a texture in order to characterise and describe plant leaves. Their system achieved a classification rate of 89.6 % for 10 species of plants.

2.3 Application of Sinuosity Measure

The sinuosity measure has been applied in many fields of science such as geography, biology and medical science, mostly as a parameter to explain other observations such as the meandering of a river.

The sinuosity measure is used to evaluate the degree of meandering of a river in [2]. It is also explained that the meandering in the case of a river is the result of a process of erosion, tending toward the most stable form in which the variability of certain essential properties such velocity and depth are minimized.

Jaekel et al. in [6] present an application of the sinuosity measure for the characterisation of the webbing on a salamander foot to demonstrate the morphological changes performed on the animal foot to adapt to a given surface.

Alain and Bernard in [3] developed a method for the analysis of spine meandering for the early detection of spine deformation.

Due to the growing interest in spacial exploration, Lazarus et al. [8] used the sinuosity measure to demonstrate that there were rivers on the surface of mars by analysing the planet surface to detect and analyse meandering shapes.

In the next section the mathematical foundation of the sinuosity measure will be discussed.

3 Image Preprocessing

If M is a leaf colour image, the features for the characterisation of M will be extracted on its three colour components (Red, Green, Blue), to explore the possibility that some specific variations exists on each components of the colour image and to increase the size of the feature set.

Let I be the binary representation of one of the component of M. Considering a set of boundary points $(x_i, y_i)_{i=1,...,n}$ of the binary image. The following elements used for the construction of the MBR(Minimum Bounding Rectangle) in [7] will be considered:

Determination of the boundary centroid. The centroid, $(\bar{x}, \bar{y})$, is the average of the coordinates of the boundary points, defined as

$$(\bar{x}, \bar{y}) = (\frac{1}{n} \sum_{i=1}^{n} x_i, \frac{1}{n} \sum_{i=1}^{n} y_i) \tag{1}$$

Determination of the image orientation and the image principal axes. The least square method applied to the coordinates of the boundary points will be used to determine the orientation of the image shape and the main axis dividing the object shape into 4 parts. The formula given in Eq. 2 is used to obtain the object orientation, θ.

$$\tan 2\theta = \frac{2 \sum_{i=1}^{n} (x_i - \bar{x})(y_i - \bar{y})}{\sum_{i=1}^{n} [(x_i - \bar{x})^2 - (y_i - \bar{y})^2]} \tag{2}$$

This formula is obtained by using the Eq. 3 of the line passing through the centroid with the angle θ.

$$x \tan \theta - y + \bar{y} - \bar{x} \tan \theta = 0 \tag{3}$$

and the perpendicular distance from that line to an edge point is defined as:

$$p_i = (x_i - \bar{x}) \sin \theta - (y_i - \bar{y}) \sin \theta \tag{4}$$

Determination of the lower and further points associated to each axis. These are the points associated to each principal axes. They are obtained using the following property:

$$\text{If } f(a,b) \begin{cases} > 0 & \text{then } (a,b) \text{ is above} \\ = 0 & \text{then } (a,b) \text{ is on} f(x,y)=0 \\ < 0 & \text{then } (a,b) \text{ is bellow} \end{cases} \tag{5}$$

Considering the principal axes equations, bounding points are organized into upper and lower further points. The equations of these axes are:

$$(y - \bar{y}) - \tan \theta (x - \bar{x}) = 0 \tag{6}$$

$$(y - \bar{y}) + \cot \theta (x - \bar{x}) = 0 \tag{7}$$

Fig. 1 Points maximizing
the distance P_i

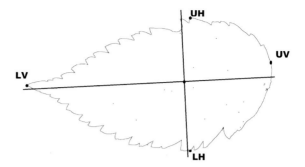

To determine the upper and lower points, the distance associated to each lower and upper point is computed using the following equation:

$$P_i = (x_i - \bar{x}) \sin \theta - (y_i - \bar{y}) \cos \theta \qquad (8)$$

The considered points are the one maximising P_i, as shown in Fig. 1 with *UH* represented as the Upper most Horizontal point, *LH* as the Lower most Horizontal point, *LV* as the Upper most Vertical point and *UV* as the Upper most Vertical point.

4 Feature Extraction

4.1 Sinuosity Measure

Lets consider $(x_i, y_i), i = 1, 2, \ldots, n$ the coordinates of points composing the curve *l*. If *l* is a continuously derivable curve having at least one inflexion point, then the sinuosity of *l* is equal to the ratio between the length of *l* and the length of the straight line joining the two end points $A(x_0, y_0)$ and $B(x_n, y_n)$ of *l*.

The sinuosity measure of the curve *l* is expressed by the following equation:

$$S_l = \frac{\displaystyle\sum_{i=1}^{n} \sqrt{(x_i - x_{i-1})^2 + (y_i - y_{i-1})^2}}{\sqrt{(x_n - x_0)^2 + (y_n - y_0)^2}} \qquad (9)$$

The values generated by Eq. (9) are from 1 (for a straight line) to infinity (closed loop where the shortest path length is zero) or for an infinitely long curve [8].

Lets consider a curve formed by 2 inverted semicircles located in the same plane, the sinuosity measure of this curve is equal to: $S = \frac{\pi}{2} \approx 1.5708$

In order to evaluate the sinuosity measure of a curve C one should make sure that C is continuous between its two ends. Generally evaluated in dimension 2, the sinuosity measure is also valid in dimension 3. The basic classification of the sinuosity is either Strong or Weak (*Strong* : $1 \ll S$ and *Weak* : $S \approx 1$).

This basic classification is the point of interest of this paper, because the sinuosity measure will provide the information we need for the classification of the leaf edge into two groups (smooth leaf = "Weak" edge and jagged leaf = "Strong" edge).

4.2 Sinuosity Coefficients

The sinuosity measure of a complete leaf boundary is infinite because leaf shape is a closed contour. In order to apply the sinuosity measure to a leaf contour, a leaf shape can be divided into 4 different parts as present in Fig. 1. To obtain the leaf shape sinuosity coefficients, the sinuosity measure of each of the following curve (UV, UH); (UV, LH); (LV, UH);(LV, LH) can be evaluated. The sinuosity coefficient of a leaf shape is a vector of sorted values of sinuosity measure of the curves composing the leaf shape. Figure 3 present some leaves with their associate sinuosity coefficients. The proposed feature extraction technique is rotation, translation and scale invariant (Fig. 2).

4.3 Translation Invariant

Let consider the curve (UV, UH) and a distance d of the line joining the two ends of the curve. Since the translation of the curve will not change the length of the curve, the expression in Eq. (9) will not change too.

Fig. 2 Sinuosity coefficients of the *Red*, *Green* and *Blue* components of some images

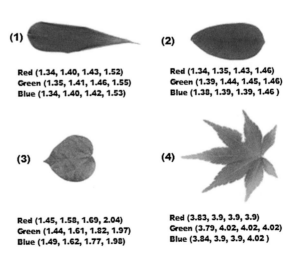

(1)
Red (1.34, 1.40, 1.43, 1.52)
Green (1.35, 1.41, 1.46, 1.55)
Blue (1.34, 1.40, 1.42, 1.53)

(2)
Red (1.34, 1.35, 1.43, 1.46)
Green (1.39, 1.44, 1.45, 1.46)
Blue (1.38, 1.39, 1.39, 1.46)

(3)
Red (1.45, 1.58, 1.69, 2.04)
Green (1.44, 1.61, 1.82, 1.97)
Blue (1.49, 1.62, 1.77, 1.98)

(4)
Red (3.83, 3.9, 3.9, 3.9)
Green (3.79, 4.02, 4.02, 4.02)
Blue (3.84, 3.9, 3.9, 4.02)

4.4 Scale Invariant

Applying a scale to an object is to multiply its dimensions by a constant k. Considering the curve (UV, UH) putting it into scale k means multiply it by the constant k, also the shortest path joining UV and UH is also multiplied by K. Taking this information into consideration the expression in Eq. (9) will not change.

5 Experimental Results and Discussions

5.1 Experimental Results

This experiment is conducted using a subset of FLAVIA leaf images database [5]. Hundred leaves of four species (25 samples per plant species) of plant are randomly selected for the experimentation as in [14], the selection is based on the representativity of the database in terms of shape. For each leaf the three colour components (Red, Green, Blue) and the associate grayscale image will be considered.

The grayscale image is generated by combining the three colour components of the image using the luminosity formula: $0.21 * Read + 0.72 * Green + 0.07 * Blue$.

A sorted vector of four sinuosity measures (Sinuosity coefficient) of each colour components and of the grey scale image are used. For each leaf in our data set 3 vectors of sinuosity measure is used to describe its shape in the colour context, a total of 12 values (Sinuosity coefficient of 12 values) will be used for leaf shape analysis in the colour context and one vector of four sinuosity measures will be used to described the shape of the graycale image.

Two experiments will be conducted to compare the recognition in the colour image context and in the grayscale image context. The vectors of sinuosity measures (sinuosity coefficients) are the input of the classifiers and the output are the plants species.

For the classification purpose 2/4 of the dataset will be used for the training process 1/4 of the rest for the testing and the remaining for the validation.

Table 1 presents the accuracy of proposed feature using WEKA classifiers. A classification rate of 92 % was obtained using the MLP(Multilayer Perceptron), 91 % using the KNN (K-nearest neighbor) and 94 % using Naive Bayes with all four sinuosity coefficients as input(12 values are used). When using grayscale images an accuracy of 84 % was obtained using the MLP, 86 % using the KNN and 91 % using Naive Bayes, with a sinuosity coefficient(4 values are used).

Figure 3 presents the comparisons between the sinuosity coefficients representing the shape of the red component of each leaves species for a subset of 25 leaves (The red component is take to illustrate the comparison). On (a) the sinuosity measure of the curve (LU, UH) on the shape of the red component of each leaf in the dataset is

Table 1 Classification rates, Recall, Precision and ROC Area of the classifier with colour and grey level images

	Colour image			Grey scale image		
	K-NN	MLP	Naive Bayes	KNN	MLP	Naive Bayes
Precision	0.91	0.93	0.94	0.86	0.84	0.91
Recall	0.91	0.92	0.94	0.86	0.84	0.91
ROC area	0.94	0.97	0.98	0.9	0.95	0.98
Classification rate	91 %	92 %	94 %	86 %	84 %	91 %

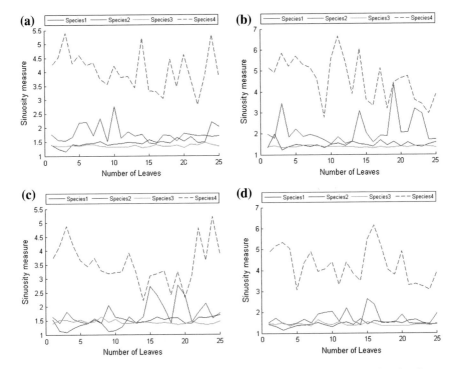

Fig. 3 Comparison of the sinuosity measures of each curve of the *red* component of a colour leaves image. **a** Sinuosity measures of (LV, UH). **b** Sinuosity measures of (UH, UV). **c** Sinuosity measures of (UV, LH). **d** Sinuosity measures of (LH, LV)

present. On the *x axis* the number of leaves and on the *y axis* the sinuosity measure. On Fig. 4. The classification rate per-species is present. On (a) the classification rate in the colour context is present, an accuracy of 88 % was obtain with the Naive Bayes, 80 % with KNN and 72 % with MLP classifier for species 1.

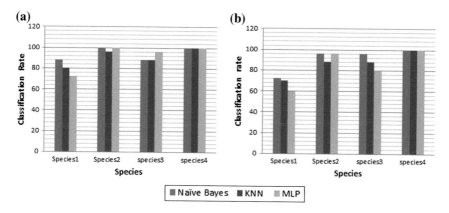

Fig. 4 Classification rate per-species. **a** With colour images. **b** With grey level image

5.2 Discussion

These results are the proof that the proposed feature is a valid feature for shape analysis and using a colour image improves the recognition accuracy, by improving the quality of the shape description since more features are used to describe a given leaf.

It is impossible to claim that the sinuosity measure of a given shape is rotation invariant, because it is obtained based on the initial angulation of the shape.

Compared to the approach in [14] this approach is based on a single feature to described a leaf shape, it is translation invariant and scale invariant and can be used to recognized deformed leaves. In addition it can be used to extract features from leaf shape regardless of the original orientation and to recognized deformed leaves.

On each graph in Fig. 3 it is clearly shown that the lines representing the sinuosity measures of the curves $(LV, UH), (UH, UV), (UV, LH), (LH, LV)$ of species 1, 2 and 3 are relatively close and the line representing the sinuosity measure of species 4 is completely separate. This shows that the sinuosity measure provide a good discrimination between the leaf species.

In Fig. 4 the misclassification of the species 1, 2 and 3 are due to the three species being very close in terms of shape. The higher classification rate observed for species 4 is proof that species 4 is not close to the other species in terms of shape.

6 Conclusions

This paper presents a new feature extraction technique for leaf shape analysis: the sinuosity coefficients. The proposed feature extraction technique for leaf shape characterisation is translation and scale invariant. The experiments show that the sinuosity coefficients are a valid and efficient features for leaf shape recognition.

A classification rate of 92 % was achieved using the Multilayer Perceptron on the dataset and very good results were also obtained with other classifiers like KNN(91 %) and Naive Bayes (94 %). The classification rate obtained prove that the proposed feature is a valid element for shape analysis in general. Further analysis of this feature for the characterisation of the variations on leaf shape, the analysis of the rotation invariant and the combination of this feature to other shape features are suggested future works.

References

1. Cope, S.J., Clark, C.D., Jonathan, Y., Remagnino, P., Wilkin, P.: Plant species identification using digital morphometrics: a review. Expert. Syst. Appl. **39**, 7562–7573 (2012)
2. Langbein, W.B., Leopold, L.B.: River Meanders-Theory of Minimum Variancew. United States Government Printing Offic, Washington (1966)
3. Alain, T., Bernard, P.: Research into spinal deformities. Technology and Informatics. IOS Press, Amsterdam (2002)
4. Cerutti, G., Tougne, L., Coquin, D., Vacavant, A.: Curvature-scale-based contour understanding for leaf margin shape recognition and species identication. In: 2013 International Conference on Computer Vision Theory and Applications, pp. 277–284. VISAP (2013)
5. Wu, S., Bao, F., Xu, E., Wang, Y., Chang, Y., Xiang, Q.: A leaf recognition algorithm for plant classification using probabilistic neural network. In: 7th IEEE International Symposium on Signal Processing and Information Technology (2007)
6. Jaekel, M., Wake, D.B.: Developmental processes underlying the evolution of a derived foot morphology in salamanders. In: Proceedings of the National Academy of Sciences of the United States of America, pp. 20437–20442. National Academy of Sciences (2014)
7. Kala, J.R., Viriri, S., Tapamo, J.R.: An approximation based algorithm for minimum bounding rectangle computation. In: 2014 IEEE 6th International Conference on Adaptive Science Technology (ICAST), pp. 1–6 (2014)
8. Lazarus, E., Constantine, J.: Generic theory for channel sinuosity. In: Proceedings of the National Academy of Sciences of the United States of America, pp. 8447–8452. National Academy of Sciences (2013)
9. Hussein, A.N., Mashohor, S., Iqbal, M.: A Texture based approach for content based image retrieval system for plant leaves images. In: IEEE 7th International Colloquium on Signal Processing and its Applications (2011)
10. Aickelin, U., Bentley, P., Cayzer, S., McLeod, J.: Danger theory: the link between AIS and IDS. Lecture Notes in Computer Sciences (LNCS), pp. 147–155 (2003)
11. Backes, A.R., Casanova, D., Martimez, O.: Plant leaf Identification based on volumetric fractal dimension. In: International J. Pattern Recogn.Artifi. Intell. 1145–1160 (2009
12. Bendiab, E., Kheirreddine, M.: Recognition of plant leaves using the dendritic cell algorithm. Int. J. Digit. Inf. Wirel. Commun. (IJDIWC) (2011)
13. Neto, J.C., Meyer, G.E., Jones, D.D., Samal, A.K.: Plant species identification using elliptic fourier leaf shape analysis. Comput. Electron. Agric. (2006)
14. Panagiotis, T., Stelios, P., Dimitris, M.: Plant leaves classification based on morphological features and a fuzzy surface selection technique. In: Fifth International Conference on Technology (2005)

A Study of Genetic Programming and Grammatical Evolution for Automatic Object-Oriented Programming: A Focus on the List Data Structure

Kevin Igwe and Nelishia Pillay

Abstract Automatic programming is a concept which until today has not been fully achieved using evolutionary algorithms. Despite much research in this field, a lot of the concepts remain unexplored. The current study is part of ongoing research aimed at using evolutionary algorithms for automatic programming. The performance of two evolutionary algorithms, namely, genetic programming and grammatical evolution are compared for automatic object-oriented programming. Genetic programming is an evolutionary algorithm which searches a program space for a solution program. A program generated by genetic programming is executed to yield a solution to the problem at hand. Grammatical evolution is a variation of genetic programming which adopts a genotype–phenotype distinction and uses grammars to map from a genotypic space to a phenotypic (program) space. The study implements and tests the abilities of these approaches as well as a further variation of genetic programming, namely, object-oriented genetic programming, for automatic object-oriented programming. The application domain used to evaluate these approaches is the generation of abstract data types, specifically the class for the list data structure. The study also compares the performance of the algorithms when human programmer problem domain knowledge is incorporated and when such knowledge is not incorporated. The results show that grammatical evolution performs better than genetic programming and object-oriented genetic programming, with object-oriented genetic programming outperforming genetic programming. Future work will focus on evolution of programs that use the evolved classes.

Keywords Automatic programming, Genetic programming · Grammatical evolution · Object-oriented programming

K. Igwe (✉) · N. Pillay
School of Mathematics, Statistics and Computer Science, University of KwaZulu-Natal, Pietermaritzburg, South Africa
e-mail: igwekevin@gmail.com

N. Pillay
e-mail: pillayn32@ukzn.ac.za

© Springer International Publishing Switzerland 2016
N. Pillay et al. (eds.), *Advances in Nature and Biologically Inspired Computing*,
Advances in Intelligent Systems and Computing 419,
DOI 10.1007/978-3-319-27400-3_14

151

1 Introduction

The term automatic programming has been used and understood from different perspectives. In the late 1980s, Rich and Waters [1] have categorized programs such as Microsoft Excel as automatic programming systems. Today, the Excel program is not considered an automatic programming system. This implies that the term automatic programming is relative to the current technologies available [2].

This work considers automatic programming from evolutionary perspectives. O'Neil [2] as well as Koza [3] indicated that GP is an automatic programming technique because it involves telling the computer what to do and not how to do it [4]. Strictly speaking, GP generates programs by taking an analogy from human programmers. Whereas previous work on object-oriented programming and genetic programming has involved taking an analogy from object-oriented programming to improve the scalability of genetic programming, this research forms part of an initiative to investigate the use of genetic programming (and variations) to perform automatic programming with the aim of producing software. Such systems could replace a human programmer, and as such, strictly obeying the quote "tell a computer what to do, not how to do it" [4].

Grammatical evolution (GE) [2] is a variation of GP. It provides a more flexible encoding of programs thus allowing for programs to be generated in any language [2]. GE represents a program as a population of binary strings. A program execution involves converting the binary string into an integer which is then mapped onto a grammar defined by production rules. This results in a production rule of the grammar being executed [2]. Previous research on object-oriented programming and genetic programming can be divided into two categories. The first aims to improve the scalability of genetic programming by either incorporating human knowledge of how to solve the problem, i.e., supplying GP with the functions and terminals that would rather be used by an expert programmer, or taking an analogy from object-oriented programming (OOP). The later results in an extension of GP to object-oriented genetic programming (OOGP) [5] which evolves object-oriented programs. The second focuses on using genetic programming for automatic object-oriented programming with the aim of producing software.

Work in the area of using genetic programming for automatic programming includes [6–8]. In [7], the automatic generation of solution algorithms for problems involving memory, iteration, conditional statements and modularization was investigated. As an advancement in automatic programming, the work in [6] induced an object-oriented design (OOD) from a program specification. A rule-based expert system is used to induce the OOD which forms input to a genetic programming component. Genetic programming then automatically evolves the methods for the program sequentially. This allows function calls between methods.

The research presented in this paper forms part of a larger initiative aimed at investigating the use of genetic programming and variations thereof for automatic programming with the aim of software development. The objective is to evaluate genetic programming and grammatical evolution for producing the code for classes and programs that use the classes. The research in this paper focuses on the former. Genetic programming and grammatical evolution are evaluated on object-oriented programming problems of easy, medium and hard levels. The domain of abstract data types (ADTs) are used for this purpose. From the study conducted in [9] it is clear that evolving the stack, queue and list data structure classes correspond to easy, medium and difficult object-oriented programming problems. In previous work we have compared the performance of GP and GE for evolving classes for problems of easy and medium difficulty. The current study extends the work by testing the automatic OOP capabilities of GP and GE on a problem of hard difficulty level, i.e., the list.

The rest of the paper is presented as follows: Sect. 2 provides the OOGP approach for automatic algorithm induction. Section 3 provides GE approach for the same purpose. The experimental setups for both the approaches are discussed in Sect. 4. Section 5 presents the result of the comparison done on both approaches. Conclusion drawn from the study and the future work are presented in Sect. 6.

2 Object-Oriented Genetic Programming (OOGP) Approach for Automatic Programming

The genetic programming approach implemented for automatic programming is described in this section. The study implements a generational genetic programming algorithm outlined in Fig. 1. The algorithm creates an initial population and iteratively refines the population by applying evolutionary strategies, namely, selection, crossover and mutation until one or more termination criteria are met. The termination criteria could be the maximum number of generations reached or the solution found. The following sections describe the process in the algorithm.

```
Create an initial population
Do {
   Evaluate the population
   Select parent
   Apply genetic operators
} While (termination criterion not met)
```

Fig. 1 Generational genetic programming algorithm

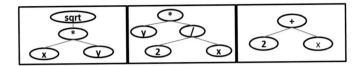

Fig. 2 An example of a chromosome with n = 3 genes

2.1 Initial Population Generation

The current study adopts the representation by Bruce [10] and Langdon [9]. Each element of the population i.e., a chromosome represents a class. A chromosome is represented as indexed memory of *n* indices with each index a parse tree as shown in Fig. 2. Each parse tree i.e., a gene represents one method of the class and is expected to perform a unique function when executed. An internal representation language is defined for the algorithm. This language is independent of any programming language and allows for the generated solution algorithms to be converted to any programming language.

The function and terminal sets are subsets of the internal representation language. Each parse tree is created by randomly choosing elements from the function and terminal set. A complete parse tree is created when a preset maximum depth is reached. At this depth, a terminal element must be randomly chosen. The grow method described in [3] is used for this purpose. The function and terminal set for the problem is a subset of the internal representation language described in Table 1.

The initial population is created using two methods. The first is the random creation of the initial population as in the standard genetic programming. This is referred to as the OOGP approach. The second is the use of the greedy approach, namely, GOOGP described in [8]. The approach involves randomly creating a population of *m* parse trees for each gene position. Each of the created parse trees is evaluated as described in Sect. 2.2 to determine the fitness of the tree. The fittest tree in the population is stored as the gene for that position. The next section describes the fitness evaluation.

2.2 Fitness Evaluation

The fitness of a chromosome is better for a higher value, i.e., the fitness is maximized. For each run, 15 fitness cases are randomly generated. Each method in the chromosome is evaluated on these cases. Each fitness case is a list of length in the range 1–15. Also the elements of the list are randomly generated integer values in the range 1–999. In order to calculate the fitness of each method, a set of problem dependent criteria must be defined. The fitness of each method is scored based on the number of the criteria the method has met. Table 2 shows the problem dependent criteria for the list data structure. The five methods in Table 2 are

Table 1 The elements of the internal representation language

Elements of the Internal representation language	Description
Arithmetic operators	+, −, *, and / operators. The / is protected so that it returns 1 if the denominator is zero
Indexed memory operators	*Read*: The *read* operator takes a single argument which must evaluate to an integer. It returns the content in the memory indexed by the value its argument evaluates to *Write*: The *write* operator takes two arguments, first the content to be written and second the index in the memory to which it should be written
Named memory operators [9]	A named memory location, *aux*, is maintained by the system and serves as a temporary memory location that can be used by a program. The following can be performed on *aux*; The *set_aux* operator takes a single integer argument and sets *aux* to the value of evaluating this argument. The *dec_aux* decrements the value stored at aux by one while the *inc_aux* increments the value stored at aux by one
Multiple statement operators	*Block2* and *block3* are used to combine two and three programming statements respectively. The *block2* as well as the *block3* operator returns the value that the last argument evaluates to
Iteration operator [11]	*For*: the *for* operator takes three arguments. The first two arguments, which must each evaluate to an integer, represent the loop bounds. The third argument represents the body of the loop. The operator returns the value the body evaluates to on the last iteration of the loop
Named memory locations [11]	In addition to *aux*, *Ivarα* and *Cvarα* are named memory locations that are added to the terminal set when creating an instance of the *for* operator. The symbol α is an integer indicating which *Ivar* or *Cvar* belongs to a particular instance of the *for* operator
Variable Names	Names of Automatically Defined Functions (ADFs): for example, ADF0 Input variables: for example, *i*, representing input to the problem, arg1 and arg2 representing ADF0 arguments
Constants	Zero and one

considered suitable to represent the list operations [12]. Each of the methods can attain a maximum score of 1, 4, 4, 3, and 3 respectively. Hence a maximum score of 15 per fitness case and a total of 225 over all the 15 fitness cases are obtainable per chromosome. The next section describes the selection process.

2.3   Selection

The selection process is based on the relative fitness described in [8] which is calculated relative to the best performing method in the current population. For instance, given the best fitness for each of the five methods in Table 2 as BF_1, BF_2, BF_3, BF_4 and BF_5 respectively, the relative fitness (RF) of a chromosome with fitness F_1 to F_5 is calculated as: $RF = \Sigma(F_i/BF_i) * 100$ $i = 1, ..., $ no. of methods.

The methods described in Table 2 have a different number of criteria. This could result in an individual which satisfies a few methods being selected because these methods have a large number of criteria. For example, the fitness of individual A which passed the criteria for the *getElement()* and the *insertAt()* methods is 8, while the fitness of individual B which passed the criteria for the other 3 methods is 7. Individual A satisfies 2 methods and appears to be better than individual B which satisfies 3 methods. Relative fitness is used to ensure that individual B is selected over individual A.

Tournament selection [3] is used to choose the parent for the purpose of regeneration. This selection process chooses, from the population, a number of chromosomes equal to a preset size, say t, of the tournament. The relative fitness of each of the chromosomes is then calculated and the one with the highest relative fitness is returned as the parent. The selection process is with replacement, thus a chromosome can be selected more than once. The following section describes the regeneration process.

Table 2 The problem specific criteria for the list data structure

Methods	Function
makeNull()	• list pointer must be set to −1
getElement(p)	• No change in pointer value • Elements in the list should not be altered • The value returned must be the value in the position indexed by p in the list • Only one value must be returned
insertAt (p,x)	• Pointer must be updated correctly • The position of the elements in the list should be correctly updated • The inserted value must be at the position indexed by p in the list • The value must be inserted only once
removeElement (p)	• Pointer must be updated correctly • The position of the elements in the list should be correctly updated. Thus, the removed Element must be that previously at the position indexed by p in the list before the execution of the method • The removed value is returned
empty()	• No change in pointer value • Elements on the stack should not be altered • The correct position of the pointer must be returned. A negative value of the pointer implies an empty list while a value $>= 0$ implies that the list is not empty

2.4   Regeneration

Two phases of crossover [8], namely, external and internal crossover are implemented. This is done to ensure a proper mixing of the genetic materials from the two selected chromosomes. Whereas the external crossover allows exchange of genes between two chromosomes, the internal crossover allows exchange of sub trees between two genes from two selected chromosomes. External crossover is implemented followed by internal crossover. External crossover swaps one or more genes between two chromosomes. Two chromosomes are selected using tournament selection and external crossover is applied as follows: For each gene position, the genes are swapped between the two selected chromosomes if the randomly generated probability in the range 1–100 is less than the preset probability. For instance, assume $G_{11}G_{12}G_{13}G_{14}$ and $G_{21}G_{22}G_{23}G_{24}$ to be chromosomes on which the external crossover is to be applied. Each gene G_{ij} is a parse tree representing a method of the class. Assume the preset external crossover probability as 40 %. If the randomly generated probabilities for each gene position are 64 %, 35 %, 80 %, and 45 % respectively, the resulting offspring will be $G_{11}G_{22}G_{13}G_{14}$ and $G_{21}G_{12}G_{23}G_{24}$.

Internal crossover is applied once the external crossover is completed. It starts by randomly generating a number in the range 1–100. This operator is applied if this number is less than the preset internal crossover probability. The operator is applied as follows: A chromosome index is randomly chosen and the standard genetic programming crossover operator [3] is applied to the parse trees at the selected index in both parents. The resulting trees replace the genes at the selected index in both parents. The fitter of the two offspring is added to the population of the next generation. The trial runs done with the mutation operator did not improve the OOGP and GOOGP solutions. Thus, the operator is not used in the final runs presented in this paper.

3   Grammatical Evolution Approach for Automatic Programming

The grammatical evolution approach is a generational algorithm and uses the same approach as OOGP for fitness evaluation, selection and termination criteria. In grammatical evolution, a chromosome is a binary string. Program execution involves mapping from the space of binary string onto an integer space. The integer space is then mapped onto a grammar from which one or more programming statements are formed. The following section describes the initial population generation for the grammatical evolution.

3.1   The Initial Population Generation

The OOGP approach represents each chromosome as indexed memory with each
gene a parse tree stored at each index. Again each chromosome is indexed memory
but with each gene a binary string representing a method of a class or an ADF. Each
gene is made up of codons each of which comprises n alleles. Execution starts by
converting each codon (binary string) to a denary value. Each denary value is then
mapped to a production rule in the grammar. The grammar used for producing the
code for the list data structure class is given in Fig. 3. The grammar is structured to
ensure that the generated programs are syntactically correct. Using the grammar as
an example, the 2 production rules for the start symbol <stmts> are:

$$< stmts > :: = < stmt > \tag{0}$$

$$< stmt > ; < stmts > \tag{1}$$

The mapping process is performed from left to right of a chromosome. The modulus of the denary value formed from the first codon and the number of production
rules is taken. For instance, 13 mod 2 will result in the production rule number (1).
This process is continued until all the non-terminal variables have been expanded.
Each non-terminal variable is enclosed inside the <> bracket. Once the codon has
been processed and there are more non-terminals to be expanded, the process
begins at the beginning of the gene again with the first codon. Cyclic calls to
non-terminals are prevented by setting a limit on the number of times a
non-terminal can be called in the program. The symbol α in the grammar is replaced
by a positive integer during the mapping process.

```
<stmts> ::= <stmt>; | <stmt>;<stmts>
<stmts> ::= <set_aux(<expr>) | write(<expr><expr>)
       | read (<expr> ) | write(<element><expr>) | <expr>
       | forα(<expr><expr><loopstmt>) | ADF0 (<expr><expr>)
<expr> ::= <var> | + <expr><expr>  | - <expr><expr>
<element> ::= i
<loopStmt> ::= write (<loopElement><loopExpr> | <loopExpr>
       | read(<loopExpr>) |forα(<expr><expr><loopstmt>)
<var> ::= one |zero|aux|dec_aux  |inc_aux   |p
<loopExpr> ::=  <loopVar>| + <loopExpr><loopExpr>
       |- <loopExpr><loopExpr>
<loopVar> ::= one |zero| aux |dec_aux  |inc_aux |p| Cvarα
```

Fig. 3   The grammar for evolving the methods for the list ADT

3.2 Regeneration

The regeneration process implements both the crossover and mutation operators. Again the crossover operator is applied in two phases—the internal and external. External crossover operators swap genes between parents if the randomly generated probability for the gene is less than the preset probability. Internal crossover is only applied if the random number generated is less than the preset probability. Internal crossover is basically one-point crossover [13] which produces two offspring. As with the OOGP, the fitter of the two offspring is returned as the result of the operation.

The mutation operator uses two preset probabilities, namely, the mutation probability and the bit flip probability. This operator is applied to the offspring created by crossover depending on the preset probabilities. As in the case of the crossover operator, a random number between 1 and 100 is generated for each gene in the chromosome. If this value is less than the mutation probability, bit mutation [2] is performed. The bit mutation is performed as follows: A random number between 1 and 100 is generated for each bit in the codon. The bit is flipped if this number is less than the bit flip probability. A bit flip implies that the bit becomes 1 if it is 0 and vice versa.

4 Experimental Setup

This section describes the experimental setup for testing the implemented algorithms. Table 3 lists the function and terminal sets used for the list ADT. The parameter values used for OOGP, GOOGP and GE are listed in Table 4. These values were obtained empirically by performing trial runs. The parameters that give the best result were chosen from various tested parameters. Maximum depth and maximum offspring depth are set differently for each method.

Table 3 Function and terminal set for OOGP and GOOGP approach

List ADT methods	Primitives
makeNull()	+, −, set_aux, block2, write, read, for, zero, one, aux, inc_aux, dec_aux, Cvar, Ivar
insertAt(p,x)	+, −, set_aux, block2, write, read, for, zero, one, aux, inc_aux, *p*, *i* (an integer to be added to the list), Cvar, Ivar
getElement(p)	+, −, set_aux, block2, write, read, for, zero, *p*, aux, inc_aux, dec_aux, Cvar, Ivar
removeElement (p)	+, −, set_aux, block2, write, read, for, zero, one, aux, *p*, dec_aux, Cvar, Ivar
empty()	+, −, set_aux, block2, write, read, for, zero, one, aux, inc_aux, dec_aux, Cvar, Ivar
ADF0	+, −, *, /, block2, for, write, Cvar, Ivar, zero, one, arg1, arg2

Table 4 Parameter values

Parameter	OOGE	GOOGE	Parameter	GE
Population size	1000	100	Population size	1000
Maximum depth range	3–5	3–5	Codon length	10
Tournament size	2	2	Allele length	8
External crossover probability	50	50	Tournament size	4
Internal crossover probability	50	50	Mutation probability	40
Maximum offspring depth range	10	10	Bit flip probability	70
Number of generations	100	100	External crossover probability	50
Population size (n) for GOOGP	N/A	500	Internal crossover probability	80
			Number of generations	100

```
<stmts> ::=    <stmt>;|<stmt>;<stmts>
<stmts>::= forα(<expr><expr><loopstmt>)|<expr>|write(<expr><expr>)
<expr> ::= <var> | + <expr><expr> | - <expr><expr>
<var> ::= arg1 |arg2|zero|one
<loopStmt> ::= write(<loopElement><loopExpr>)
               |forα(<expr><expr><loopstmt>)| <loopExpr>
<loopElement> ::= Ivarα
<loopExpr> ::= Cvarα|zero|one
```

Fig. 4 The grammar for evolving the ADF

An ADF, namely, ADF0 is implemented which takes 2 arguments and returns no value. This ADF is added to the function set (and the grammar for the GE) when creating the *insertAt(p,x)* and the *removeElement(p)* methods. The body of the ADF is automatically determined during evolution. The inclusion of an ADF allows GP and GE to generate code which can be reused by different list methods. The grammar for the ADF is given in Fig. 4. The system was implemented in Java using Netbeans IDE 7.2.1 with JDK 1.7.2_25. Thirty runs, each with a different seed, were performed for each approach. This was done due to randomness of the approaches. These runs were performed on an Intel Core 3.1 GHz machine with 8192 MB of RAM.

5   Results and Discussion

The performance of OOGP, GOOGP and GE approaches are discussed in this section and are based on their success rates. Thirty runs were performed for each approach. The success rate listed in Table 5 is the number of the 30 runs that

Table 5 Success rates (SR) and average fitness (AF)

List ADT tests	OOGP		GOOGP		GE	
	SR	AF	SR	AF	SR	AF
A	0	176.7333	0	208.5667	3	207.8
B	0	195.96	4	211.83	2	207.53
C	0	196.65	7	217.73	16	218

produced a solution i.e., a chromosome which when executed produces a solution for all methods of the list ADT. We performed three set of tests. (a) We test the ability of each approach to automatically and simultaneously produce the code for the list data structure class. (b) Then based on the analysis of the functions of each method to be evolved, an ADF is implemented and tested for each approach. (c) Finally, for each method to be evolved, human knowledge of the solution is incorporated into the approaches.

In 30 runs performed, OOGP and GOOGP were not able to find a solution. GE was able to automatically evolve 3 solutions. The presence of the ADF improved the performance of OOGP and GOOGP but degrades the performance of GE as shown by the average fitness in Table 5. OOGP was not able to find any solutions. GOOGP and GE were able to find 4 and 2 solutions respectively. Finally, we observed further improvements with the incorporation of human knowledge into the architecture of the approaches. Again, OOGP was not able to find any solution algorithm. GOOGP and GE found 7 and 16 solutions respectively.

Hypothesis testing was conducted to test statistical significant of the results obtained with C. The hypothesis is that GE performs better than GOOGP when human knowledge of the solution is incorporated into the architecture of the approaches. The hypothesis was tested using the success rate and is significant at all level of significance with Z value of 2.51. The average fitness found with the GOOGP approach for test A and B are better than that found with the GE approach. However, there is no significant improvement of GOOGP over GE in terms of the number of solutions generated.

The use of an ADF was found to improve the GP performance but not the GE performance for this problem. This result corresponds to the work in [14] which states that an ADF improves GP performance on complex problems. Again, this study corresponds to the suggestion made in [15] that the performance of GE using ADFs depends on size and type of problem.

Figure 5 shows the convergence graph obtained from run 2 of the approaches. OOGP started with a low fitness compared to GOOGP and GE. It converges to a local optimum at generation 6. This indicates that OOGP requires additional techniques to escape local optima, thus GOOGP is implemented. The initial population of the GOOGP is informed and hence the GOOGP started with a very high fitness. The fitness slightly increased at generation 3 before it converges to a local optimum. GE fitness started with a considerably high fitness but not as high as the GOOGP fitness. The fitness fluctuates but gradually converges as the generation progresses. This suggests the reason for its better success rates than the OOGP and GOOGP.

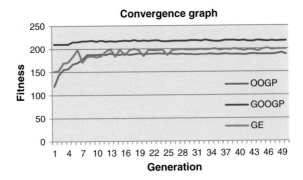

Fig. 5 The convergent graph from run 2 of the OOGP, GOOGP and the GE

Table 6 An example solution generated by GE

Method	Generated solution algorithm
makeNull()	set_aux (+−− aux ++ inc_aux aux zero zero zero);
inserAt(p,i)	for 1{+− zero inc_aux p, aux, write (Ivar1, Cvar1)}; write (i, p);
getElement(p)	read (p);
removeElement()	for 18{aux, + dec_aux p, write (Ivar18, Cvar18)};
empty()	aux;

The GOOGP and GE implemented in the current study for test B and C performs competitively well with the OOGP reported in [9]. The study reported that OOGP was able to find 2 solutions in 56 runs for the list ADT implemented. One of the solutions evolved by GE in the current study of test A is shown in Table 6. Compared to the results obtained in [8] for the stack and queue ADTs, we found out that the ability of GP and GE to be used for automatic programming decreases with an increase in the complexity of the problem.

Each program statement is represented in prefix notation. The functions *inc_aux*, *dec_aux* and *set_aux* are defined to return 0. The generated code satisfies the criteria described in Table 2 and hence correctly implements the list data structure class.

6 Conclusion and Future Work

This paper aimed at comparing the performance of genetic programming and grammatical evolution for automatic object oriented programming. OOGP, GOOGP and GE were tested on the list data structure—a more difficult problem than the stack and queue tested in [8]. We found out that GE is more successful at automatic programming than OOGP and GOOGP as it scales better than both these approaches without incorporating programmer knowledge of how to solve the problem. The ability of both GP and GE to be used for automatic programming decreases with an increase in the complexity of the problem.

Future work will include an analysis of the GOOGP algorithm to identify the reasons for their poor performance compared to the GE. Evolution of problems that make use of the evolved class will also be investigated.

References

1. Rich, C., Waters, R.C.: Automatic programming: myths and prospects. IEEE Comput. **21**, 40–51 (1988)
2. O'Neil, M., Ryan, C.: Grammatical Evolution, Evolutionary Automatic Programming in an Arbitrary Language. Kluwer, Norwell (2003)
3. Koza, J.R.: Genetic Programming 1: On the Programming of Computers by Means of Natural Selection. MIT press, Cambridge (1992)
4. Samuel, A.L.: Some studies in machine learning using the game of checkers. IBM J. Res. Dev. **3**, 210–229 (1959)
5. Abbott, R.: Object-oriented genetic programming, an initial implementation. In: International Conference on Machine Learning: Models, Technologies and Applications, pp. 26–30 (2003)
6. Pillay, N., Chalmers, C.K.: A hybrid approach to automatic programming for the object-oriented programming paradigm, In: 2007 Conference of the South African Institute of Computer Scientists and Information Technologists, pp. 116–124 (2007)
7. Igwe, K., Pillay, N.: Automatic Programming Using Genetic Programming. In: World Congress on Information and Communication Technologies, pp. 339–344. Hanoi, Vietnam (2013)
8. Igwe, K., Pillay, N.: A Comparative study of genetic programming and grammatical evolution for evolving data structures. In: 2014 PRASA, RobMech and AfLaT International Joint Symposium, pp. 115–121. Cape Town, South Africa (2014)
9. Langdon, W.B.: Genetic Programming and Data Structures: Genetic Programming + Data Structures = Automatic Programming! Kluwer, Boston (1998)
10. Bruce, W.S.: The Application of genetic programming to the automatic generation of object-oriented programs. Ph.D. thesis, Nova Southeastern University (1995)
11. Pillay, N.: Evolving solutions to ASCII graphics programming problems in intelligent programming tutors. In: International Conference on Applied Artificial Intelligence (ICAAI'2003), pp. 236–243 (2003)
12. Goodrich M.T., Tamassia, R.: Data Structures and Algorithms in Java. Wiley (2008)
13. Goldberg, D.E.: Genetic Algorithms in Search, Optimization and Machine Learning. Addison-Wesley Longman Publishing Company, Boston (1989)
14. Koza, J.R.: Genetic Programming II, Automatic Discovery of Reusable Programs. MIT Press, Cambridge (1994)
15. Hemberg, E., O'Neill, M., Brabazon, A.: An investigation into automatically defined function representations in grammatical evolution. In: 15th International Conference on Soft Computing, Mendel (2009)

Evolving Heuristic Based Game Playing Strategies for Checkers Incorporating Reinforcement Learning

Clive Frankland and Nelishia Pillay

Abstract The research presented in this paper forms part of a larger initiative aimed at creating a general game player for two player zero sum board games. In previous work, we have presented a novel heuristic based genetic programming approach for evolving game playing for the board game Othello. This study extends this work by firstly evaluating it on a different board game, namely, checkers. Secondly, the study investigates incorporating reinforcement learning to further improve evolved game playing strategies. Genetic programming evolves game playing strategies composed of heuristics, which are used to decide which move to make next. Each strategy represents a player. A separate genetic programming run is performed for each move of the game. Reinforcement learning is applied to the population at the end of a run to further improve the evolved strategies. The evolved players were found to outperform random players at checkers. Furthermore, players induced combining genetic programming and reinforcement learning outperformed the genetic programming players. Future research will look at further application of this approach to similar non-trivial board games such as chess.

Keywords Genetic programming · Heuristics · Reinforcement learning · Game playing · Board games · Checkers

1 Introduction

Since the first strong checkers program presented by Arthur Samuel [1] in the 1950s, developing artificial intelligence (AI) players for the game has become an integral part of AI and machine learning research [2]. In 1989, the 'Chinook'

C. Frankland (✉) · N. Pillay
School of Mathematics, Statistics and Computer Science, University of KwaZulu-Natal, Durban, KwaZulu-Natal, South Africa
e-mail: clivefrankland@gmail.com

N. Pillay
e-mail: pillayn32@ukzn.ac.za

© Springer International Publishing Switzerland 2016
N. Pillay et al. (eds.), *Advances in Nature and Biologically Inspired Computing*,
Advances in Intelligent Systems and Computing 419,
DOI 10.1007/978-3-319-27400-3_15

project began with the goal of producing a program that could challenge world checkers champions. These early checkers programs relied on programmed human expertise and brute force search techniques for their intelligence [3, 4]. In 1999, Chellapilla and Fogel's 'Blondie 24' program became the first example of a program that was able to learn from scratch without any human guidance, in a knowledge-free regime [5, 6]. AI research into game playing has now turned to heuristic based evolutionary algorithms that allow machines learning abilities without using features that would normally require human expertise [4, 7]. Genetic algorithms [8] and more recently genetic programming (GP) [8–10] for evolving evaluation functions are used with standard game playing algorithms such as the alpha-beta algorithm or Monte Carlo tree search, to produce grand master level computer players for games such as chess [8], Othello [10, 11], lose checkers [13], checkers [10] and give away checkers [9]. In previous work on applying genetic programming to checkers thus far, GP has generally been used to evolve a heuristic to assess how good a board state is [9, 11]. The heuristic has been used with the minimax algorithm or alpha-beta pruning to generate game playing strategies. The research presented in this paper differs from previous work in that genetic programming is used to evolve the actual game playing strategies comprised of heuristics in real time during game play. This provides an advantage over offline induction of game playing strategies in that the player evolved offline may not cater for the current scenario in the game play We have previously shown the effectiveness of this approach for Othello [12]. In this study we extend this work to firstly evaluate the approach further by applying it to a different two player zero sum board game, namely, checkers. The second objective of this research is to investigate the effect of using reinforcement learning to improve the strategies evolved by genetic programming.

The usefulness of this approach was validated by employing it to evolve strategies for the 8 × 8 variation of British and American tournament checkers [13]. The results show that this approach is able to evolve good playing strategies for checkers and that reinforcement learning significantly enhanced heuristics, resulting in improved performance by the evolved players. Section 2 describes the heuristic based genetic programming approach for checkers and the incorporation of reinforcement learning. The experimental setup used to evaluate the approach is presented in Sect. 3. Section 4 discusses the performance of the genetic programming approach with and without reinforcement learning. The findings of the study and future work are presented in Sect. 5, which concludes the paper.

2 Evolution of Heuristics Based Strategies

The genetic programming algorithm implemented in this study to evolve game playing strategies is that presented by Koza [14]. The algorithm is illustrated in Fig. 1. The algorithm begins with an initial population, which is iteratively refined over g generations via the processes of evaluation, selection and regeneration. The

```
Create an initial population
Repeat
  Evaluate the population
  Select parents
  Apply genetic operators to create next generation
Until termination criteria are met
```

Fig. 1 Genetic programming algorithm [14]

process of creating an initial population and evolving it over a set number of generations is referred to as a run. A new run is performed for each move made by the player.

At the end of each run, the population for the move is preserved and is used as the initial population of the next move. The strongest player of each run, which we refer to as the alpha player, is maintained throughout the run and is used to evaluate the population. The fitness of each game playing strategy, i.e. player, is evaluated by playing against the alpha player. The alpha player can be randomly selected at the beginning of a run from the initial population or if a move has already been made the alpha player from the previous run can be used. The following section describes the heuristic based strategies evolved by the genetic programming approach and how these are applied. This is followed by an overview of the processes of initial population generation, fitness evaluation, selection and the genetic operators. The section finally describes how reinforcement learning is applied to improve the evolved game playing strategies.

2.1 Heuristic Based Strategies

A strategy is applied by considering all legal moves and choosing the piece in the position with the highest heuristic value to make the move. That piece is moved to a corresponding legal position with the highest heuristic value. The following example illustrates the use of these heuristics in applying a strategy. Figure 2 represents a strategy and the corresponding board heuristic values allocated by the strategy are illustrated in Fig. 3. In this example, the third terminal in the strategy assigns a heuristic value of 54 to position 21 on the board. The piece occupying this legal move position will be selected to make the move as it has the highest heuristic value of all legal move positions. The sixth terminal in the strategy assigns 23 to position 28 and in applying this strategy the player will move the selected piece (54) to that position as it has the highest heuristic value of available 'move to' positions.

If there are pieces in a position to jump, then it is compulsory that that piece be moved (regardless of heuristic value). In a case where there are multiple pieces able to jump, the piece in the position with the highest heuristic value will be chosen to make the move. If two pieces have the same heuristic value, one piece will be randomly selected to make the move. If there are a number of legal positions

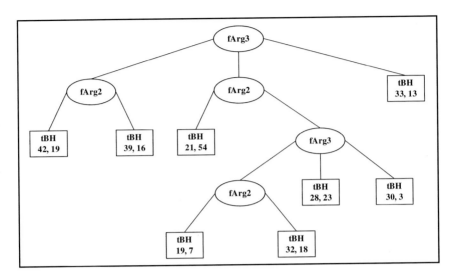

Fig. 2 Example of an element in population representing one strategy

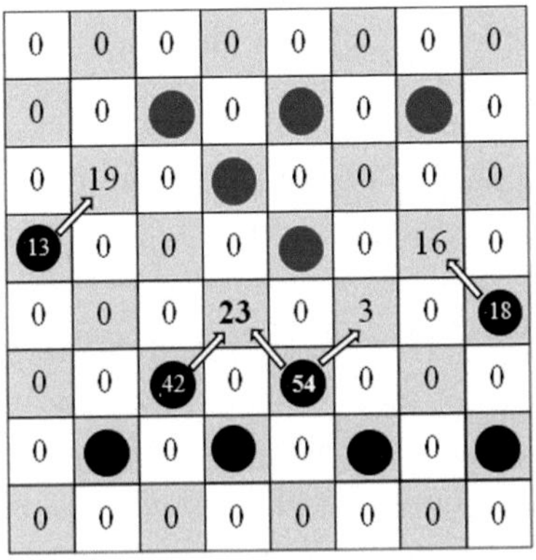

Fig. 3 Heuristic values
produced by the game playing
strategy in Fig. 2

available to move to with the same heuristic value, one position will be randomly
selected as the 'move to' position.

The following section describes how the game playing strategies are created
during initial population generation of the genetic programming algorithm.

2.2 Initial Population Generation

The initial population is created using the ramped half-and-half method [14]. Each
element of the population represents a player and defines a game playing strategy in

terms of heuristics as illustrated in Fig. 2 in the previous section. Each heuristic represents a board position. The elements of the function set are used to combine heuristics to form game playing strategies. The function set contains two operators, namely, fArg2 and fArg3, which takes 2 and 3 arguments respectively. The terminal set is represented by one terminal, tBH, taking 2 integer arguments representing the board position and its heuristic value. During initial population generation a heuristic value is randomly generated for each heuristic. A range of 1 to 64 was used to test the GP. This range of values was selected simply from observation, as a greater or lesser range did not influence the computer player's playing ability. In this setup, a strategy need not necessarily contain all the heuristics representing each position on the board. If a board position is not represented in a strategy, that board position is assigned a value of zero. If a terminal representing the same board position appears more than once in a strategy but with different values, the first value is chosen. The game playing strategy is evaluated by its performance in game play against the alpha player. The following section explains this further.

2.3   Fitness Evaluation

Fitness evaluation begins with each player in the population competing against the alpha player over a preset number of games. The number of games played is a parameter value. The alpha player is randomly selected from the initial population or if a move has already been made, the alpha player of the previous run may be used. At the end of each game 1 point is awarded to a single piece and 2 points to a king. The fitness of the player is determined as the sum of points accumulated over a preset number of games. The fitness of the alpha player is divided by the number of elements in the population as a better reflection of its ability as it plays as many game sets as the number of elements in the population, while each element of the population plays only one set. If an element of the population has a higher fitness than the alpha player, that element replaces the alpha player in the next generation.

2.4   Selection and Regeneration

Tournament selection [14] is used to choose parents to create offspring of successive generations. This method randomly chooses t elements from the population. The fittest element is returned as the winner of the tournament and a parent. Selection is with replacement, so an element of the population can serve as a parent more than once. The reproduction, mutation and crossover operators are used for regeneration. The reproduction operator copies the parent chosen using tournament selection into the next generation. Mutation randomly selects a mutation point and replaces the subtree rooted at the point with a new subtree. Hence, mutation is the only operator that can change the heuristic value of a terminal. The crossover operator randomly selects a crossover point in each of the parents and swaps the subtrees rooted at the crossover points to create two offspring. The actions of both the mutation and crossover operators could result in reproduction if the resultant offspring is identical to another element in the population. Offspring that exceed the preset size limit are pruned by replacing subtrees with terminals nodes. The alpha player from a generation can be used in the successive generation if the parameter value to use it is set to true. In the case of crossover, the alpha player and a parent selected from the population are chosen to produce two offspring. In the case of mutation, it is randomly decided whether to mutate the alpha player or not.

2.5 Reinforcement Learning

Reinforcement learning [3, 6] is often incorporated into computational intelligence techniques to evolve game playing strategies. Reinforcement learning is based on the surmise that a series of actions lead to success or failure, and some form of reward or penalty for the final result can be apportioned back to each action. Correct actions are rewarded and incorrect actions are penalized. Reinforcement learning seeks to decompose these series of actions, into meaningful smaller bits that can indicate whether or not specific actions should be favoured or disfavoured [5].

The reasoning behind reinforcement learning in the context of this research is to capture the performance of the heuristics in strong and weak strategies during a run and update the heuristic values accordingly on the board at the end of a run. Two matrices representing the game board for strong (SPMatrix) and weak (WPMatrix) players are maintained during a run. The matrices store the average heuristic value produced for each board position by the strong and weak players. The strong players or strategies are those that were promoted to alpha players during the run. The weak players are those with a raw fitness less than 30 % of the alpha player's raw fitness during a generation. The heuristic values produced by the alpha player (AMatrix) in the last generation together with the components used in the reinforcement learning (SPMatrix) and (WPMartix) are combined to generate a learned matrix (RIMatrix) which represents the final strategy that is used to make a move. The 'learned matrix' is produced using the following formula: $RIMatrix(x, y) = AMatrix(x,y) * SPMatrix(x,y)/WPMatrix(x,y)$ where, RIMatrix is the resulting reinforcement learning matrix, AMatrix is the matrix storing the heuristic values produced by the alpha player, SPMatrix is the matrix storing the average heuristic values for the strongest players and WPMatrix the average heuristic values for the weakest players. The resulting reinforcement learning matrix (RIMatrix) is carried through to the next move and is used as the alpha matrix during the fitness evaluation process (Sect. 2.4). An example of this applied to four cells of a matrix is illustrated in Fig. 4.

In this example, two cells in the Alpha matrix, have equal values (7) and are potential move positions, as they are also the highest values in the matrix. The

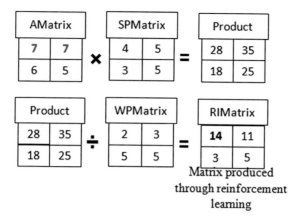

Fig. 4 Application of the reinforcement learning formula

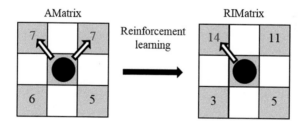

Fig. 5 Reinforcement learning

proposed reinforcement learning technique is applied to the AMatrix, generating a learned matrix. The resulting value of the first cell, in this example, is now the highest. The corresponding position on the board will be selected to make the move (Fig. 5).

3 Experimental Setup

The aims of the research presented in this study were to evolve good heuristic based strategies for checkers using GP and to investigate the effectiveness of incorporating reinforcement learning to enhance the performance of the evolved players. Two experiments were performed to ascertain this:

- Experiment 1—This experiment compares the performance of the heuristic based genetic programming approach, without reinforcement learning, to a random move player. Computational intelligence in board game simulations (also known as playouts) [15] often incorporate random move, or better, pseudo-random move players [16–18] to demonstrate improved levels of play

Table 1 GP algorithm
parameters

Parameter	Value
Population size	50
Maximum depth	6
Generation method	Ramped half and half
Selection method	Tournament
Tournament size	4
Reproduction	0 %
Crossover	80 %
Mutation	20 %
Generations	25
Preserve alpha	True
Game moves	150

for the artificial intelligence controlled player. The random move player in this experiment, referred to as RPlayer, randomly selects a legal position to make a move. The heuristic based genetic programming approach will be referred to as HGP. The performance of the HGP will be compared to the RPlayer simply to determine improved levels of playability.

- Experiment 2—This experiment compares the performance of the HGP to the heuristic based genetic programming approach incorporating reinforcement learning. This player will be referred to as RHGP.

In both experiments 10 games were played, with each player alternating between black and white per game. Five games as black, thus moving first, and 5 games as white. Ten games proved to be sufficient for purposes of evaluation. The score and number of single pieces promoted to kings during play, for each game were recorded for each experiment. The score was determined to be 1 point for a single piece and 2 points for a king. The following parameter values, depicted in Table 1, were used for the GP algorithm:

- A population size of 50 evolved over 25 generations was used. A population size greater than 50 increased the time it took the GP to find a solution and did not improve performance. Twenty-five generations allowed for sufficient convergence without creating high runtimes for real time play. Tests on runs with more generations did not produce fitter individuals.
- The ramped half-and-half method with a depth limit of 6 was used to create the initial population. This was done to establish a wide range of heuristic values at the start.
- Mutation is a global search operator, resulting in high genetic variety during the evolutionary process and so was given a typically low probability of 20 %.
- Crossover is a local search operator [13], which limits variety thus preserving genes, was given a probability of 80 %, higher than the standard value of 60 % suggested by Koza [14]. This was considered important as one of the main objectives of the experiment was to evolve a population that was highly fit,

combining several individuals for a final result. By sharing more genetic code through increased crossover, it was anticipated that the population as a whole would stay more tightly bunched, resulting in better individuals at the end of each run.

- The probability used for reproduction was set to 0 % as both crossover and mutation could result in producing an offspring identical to another element in the population.
- A maximum offspring node limit of 1000, to cater for the more general representation, was used in this study.
- The parameter 'Preserve Alpha' was included to specify whether the alpha player from a previous run should be used as the alpha player at the beginning of a new run or a new 'Alpha' should be created by randomly selecting an element of the initial population. It also specifies whether the alpha player is selected as a parent for a single crossover operation and it is randomly decided whether the alpha player should be mutated to create an offspring or not. For both experiments, this value was set to true.
- A game was determined to be at an end when one of the following conditions was reached:

  - No more pieces of one colour were left on the board.
  - One colour was unable to move.
  - No pieces were taken or promoted to kings in the preceding 50 moves.
  - The number of moves in the game, set by the parameter 'Game moves', exceeded the set value.

The game of checkers and the genetic programming approach were written in Java 8 (build 1.8.0_45-b15). The hardware and software used was Windows 8.1 operating on a Toshiba© Intel© Core i7 4700MQ (2.40 GHz) laptop with on-board Intel HD Graphics 4600 graphics card and 4 GB DDR3 RAM.

4 Results and Discussion

This section reports on the performance of the proposed GP approach in inducing game playing strategies for checkers and compares the performance of HGP and RHGP in evolving these strategies. A discussion of the performance of the approach is firstly presented followed by an analysis of some of the game playing strategies typically evolved by RHGP.

4.1 Performance of the Genetic Programming Approach

The proposed approach was able to successfully evolve game playing strategies for checkers. One of the differences between this study and previous work in this field

Fig. 6 Games won

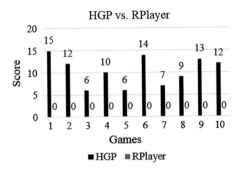

is that strategies are evolved in real time during game play, instead of offline. Using the parameters in Table 1, HGP was able to evolve sound strategies in under a minute. Experiment 2 has revealed that game playing strategies were improved through the incorporation of reinforcement learning:

- **Experiment 1** (HGP vs. RPlayer)—HGP was found to outperform the random move player, winning 10 out of the 10 games as shown in Fig. 6. All of the games resulted in the HGP capturing every RPlayer piece. This is anticipated as the heuristic representation used for HGP is aimed at providing sound strategies for the checkers player. HGP promoted a significantly greater number of single pieces to kings in each game (Fig. 7), 39 kings to the HGP and 6 kings to the RPlayer. Each game lasted no more than 50 moves.
- **Experiment 2** (RHGP vs. HGP)—RHGP performed better than HGP winning 7 of the 10 games and drew 1 game (Fig. 8). RHGP promoted 33 single pieces to kings and HGP promoted 19 single pieces to kings (Fig. 9).

It is important to note that 6 of the 10 games ended before one colour captured all of the pieces of the other colour, indicating that both the RHGP and HGP were able to evolve defensive strategies, preserving their kings. These games in a tournament environment would normally constitute a draw but for the purpose of this experiment, a win was given to the player with the highest score at the end of a game. The RHGP gained a sum of 58 points over 10 games and the HGP gained 28 points.

Fig. 7 Kings captured

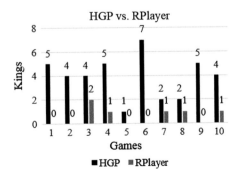

Fig. 8 Games won

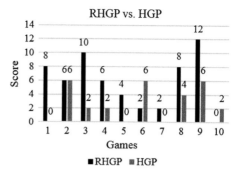

Fig. 9 Kings achieved

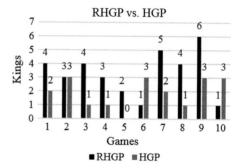

4.2 Analysis of Game Playing Strategies

This section outlines 4 important strategies that contribute significantly to the overall game play of the RHGP. In these examples, the RHGP is playing black and the HGP is playing white. The arrows indicate the direction in which a selected piece will move.

- **Strategy 1**—The start game in checkers is usually considered 'the battle for mobility' [11]. Advancing pieces into strategic positions from the start is guaranteed to result in the promotion of single pieces to kings and the prevention of your opponent gaining kings. Figure 10 illustrates a typical example of a strategy evolved by the RHGP in which each rank (row of four pieces) is sequentially advanced in unison. The back rank (B) in particular is noted to form part of the maneuver, unlike in the case of the HGP, in which the back rank (W) has remained in position resulting in fragmentation of the front ranks (Fig. 11).
- **Strategy 2**—During the start and early middle games, the RHGP is noted to adopt a powerful strategy defending territory and pieces. In Fig. 11 the X's form central points around which the RHGP builds a 4 man formation. In the game of checkers, this is an almost impenetrable formation, and if successfully maneuvered to the king side of the board, will result in the promotion of single pieces to kings.

Fig. 10 RHGP evolved
strategy 1

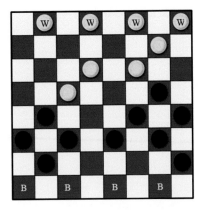

Fig. 11 RHGP evolved
strategy 2

- **Strategy 3**—Both the RHGP and HGP evolved strategies that favoured the promotion of single pieces to kings. Kings however often remained on the 'King Rank', and only moved to allow passage for a single piece onto the rank. This is depicted in Fig. 12.

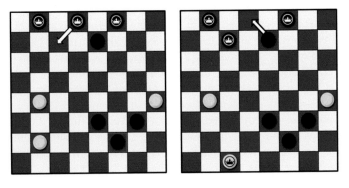

Fig. 12 RHGP evolved strategy 3

Fig. 13 RHGP evolved
strategy 4

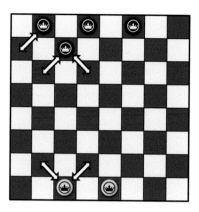

- **Strategy 4**—A number of end games between the RHGP and HGP resulted in kings from both sides simply moving backwards and forwards between ranks (Fig. 13). Unlike strategies adopted by humans, in which the aim is to gain the upper hand resulting in a potential win, this defensive strategy seems intuitive of the evolved learning process to preserve kings disregarding the fundamental aim of winning. In this case, games were ended based on the 'no take rule' [13].

5 Conclusion

The aim of the study presented in this paper is twofold. Firstly, we evaluate a novel heuristic based approach for evolving game playing strategies for checkers. Secondly, we investigate the effectiveness of improving the evolved strategies using reinforcement learning. The evolved strategies were tested against random players and were found to outperform the random players. Furthermore, the use of reinforcement learning to improve the evolved strategies resulted in generation of better game playing strategies which performed better than the evolved players applied as is. This study has revealed the potential of both the heuristic based approach and the incorporation of reinforcement learning. Future work will look at applying it to similar non-trivial mind games such as chess and investigating other options for incorporating reinforcement learning into the evolutionary process. In addition to this other methods of evaluating the game playing strategies instead of play against a random player will also be investigated.

References

1. Schaeffer, J., Burch, N., Bjornsson, Y., Kishimoto, A., Muller, M., Lake, R., Lu, P., Sutphen, S.: Checkers is solved. Science **317**(5844), 1518–1522 (2007)
2. Duch, W.: What is computational intelligence and where is it going? In: Duch, W., Mandziuk, J. (eds.) Challenges for Computational Intelligence. Springer Studies in Computational Intelligence, vol. 63, pp. 1–13. Springer, Heidelberg (2007)
3. Waledzik, K., Mandziuk, J.: The Layered Learning method and its application to generation of evaluation functions for the game of checkers. In: 11th International Conference, pp. 543–552. Kraków, Poland, (2010)
4. Lucas, S.: Computational intelligence and games: challenges and opportunities. Int. J. Autom. Comput. **5**(1), 45–57 (2008)
5. Chellapilla, K., Fogel, D.: Evolving an expert checkers playing program without using human expertise. IEEE Trans. Evol. Comput. **5**(4), 422–428 (2001)
6. Kusiak, M., Waledzik, K., Mandziuk, J.: Evolutionary approach to the game of checkers. 8th International Conference on Adaptive and Natural Computing Algorithms, ICANNGA 2007, pp. 432–440. Warsaw, Poland (2007)
7. Pell, B.: Strategy Generation and Evaluation for Meta-Game Playing. Ph.D. Dissertation, University of Cambridge (1993)
8. David, O., van den Herik, J., Koppel, M., Netanyahu, N.: Genetic algorithms for evolving computer chess programs. IEEE Trans. Evol. Comput. **18**(5), 779–789 (2001)
9. Mukherjee, A.: A Genetic Programming Approach to. Modified Chinese Checkers. https://cs4701.wikispaces.com/file/view/4701+final+report.pdf (2013)
10. Kusiak, M., Waledzik, K., Mandziuk, J.: Evolutionary-based heuristic generators for checkers and give-away checkers. Expert Syst. **24**(4), 189–211 (2007)
11. Benbassat, A., Sipper, M.: Evolving lose-checkers players using genetic programming. In: Proceedings of the 2010 IEEE Symposium on Computational Intelligence and Games (CIG), pp. 30–37. Dublin (2010)
12. Frankland, C., Pillay, N.: Evolving game playing strategies for othello. In: Proceedings of the 2015 IEEE Congress on Evolutionary Computation (CEC 2015), pp. 1498–1504. Sendai, Japan, 25–28 May 2015
13. Rules of Draughts (Checkers), http://www.wcdf.net/rules/rules_of_checkers_english.pdf
14. Koza, J.: Genetic Programming: On the Programming of Computers by Means of Natural Selection. MIT Press (1992)
15. Finnsson, H., Björnsson, Y.: Simulation control in general game playing agents. In: Proceedings of the AAAI conference on Artificial Intelligence, Atlanta, pp. 954–959. Atlanta (2010)
16. Van Lishout, F., Chaslot, G., Uiterwijk, J.: Monte-Carlo tree search in backgammon. In: Proceedings of the Computer Games Workshop (GCW 2007), pp. 175–184. Amsterdam (2007)
17. Chaslot, G., Bakkes, S., Szita, I., Spronck, P.: Monte-Carlo tree search: a new framework for game AI. In: Proceedings of the 4th Artificial Intelligence and Interactive Digital Entertainment Conference, pp. 216–217. AAAI Press, Menlo Park (2008)
18. Benbassat, A., Sipper, M.: Evolving board-game players with genetic programming. In: Proceedings of the 13th Annual Genetic and Evolutionary Computation Conference (GECCO '11), pp. 739–742 (2011)

PARA-Antibodies: An Immunological Model for Clonal Expansion Based on Bacteriophages and Plasmids

Mark Heydenrych and Elizabeth Marie Ehlers

Abstract This paper presents a novel method for modelling the growth of agents undergoing clonal expansion. Using the concept of plasmids, clonal agents are broken into components which can be judged independently of the agents themselves. These components can be redistributed by objects analogous to bacteriophages. Full algorithms for the model are given, and results of an implementation provided. The work presented in this paper has significant implications for the field of clonal expansion.

Keywords Immunological computation · Clonal expansion · Horizontal gene transfer

1 Introduction

This paper discusses a model for implementing component-based clonal expansion. Clonal expansion, even without any supplementation, is an incredibly useful tool for distributed problem solving. This is due to the fact that, when faced with a large search space, traditional agents may struggle to find an optimum. Clonal expansion, however, allows cooperation: each B cell explores only a small portion of the search space throughout its life, but the entire population of agents will explore much of the search space together. However, by augmenting clonal expansion it is possible to gain further improvements.

The improvement suggested by this paper will transform clonal agents into component based agents; that is, the antibody sequence which composes these agents will not be seen as a indivisible DNA sequence but rather a combination of smaller sequences of antibodies. These sequences are neither predefined nor fixed, and can

M. Heydenrych (✉) · E.M. Ehlers
Academy of Computer Science and Software Engineering,
University of Johannesburg, Johannesburg, South Africa
e-mail: mheydenrych@uj.ac.za; swinefish@gmail.com

© Springer International Publishing Switzerland 2016
N. Pillay et al. (eds.), *Advances in Nature and Biologically Inspired Computing*,
Advances in Intelligent Systems and Computing 419,
DOI 10.1007/978-3-319-27400-3_16

179

outlive the individuals in which they are found. The full nature of these sequences will be discussed.

Section 2 gives a brief introduction to clonal expansion to those unacquainted with the field. An understanding of clonal expansion is essential to understanding the contribution of this paper.

Section 3 will discuss the important concepts of horizontal gene transfer, bacteriophages and plasmids. It will discuss the concepts in the context of the human immunological system as well as an analogy for these items in clonal expansion.

Section 4 will elaborate on the PARA-Antibodies model which is the major contribution of this paper. It will discuss the use of phages, the use of plasmids, and discuss core algorithms for the system.

Section 5 will describe the results of a prototype implementation, while Sect. 6 concludes the work presented in this paper.

2   An Introduction to Clonal Expansion

This section covers clonal expansion, the key technique used in this paper. Horizontal gene transfer has recently been recognised as an important mechanism in adaptability [8], often contributing to drug resistance in bacteria [2]. However, to the best of the authors' knowledge there is little work examining the utilisation of horizontal gene transfer in clonal expansion. Research using horizontal gene transfer with genetic algorithms has been performed by Monteiro, Goldbarg and Goldbarg [6] as well as Rocha, Goldbarg and Goldbarg [7]; little further research appears to have been done since. However, both of these papers focus on using genetic algorithms to solve the minimum spanning tree problem while the algorithm in the PARA-antibodies model is a general purpose clonal expansion algorithm. It is in this that the PARA-Antibodies model separates itself from the related works.

Clonal expansion is a population based optimisation strategy based on the working of the vertebrate immune system. Clonal expansion typically maintains a population of B cells (common immune cells), each of which has a set of antibodies. These antibodies are intended to protect the organism from a variety of antigens. When an antibody matches with an antigen, the B cell with that antibody is stimulated and the immune system can neutralise the antigen. B cells which match many antigens should be maintained while B cells which match few antigens should be replaced. The number of antigens matched by a specific B cell is called its affinity [3].

However, the antigens in a population may not be sufficient for full protection of an organism. For this reason, simply keeping good B cells is not enough; these good cells must be further improved. The clonal expansion algorithm also introduces a mutation operator to change the antibodies found in a B cell. It is not uncommon for these changes to be detrimental to a B cell, but sometimes this mutation will result in an improvement. It is this combination of selecting the best B cells and mutating that leads to the necessary optimisation. (For those familiar with genetic algorithms, clonal expansion can be thought of as a genetic algorithm with no crossover and a higher than normal mutation rate). From a computational point of view, the set of antibodies in a B cell represents a possible solution to a problem, the set of antigens represent the parameters of the problem and a cell's affinity indicates the quality of the solution it represents.

3 An Introduction to Horizontal Gene Transfer

This section discusses the concept of horizontal gene transfer, its importance to clonal expansion, and the proposed method by which it can be achieved.

Horizontal gene transfer occurs when a sequence of genes is transferred from one individual to another unrelated individual. An advantage of horizontal gene transfer is that useful genes from one individual can be passed on to another, increasing the frequency of those genes in the population, and as such increasing the overall fitness of the population.

It is important to note that horizontal gene transfer allows a number of genes, usually contiguous, to be transferred from one individual to another. This is important in a number of cases, both the order and position of antibodies in a B cell is important to its affinity. This is common in both organisms and clonal expansion algorithms.

There are three common ways in which horizontal gene transfer can occur between different cells [4]. These three are:

- Transformation: DNA without any carrier is taken from one individual to another. Typically this can only transfer short DNA segments, and only between particular types of bacteria.
- Conjugation: By direct contact, gene segments can be transferred from one organism to another. This allows fairly long gene sequences to be transferred. Conjugation can occur between individuals which are distantly related—even occasionally between different species.
- Transduction: An infectious agent such as a virus carries genetic material from one individual to another. Since both individuals must be susceptible to the virus, this can only occur between similar species.

Only transduction is of interest in this paper. An advantage of transduction is the use of a virus (typically a bacteriophage, which will be shortened as 'phage' from this point on). This phage, since it acts as a carrier, is useful from a programming point of view. The process of transduction can be seen in Fig. 1 (adapted from the description given in [5]).

Since B cells in clonal expansion do not have a genetic sequence, infections will instead carry antibody sequences horizontally. It is essential to identify which antibody sequence in a B cell will be copied when horizontal gene transfer occurs. In the ideal situation, these antibody sequences would be independent, self-sustained objects rather than just segments from the agent itself. Fortunately, biology again provides inspiration, with such a structure already occurring in nature—plasmids. Plasmids are useful because they are completely independent, and have the ability to

Fig. 1 The process of transduction: **a** A bacteriophage infects a cell. **b** The bacteriophage copies a portion of the cell's DNA, and becomes a carrier. **c** The carrier bacteriophage infects a new cell. **d** The DNA which the bacteriophage carried is copied into the new cell. (Adapted from [5])

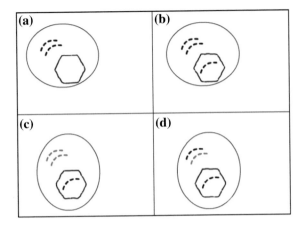

replicate without any external aid. An advantage of plasmids is that they can move easily from one cell to another, a useful feature when dealing with antibody segments that are expected to move between individuals in an environment. Finally, plasmids have the ability to insert themselves into the genetic sequence of cells [1]—that is, a plasmid can become part of the make-up of an individual, and the changes caused by these plasmids will be inherited by future generations. In the proposed model, plasmids will be used to carry antibodies rather than genes.

Now that the inspirations for the model have been explained, the next section will detail the workings of the model itself in order to provide readers with an understanding of how plasmids and transduction could be used to improve clonal expansion.

4 PARA-Antibodies Model

Section 4 will introduce the major contribution of this paper, a proposed model for adding horizontal gene transfer to clonal expansion. This section will focus on the following:

- The use of phages, which will be discussed in Sect. 4.1
- The use of plasmids as an extension to the phage managed model, which will be discussed in Sect. 4.2
- An initial set of algorithms for the proposed model, which will be discussed in Sect. 4.3

4.1 Phages

The first item for consideration is phages: it is essential to have some kind of infectious agent through which transduction occurs. These phages are expected to be fairly simple classes, focused on being carriers and transmitters for antibodies. However, the antibodies carried by a phage must be acquired through some infection—initially a phage will have no antibodies to carry.

That said, once a phage has acquired antibodies, its job is to infect as many other individuals as possible. With each individual it infects, there is a chance that the antibodies it carries will be copied into the antibodies of the individual it is currently infecting. However, since it is undesirable to have poor antibodies being copied into strong individuals, it is proposed that each clonal agent be given an immune system. This immune system will abstract away all the complexity of a true immune system, and replace it instead with the agent's affinity—the better an individual is, the less likely it is to be infected. It is expected that such a system will encourage strong antibodies to be copied into weaker individuals.

However, if this same immune system is used when determining the infection which provides a new phage with antibodies, then an unfortunate result is that the phage is more likely to acquire poor antibodies rather than strong antibodies, since it is more likely to infect a weak individual. Therefore the authors proposed two types of infections:

1. Carrier infections: In the case of a carrier infection, a phage with no antibodies infects an individual in order to gain an set of a B cell's antibodies
2. Transmission infections: In the case of a transmission infection, a phage with antibodies infects an individual with the chance of copying its antibodies into that individual

Carrier infections are the first step in a phage's life cycle. In a carrier infection, an individual in the population is stochastically chosen for the phage to infect. Once the infection has taken place, the phage copies a portion of the antibodies of the individual to become the segment which it transmits. When selecting an individual to infect, the likelihood of infection should be directly proportional to the individual's affinity.

There are also important considerations when choosing the segment of antibodies to copy from the individual into the phage most importantly, the length of the segment. The segment should have a length between some minimum and maximum, designated ρ_{min} and ρ_{max}. During prototype implementation, the following values were found to give good results:

- $\rho_{min} = 3$
- $\rho_{max} = \frac{1}{10}$ of the length of an antibody sequence

Once a phage has attained antibodies, it begins to infect other agents throughout its lifetime. In each generation, each phage may infect at most one agent. This process should be pseudo-random, but should be set up such that there is a high probability that at least one phage will infect at least one agent each generation. This will allow the antibodies the phages contain to be continually spread among the population. Since these are transmission infection, the likelihood of infection should be inversely proportional to an agent's affinity.

The number of phages extant in a system is left entirely to the discretion of the programmer. The more phages there are, the more horizontal gene transfer will occur, and the more good antibodies will be redistributed. As such, the number of phages in an environment will be dependent entirely on what the programmer's needs are.

Unfortunately, it is impossible at this point to know whether antibodies are helpful or detrimental to the clonal agents into which it is copied. A fitness value should be attached to the antibody segment itself—an idea that will be expanded in Sect. 4.2.

4.2 Plasmids

As discussed in Sect. 3, a plasmid is a small sequence of self-contained DNA which is able to reproduce independently. As such, plasmids are a clear analogy for the antibody segments used so far in this model. Plasmids will have three major components in the proposed model:

- The antibodies themselves. This can be represented in any way that is appropriate for the problem at hand.
- Where the sequence begins. It is important to keep track of where a plasmid begins in an antibody sequence, since in many problems this position can have important effects on the phenotype.
- The fitness of the plasmid. This is the major contribution of plasmids to the PARA-Antibodies model.

The fitness of a plasmid is simply an indicator of how useful the plasmid has been in previous interactions. This is an important consideration, since the aim of the proposed model is to create a *useful* supplementation to clonal expansion.

After a transmission infection, a phage has a plasmid, but that plasmid has not yet been judged. This fitness will be determined as infections occur, and will be fairly simple to calculate. Whenever a transmission infection occurs and the plasmid is copied into a B cell, it makes sense that the agent which was affected must have its affinity re-evaluated, since its antibodies have been modified. When the affinity is re-evaluated, the fitness of the plasmid will also be updated. The value of $affinity_{after}$—$affinity_{before}$ will be added to the fitness of the plasmid.

This value will be added as is, whether it is positive or negative. As such, if the plasmid causes an improvement in the agent to which it is added, then its fitness will be increased by the same value. If, however, the plasmid is detrimental to that agent, then its fitness will be decreased. By this method, it becomes trivial to tell how useful a plasmid is to the population—those with high values are highly useful, while those with low values are less helpful. Plasmids with negative fitness are known to be generally detrimental to the system.

However, one does not wish to have detrimental plasmids in the system. While it is possible to use plasmid fitness to modify the chances of transmission infection by phages, this does not hold with the analogy presented thus far—the quality of a plasmid is meaningless to the effectiveness of a phage. As such, rather than affect how likely phages are to infect, plasmids will be removed if they are not helpful to the system.

This can be achieved by setting a threshold value, τ. If the fitness of a plasmid falls below τ, then the plasmid dies and the phage which carries it no longer has a plasmid. In this case, the phage will be treated as a new phage and perform a carrier infection on the next generation.

The use of plasmid fitness, then, provides a significant advantage over a phage driven model—self regulation. Since plasmids die when they are no longer useful, there is no need for any central authority to handle the death of plasmids. As well as this, even a plasmid that was incredibly useful to the population of agents at one point may become detrimental at a later stage. In this case, such a plasmid would eventually obtain a fitness lower than τ, and so no longer be in the system.

Finally, one must consider the change in fitness over time. Usefulness in the past is a good predictor of success in the present, but the more distant that past, the less useful the fitness is as a predictor. As such, the proposed model proposes diminishing fitness of plasmids. It is proposed that the fitness of a plasmid in generation n be calculated as shown in Eq. 1.

$$fitness_n = \omega \times fitness_{n-1} + \Delta A \qquad (1)$$

where

$$fitness_n = \text{the fitness of the plasmid in generation } n$$
$$\omega = \text{a depreciation constant}$$
$$\Delta A = \text{the change in affinity of an agent}$$

Now that Eq. 1 has been defined, it can be seen that even a plasmid that was useful at some time in the past will eventually be eliminated if it is no longer useful, even if it is not detrimental but simply unhelpful. This will occur because the constant multiplication by ω will eventually bring the plasmid's fitness below τ. Furthermore, the depreciation constant will increase the speed at which detrimental plasmids will be eliminated from the system.

As a final note, the author suggest that plasmids be given a short evaluation period. During this period, plasmids will not be eliminated regardless of their fitness in order to ensure that all plasmids are sufficiently evaluated before they can be removed from the system.

Now that the PARA-Antibodies model has been briefly described, Sect. 4.3 will provide a set of core algorithms which form the basis of the prototype implementation of the model.

4.3 Initial Algorithms

Section 4.3 shall provide algorithms for the proposed model detailed in the previous sections. These algorithms have been implemented, and the results of this implementation will be recorded in the next section.

Two major algorithms are necessary for the proposed model: the carrier infection and transmission infection performed by the phages. These are the most important algorithms proposed by this paper, because these are the algorithms which allow horizontal gene transfer to occur.

Algorithm 1 is fairly simple: the length of the plasmid to be acquired is determined as a pseudo-random number between ρ_{min} and ρ_{max}. The starting point is then chosen, and the plasmid's antibodies are copied from the agent's antibody sequence. Lines 4 and 5 complete the plasmid's initialisation.

Input: A Phage p; A ClonalAgent a; Limits ρ_{min} and ρ_{max}
Result: p will acquire a plasmid
1   length $\leftarrow U(\rho_{min}, \rho_{max})$
2   start $\leftarrow U(-, a.\text{sequence.length - length})$
3   p.plasmid.antibodySegment $\leftarrow a$.antibodySequence.subSquence(start, length)
4   p.plasmid.start $\leftarrow$ start
5   p.plasmid.fitness $\leftarrow 0$

Algorithm 1: An algorithm for a Carrier Infection

Algorithm 2 details a transmission infection. This algorithm is fairly self-explanatory. Note that line 7 updates the fitness according to Eq. 1. Because of the for loop, this algorithm is expected to run in $O(n)$ time, linear in the length of the gene segment.

Input: A Phage p; A ClonalAgent a; Constants ω and τ
Result: The structure of a had been modified; The fitness of p's plasmid has been updated
1   oldAffinity $\leftarrow a$.affinity
2   **for** $i \leftarrow 0$ *to p.plasmid.antibodySegment.length* **do**
3   a.antibodySequence($i + p$.plasmid.start) $\leftarrow p$.plasmid.antibodySegment(i)
4   **end**
5   a.updateAffinity()
6   newAffinity $\leftarrow a$.fitness
7   $\Delta A \leftarrow$ newAffinity - oldAffinity
8   p.plasmid.fitness $\leftarrow \omega \times p$.plasmid.fitness + ΔA
9   **if** *p.plasmid.fitness $< \tau$* **then**
10   p.plasmid $\leftarrow$ NULL
11   **end**

Algorithm 2: An Algorithm for Transmission Infection

The discussion of these algorithms concludes the exposition of the proposed model. Section 5 will discuss the results of a prototype implementation of the proposed model.

5 Results

A C# implementation of the proposed algorithms was produced, solving the bounded Knapsack problem. The results of testing with that implementation shall briefly be described now. The proposed algorithm was tested against traditional clonal expansion. Ten tests were run for both PARA-Antibodies and traditional clonal expansion on problems of varying size, and the random seed was kept the same across parallel tests to ensure that differences of performance were not due to stochastic effects. For each problem size, the average solution quality was recorded for both techniques.

In a number of cases, the proposed model led either to faster convergence or an improved solution, often both. In a small number of tests, the proposed model helped the agents escape a local optimum in order to reach a better solution. This strongly suggests the proposed model is no more susceptible to getting stuck in local optima than standard clonal expansion, and can sometimes be less susceptible. This is because phages may contain antibodies that are no longer extant in the population. The injection of such antibodies can push an individual into a part of the search space outside the local optimum, which may allow for escaping.

In a majority of cases, the use of plasmids leads to improved solutions. The improvement caused by plasmids is shown in Fig. 2. From these results, it is clear not only that plasmids provide appreciable improvement in most cases, but also that the improvement offered by plasmids is greater with a larger problem space.

These preliminary results suggest that the PARA-Antibodies model is a useful and powerful addition to the field of clonal expansion.

Fig. 2 The average improvement caused by plasmids, as a percentage of fitness. The horizontal axis increases with the problem size, in terms of number of items to be packed

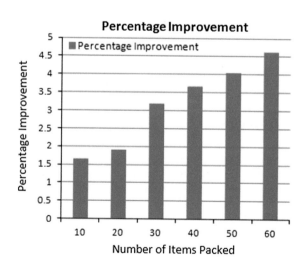

6 Conclusion

This paper has proposed a model for component-based clonal expansion. This model is based strongly on horizontal gene transfer—specifically the capability of individuals to acquire genetic material from other individuals and merge it with their own genetic code. Although clonal expansion uses antibodies rather than genes as its metaphor, the concepts remain similar.

The proposed model, as explained in this paper, focuses on the use of phages in order to redistribute antibody sequences—called 'plasmids'—among the agents. These plasmids have a fitness, which is used to judge the effectiveness of that plasmid in improving the population. Poor plasmids are eliminated from the system to ensure better overall fitness.

The results of an initial implementation showed that the proposed model is never worse than standard clonal expansion, and has the potential to be better than standard clonal expansion in some important regards, including escaping local optima.

There are many potential uses for plasmids, remembering that a plasmid has a fitness separate from the affinity of the agent from which it came. Since these plasmids represent a portion of a solution and an indication of its usefulness, plasmids may be analysed for partial solutions. Further work will be done in extracting good partial solutions from phages, especially in fields in which partial solutions are either useful, or reveal important aspects of the problem such as the travelling salesman problem.

References

1. Biology-Online: Plasmid http://www.biology-online.orgidictionary/Plasmid (2009). Accessed 4 April 2012
2. Cavanagh, J.P., Hjerde, E., Holden, M.T., Kahlke, T., Klingenberg, C., Flægstad, T., Parkhill, J., Bentley, S.D., Sollid, J.U.E.: Whole-genome sequencing reveals clonal expansion of multiresistant Staphylococcus haemolyticus in European hospitals. J. Antimicrob. Chemother. **69**(11), 2920–2927 (2014)
3. Dasgupta, D., Niño, L.F.: Immunological Computation—Theory and Application. Auerbach Publications, Boca Raton (2009)
4. Hartl, D.L., Jones, E.W.: Genetics: Principles and Analysis (1998)
5. Maloy, S.: Horizontal gene transfer, http://www.sci.sdsu.edu/smaloy/MicrobialGenetics/topicsigenetic-exchange/exchange/exchange.html (2002). Accessed 4 April 2012
6. Monteiro, S.M., Goldberg, E.F., Goldberg, M.C.: A plasmid based transgenetic algorithm for the biobjective minimum spanning tree problem, pp. 49–60 (2009)
7. Rocha, D.A.M., Goldberg, E.F.G., Goldberg, M.C.: A new evolutionary algorithm for the biobjective minimum spanning tree, pp. 735–740 (2007)
8. Soucy, S.M., Huang, J., Gogarten, J.P.: Horizontal gene transfer: building the web of life. Nat, Rev. Genet. **16**(8), 472–482 (2015)

Bioinspired Tabu Search for Geographic Partitioning

María Beatríz Bernábe-Loranca, Rogelio González Velazquez,
Martín Estrada Analco, Jorge Ruíz-Vanoye, Alejandro Fuentes Penna
and Abraham Sánchez

Abstract The analytical observation of nature induces inspiration to propose new computational paradigms to create algorithms that solve optimization and artificial intelligence problems. The artificial vision allows establishing a problem with intelligent techniques from living systems. The bioinspired systems are presented as a set of models that are based on the behavior and the way of acting of some biological systems. These models can be expressed in data mining and operations research where the clustering is a recurrent technique in the P-median problem and territorial design. On this point, we have solved clustering problems using partitioning with bioinspired aspects and variable neighborhood search to approximate optimal solutions. In this work we have improved the search strategy: we present a bioinspired partitioning algorithm with optimization by tabu search (TS). This clustering problem under a bioinspired connotation has been proposed after observing some characteristics in common between clustering and human behavior in conflict situations, where some characteristics have been modeled.

Keywords Clustering · Bioinspired · Tabu search

1 Introduction

The inspired algorithms can describe and solve complex relationships from intrinsically very simple initial conditions and rules with little to no knowledge about the search space. In general, nature is the best example to propose bioinspired algorithms for optimization. If we closely examine each and every feature or

M.B. Bernábe-Loranca (✉) · R.G. Velazquez · M.E. Analco · A. Sánchez
Facultad de Ciencias de la Computación, Benemérita Universidad Autónoma de Puebla,
Puebla, Mexico
e-mail: beatriz.bernabe@gmail.com

J. Ruíz-Vanoye · A.F. Penna
Universidad Autónoma del Estado de Hidalgo, Pachuca, Mexico

© Springer International Publishing Switzerland 2016
N. Pillay et al. (eds.), *Advances in Nature and Biologically Inspired Computing*,
Advances in Intelligent Systems and Computing 419,
DOI 10.1007/978-3-319-27400-3_17

phenomenon in nature, they manage to develop an optimal strategy, still addressing complex interactions between organisms ranging from microorganisms to full-fledged human beings, balancing the ecosystem, adaptation, physiques, clouds, rain etc. Nature teaches us; its designs and capabilities are enormous and mysterious things that researchers are trying to mimic in technology. Both fields have a much stronger connection, since it seems reasonable that new or persistent problems in computer science could have a lot in common with problems nature has encountered and solved long ago. Therefore an easy mapping between nature and technology is possible. Bio inspired computing has come up as a new era in computing encompassing a wide range of applications, covering many. Heuristic approaches seem to be superior in solving hard and complex optimization problems, particularly where the traditional methods fail. Bioinspired algorithms are heuristics that mimic/imitate the strategies of nature since many biological processes can be thought out as processes of constrained optimization. They make use of many random decisions which classifies them as a special class of randomized algorithms. Formulating a bioinspired algorithm design involves choosing a proper representation of the problem, evaluating the quality of the solutions using a fitness function and defining operators to produce new sets of solutions [1].

The previous section shares excellent arguments to propose different and broad bioinspired algorithms schemes. This is the philosophy we adopted to develop our algorithm: Building a clustering algorithm capable of processing big data bases, either biological or geographical data, in reasonable amounts of time. We have combined aspects from automatic classification (classification by partitions), in a bioinspired setting, with tabu search which induces us to the following conclusion: the classic partitioning in front of geometric restrictions over distances, can be presented as a bioinspired algorithm analogous to the "reunion" behavior of human beings under critical or emerging situations in physical spaces where they must get together into groups.

2 Description of the Problem

Clustering analysis includes a number of different algorithms and methods to group objects into categories according to their similar characteristics. In recent years, considerable efforts have been made to improve the performance of such algorithms. On this sense, this paper explores three different bioinspired metaheuristics for the clustering problem: Genetic Algorithms (GAs), Ant Colony Optimization (ACO), and Artificial Immune Systems (AIS) [2, 3].

General purpose algorithms have been designed that are inspired by processes observed in nature. Such algorithms from the field of bioinspired computation, which cover many algorithmic approaches inspired by processes observed in nature, include well known approaches such as evolutionary algorithms (EAs) ant colony optimization (ACO) and particle swarm optimization (PSO) [4].

We discuss a bioinspired algorithm that generates good results for geographic data but it can be used with other kinds of data as well. Designing it involves the following three steps:

1. Choose a representation of possible solutions.
2. Determine a function to evaluate the quality of a solution.
3. Define the neighborhood mechanism and strategic memory that produces a new set of elite solutions from a set of current solutions.

Considering these aspects, we propose a partitioning algorithm that optimizes the distance minimization criterion under a bioinspired philosophy about human behavior in times of disaster. The challenge consists in achieving an accurate evaluation function value for the implicit criterion, interpreted from a statistical and biological point of view. On this stage, the approximation method for partitioning that we have built is based on variable neighborhood search whereas the searches made are stored in memory, as human beings usually do it under emergency conditions.

Our conclusions lead to propose partitioning as a bioinspired-trajectory algorithm with responsive memory that promises quasi-optimal solutions when the optimization technique is based on variable neighborhood search and tabu memory, and they solve a considerable amount of clustering problems for spatial or biological data as long as they are spatially described [5].

2.1 Bio-Inspired Partitioning

The partitioning begins with a random initial solution, k centroids represent the groups. From this point the main cycle begins to move to neighboring solutions. Neighbor solutions are generated inside a neighborhood structure. If the solution is better, it will finish the current local search to be able to change the candidate list (choose another group). When a given number of worse solutions have been found the perturbation strategy will trigger and change one of the centroid for a geographical unit (bigger change).

This idea departs from assuming that when an unexpected disaster or a necessity to get organized surprises human beings, a leader is proposed, almost stochastically, such that the rest of the people will get close to the leader most similar to them or the physically closest to them and they will try to solve the problem in a certain amount of time (t). But if an individual doesn't see results after the established time, he/she will most likely join another group. If the new leader can't satisfy the individual's expectations, the individual will keep changing groups looking for leaders similar to the previous ones because human beings aren't usually prepared for dramatic changes. We should remark that the difference here is that the objects are the ones shifting groups while the leaders are kept in place. After a while, when bigger changes for the benefit of the group aren't noticeable, then the group itself will replace its leader. On this point our algorithm is consistent with perturbation strategies commonly used in metaheuristic procedures to restart the search process from another point in the solution space.

2.2 Grouping in Human Beings

Many living beings solve a considerable amount of problems by using different forms of searching procedures in spite of being intuitive and non-systematical. However the majority of real problems require at least a search or clustering method to be solved.

In this work we propose that by observing the way and structure that most living beings use to search for something as a group, we can propose an analogy with classification by partitions in a bioinspired way. The bioinspired partitioning proposal can be stated as the one that optimizes an objective by finding the optimal structure and sequence of a minimal and efficient trajectory with the goal of easing paths, tours and search procedures among its members with a minimum cost. Since a heuristic method is demanded to solve this partitioning, we have proposed TS because it is an algorithm that employs responsive and intelligent memory and is guided by special memory strategies adopted from the artificial intelligence field. This gives TS an advantage over its randomized rivals and a more human-like search behavior [6–8].

3 Tabu Search

The origins of the tabu search method date back to works published by the end of the 70's. Officially, its name and methodology were later introduced by Fred Glover and Manuel Laguna in 1989 in the homonymous book. Its philosophy is deriving and exploiting a set of intelligent principles to solve problems. That's why we can say that TS is based on concepts that unite the artificial intelligence and the optimization fields. TS is a metaheuristic that guides a local search heuristic (our bioinspired local search method) to explore the solution space beyond local optimums [6].

The main TS strategy is imposing restrictions to guide the search process towards difficult access regions of the solution space. These restrictions are applied using the method's memory to record complete, parts or attributes of the solutions and lock them out of the search process for a certain amount of time.

Diverse applications have employed TS to attain high quality solutions; in particular, our interest is on the works that have presented good results with TS in clustering problems. As a particular case, in [9] a modified tabu search approach is proposed that consists of two phases: the constructive phase which is an initial solution using the K-means algorithm and an improvement phase, where the modified TS algorithm is employed to improve the initial solution. Furthermore their results are better than K-means and tabu search (standard) on their own.

4 Bio-Inspired Tabu Search

Considering the bioinspired aspects described on the previous sections, we have built the following TS partitioning algorithm.

4.1 Tabu Search Bio Algorithm

```
Input:
Number of centroids p
Number of iterations nit
Number of iterations for perturbation ip
Tabu tenure tt

1: pc ← 0
2: ic ← 1
3: S ← InitialSolution()
4: S* ← S
5: While ic < nit do
6:   prev_cost ← CostOf(S)
7:   S ← BioLocalSearch()
8:   If CostOf(S) >= prev_cost then
9:       pc ← pc+1
10:    end if
11:    If Costo(S) < Costo(S*) then
12:        S* ← S
13:    end if
14:    If pc > ip then
15:        S ← PerturbSolution(S)
16:        pc ← 0
17:    end if
18:    UpdateTabuList()
19:    ic ← ic+1
20: end while
31: Return S*
```

Our algorithm is a basic tabu search template. It has a main cycle which ends when the maximum number of iterations is reached, which is given by the user.

Inside the main cycle the bioinspired local search procedure is executed (line 7). Each one of the iterations stores the cost of the current solution to compare it with the cost of the new solution returned by the bioinspired local search process.

Furthermore the input parameter ip (perturbation iterations) determines number of
solutions, worse than the previous one, will be accepted before changing the centroids (lines 8–10 and 14–17). Finally the tabu list is updated and the global
iterations counter is increased (ic).

4.2 Local Search

The following procedure corresponds to the bioinspired local search routine.

```
Input
Current cost - cost
Number of iterations for the local search - lsIterations
1: c1 ← 0;
2: c2 ← 0;
3: ug ← 0;
4: S₁ ← S;
5: solutionFound ← false;
6: cont ← 1;
7: While cont < lsIterations AND solutionFound = false do
8:  UpdateTabuList();
9:      c1 ← getHighestCostCentroid();
10: ug ← getFarthestGUFrom(c1);
11:     c2 ← getNewCentroidFor(ug);
12:     localCost <-- CostOf(c1, c2, ug);
13:     BeginTabuState(c1);
14: If cost > localCost then
15:        solutionFound ← true;
16:   else
17:        cont ← cont + 1;
18:   end if
19:   end while
20: If solutionFound = true then
21: S₁ ← Shift(c1, c2, ug);
22: end if
23: Return S₁
```

The local search cycle ends when the number of iterations is reached or when the
first improvement is found.

Three elements must be chosen in this procedure: the centroid 1 (c1), centroid 2
(c2) and the geographical unit (gu). The geographical unit belongs to c1 and will be
transferred to c2.

The selection of these elements is made following the bioinspired ideas presented in the previous section, where human being will move to another groups looking for better leadership and companions. Centroid c1 represents a dissimilar groups, this is, a group with low cohesiveness, making it susceptible to ruptures. The gu is the member that is physically farther (more dissimilar) to c1 than the rest of the gu's of the group. Lastly the centroid c2 is chosen according to the physical proximity (similarity) that gu has with another geographical unit outside of its group, which corresponding centroid is c2.

Once we obtained the three elements, we calculate the cost of transferring gu from c1 to c2 (line 12). If the transfer cost surpasses the current cost then the local search is interrupted and the new better solution is returned, the current solution doesn't suffer any changes otherwise.

The tabu strategy consists in tagging c1 as tabu active for a certain number of iterations (line 13), that's why we need to update the tabu list in the local search (line 8).

Perturbation strategy: All kinds of social groups, under normal circumstances, change leaders after a period of time (sometimes willingly and sometimes due to public pressure). From government elections to coups, it's natural for a social group to change their structure and/or leadership, in this way our perturbation strategy emulates this necessary behavior.

Algorithmically this idea is carried out by using the "findOut" algorithm designed by Resende and Weneck [10], which given a candidate geographical unit to replace a centroid, it returns the best candidate centroid to be replaced. Our contribution at this stage is to use a restricted candidate list formed by the members assigned to c1 only. The perturbation is then the best exchange possible found in the candidate pairs.

4.3 Parameter Fitting

Global iterations: This parameter represents the number of time the local search procedure will be executed. Based on trial and error experiments we concluded that for instances up to 2000 elements 5000 iterations provides adequate response time without sacrificing quality.

Local search iterations: This parameter determines how many candidate elements will be evaluated in the local search. We have determined that 15 is an adequate value.

Maximum worse solution accepted before perturbation: This value determines when the search will be restarted according to the amount of solutions that don't improve the previous one. We decided that a value 10 provides the desired diversification.

Tabu-add tenure: Determines the number of iterations that a newly added centroid will be unable to be replaced. We have set this value at the half of the size of the input problem, this is, half of the number of groups inputted.

5 Experiments and Results

Gathering results: The tests were performed on a machine with an AMD E-350 CPU at 1.60 GHz and 4 GB DDR3 in RAM. The OS is Windows 7 Ultimate 64 bits. Table 1 shows our results, including a simple simulated annealing algorithm (SA), a variable neighborhood search (VNS Pure) and bioinspired VNS (VNS Bio) and tabu search (TS Bio), the last two employ the same bioinspired local search

Table 1 Test results. O number of objects, G number of groups, A algorithm, C cost attained and T time

District	O	G	A	C	T
Colima	371	15	SA	17.389	00:01
			VNS Pure	17.227	00:01
			VNS Bio	15.618	00:01
			TS Bio	14.945	00:01
		30	SA	12.735	00:01
			VNS Pure	11.874	00:01
			VNS Bio	11.826	00:01
			TS Bio	10.506	00:01
		45	SA	10.026	00:02
			VNS Pure	9.8788	00:01
			VNS Bio	9.3687	00:01
			TS Bio	8.4645	00:01
		60	SA	8.9314	00:02
			VNS Pure	8.634	00:01
			VNS Bio	7.8389	00:01
			TS Bio	6.8495	00:01
Puebla	2580	15	SA	370.83	00:05
			VNS Pure	361.42	00:17
			VNS Bio	355.65	00:47
			TS Bio	300.93	00:50
		30	SA	265.30	00:10
			VNS Pure	253.98	00:41
			VNS Bio	256.24	01:27
			TS Bio	217.36	00:18
		45	SA	223.85	00:17
			VNS Pure	215.90	01:16
			VNS Bio	216.21	02:14
			TS Bio	172.99	00:13
		60	SA	196.8	00:21
			VNS Pure	190.5	02:32
			VNS Bio	188.94	03:00
			TS Bio	154.82	00:20

procedure. Finally we have set the parameters as follows: SA—Initial temperature: 5000, Final temperature: 0.1, α: 0.98 and L(t) = 5. VNS Pure and VNS Bio—Neighborhood structures set iterations: 4 and Local search iterations: 100. TS Bio: Global iterations: 5000, Local search iterations: 15, Iterations before perturbation: 10 and Tabu-add tenure: problem size/2.

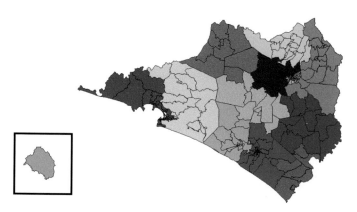

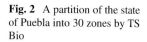

Fig. 1 A partition of the state of Colima into 15 zones by TS Bio

Fig. 2 A partition of the state of Puebla into 30 zones by TS Bio

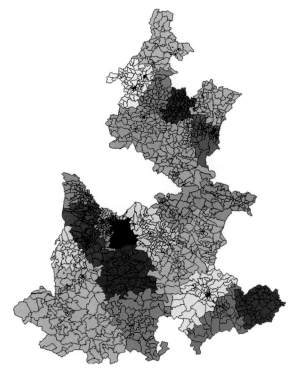

The cost attained C is the sum of the Euclidian distances between the geographic units and the centroids where they were allocated.

We can see that TS bio surpasses the other algorithms and in general achieves better computing times. It is interesting to see that for the biggest map (Puebla) it reaches solutions which costs are significantly lower than the competitors and it shows better efficiency than VNS Bio by reaching a better result for the biggest instance in 20 s in contrast to the 3 min required by VNS Bio.

Graphic results: These were generated using the same computer as above with a tool we develop aided by the GeoTools library [11] that is able to generate visual representations of the partitions on any map.

The following figures present the maps of the best configurations obtained for the selected instances (Figs. 1, 2).

5.1 Conclusions

Considering the bioinspired aspects described in the previous sections, our TS partitioning algorithm appears solid while trying to emulate human behavior making an analogy between geographic groups and social groups. A more human-like metaheuristic like TS has given better results than semi random strategies such as VNS or SA. However these are the early stages of a new approach and hence our new task is to test it on benchmark instances. Another issue left to attend is the proper calibration of three of the algorithms in Table 1 to present a fairer comparison, this is, an experiment design, which we have applied to calibrate SA so far. For the rest of the algorithms we configured the parameters based on a trial-and-error experimentation.

References

1. Binitha, S., Sathya, S.: A survey of bio inspired optimization algorithms. Int. J. Soft Comput. Eng. (IJSCE) **2**(2), 137–150 (2012) ISSN: 2231–2307
2. Thelma Colanzi, E., Klewerton Guez Assunção, W., Trinidad Ramirez, A.: Application of bio-inspired metaheuristics in the data clustering problem. CLEI Electron. J. **14**, 1–18 (2011)
3. Ruiz-Vanoye, J.A., Díaz-Parra, O., Cocón, F., Soto, A., Buenabad Arias, A., Verduzco-Reyes, G., Alberto-Lira, R.: Meta-heuristics algorithms based on the grouping of animals by social behavior for the traveling salesman problem. Int. J. Comb. Optim. Probl. Inf. **3**(3), 104–123 (2012). ISSN: 2007–1558
4. Hongbo, L., Ajith, A., Okkyungi, C., Seong, H.: Variable neighborhood particle swarm optimization for multi-objective flexible job-shop scheduling problems, SEAL 2006, LNCS, vol. 4247, pp. 197–204. Springer, Berlin, Heidelberg (2006)
5. Bernábe-Loranca, B., González, R., Olivares, E., Ramírez, J., Estrada, M.: A bioinspired proposal of clustering around medoids with variable neighborhood structures. Int. J. Comput. Inf. Syst. Ind. Manag. Appl. **6**, 45–54 (2014). [MIR Labs, Dynamic Publishers, Inc., USA]. ISSN 2150–7988

6. Glover, F., Laguna, M.: Tabu Search. Kluwer Academic Publishers (1997)
7. Leiva, S., Torres, F.: Una revisión de los algoritmos de partición más comunes de conglomerados: un estudio comparativo. Revista Colombiana de Estadística **33**, 321–339 (2010)
8. Hajmohammadi, M.S.: Graph-based semi-supervised learning for cross-lingual sentiment classification. In: Intelligent Information and Database Systems, 7th Asian Conference, ACIIDS 2015, pp. 91–110. Bali, Indonesia (2015)
9. Jain, S., Swamy, C., Balaji, K.: Greedy algorithms for k-way graph partitioning
10. Resende, M., Werneck, R.: On the implementation of a swap-based local search procedure for the p-median problem. In: Proceedings of the 5th Workshop on Algorithm Engineering and Experiments (ALENEX), pp. 119–127 (2003)
11. Project Management Committee. GeoTools The Open Source Java GIS Toolkit. http://geotools.org/. Accessed 14 Aug 2015

A Hyper-Heuristic Approach to Solving the Ski-Lodge Problem

Ahmed Hassan and Nelishia Pillay

Abstract Hyper-heuristics seek solution methods instead of solutions and thus provides a higher level of generality compared to bespoke metaheuristics and traditional heuristic approaches. In this paper, a hyper-heuristic is proposed to solve the ski-lodge problem which involves allocating shared-time apartments to customers during a skiing season in a way that achieves a certain objective while respecting the constraints of the problem. Prior approaches to the problem include simulated annealing and genetic algorithm. To the best of our knowledge, this is the first time the ski-lodge problem is approached from a hyper-heuristic perspective. Although the aim of hyper-heuristics is to provide good results over problem sets rather than producing best results for certain problem instances, for completeness and to get an idea of the quality of solutions, the results of the proposed hyper-heuristic are compared to that of genetic algorithm and simulated annealing. The hyper-heuristic was found to perform better than simulated annealing and comparatively to the genetic algorithm, producing better results for some of the instances. Furthermore, the hyper-heuristic has better overall performance over the problem set being considered.

Keywords Combinatorial optimization · Hyper-heuristics · Genetic algorithms · Simulated annealing

A. Hassan (✉)
Department of Mathematics, Taymaa Campus,
Tabuk University, 71491 Taymaa, Saudi Arabia
e-mail: ahmedhassan@aims.ac.za

N. Pillay
School of Mathematics, Statistics and Computer Science,
University of KwaZulu-Natal, 3201 Pietermaritzburg, South Africa
e-mail: pillayn32@ukzn.ac.za

© Springer International Publishing Switzerland 2016
N. Pillay et al. (eds.), *Advances in Nature and Biologically Inspired Computing*,
Advances in Intelligent Systems and Computing 419,
DOI 10.1007/978-3-319-27400-3_18

1 Introduction

In this paper, we consider a combinatorial optimization problem related to the winter tourism industry. The problem is about finding the best assignment of weeks to customers wishing to spend one week in a ski-lodge in a skiing area subject to certain constraints. The contribution of this paper is that it is the first hyper-heuristic approach to the problem since all of the previously proposed methodologies fall under the category of metaheuristics. Furthermore, the paper demonstrate the efficacy of genetic algorithms in problem solving when implemented on the hyper-heuristic level instead of being employed directly to the solution space. Unlike the usual case where hyper-heuristics are beaten by metaheuristics owing to the detailed knowledge that the latter techniques use, in this paper the proposed hyper-heuristic has a comparable performance with the metaheuristic genetic algorithm.

The paper is organized as follows. Section 2 defines the problem of interest. The technique used to solve the problem is a hyper-heuristic and thus a brief overview of hyper-heuristics is provided in Sect. 3. The details of the approach used to solve the problem is provided in Sect. 4. Since the proposed approach involves parameters, Sect. 5 discusses suitable values of these parameters and the method used to determine them. The performance of the hyper-heuristic is presented in Sect. 6 and compared with the performance of the genetic algorithm and simulated annealing. Finally, Sect. 7 concludes the paper and presents possibilities for future work and further investigation.

2 The Ski-Lodge Problem

The ski-lodge problem is originally proposed by [4]. The problem has a website [1] containing a problem instance generator, interactive program for solving the problem manually, a genetic algorithm and a simulated annealing algorithm. On the problem website there are ten instances published along with the solutions generated using the simulated annealing and the genetic algorithms. Those instances are used in this work to evaluate the hyper-heuristic.

The ski-lodge problem is as follows. There is a ski-lodge containing 4 apartments, each capable of hosting up to 8 persons. However, fire regulations require that no more than 22 persons be resident in the ski-lodge during any week. There are 16 weeks of a skiing season with the first five weeks being the most popular. Each owner (customer) has to provide a list of three choices of her favourite weeks. In addition, the owner has to state the total number of her party including herself. There are 64 owners in total (since we have 16 weeks and for each week a maximum of 4 owners can be accommodated). The operating company is contractually entitled to pay a compensation for each owner who cannot be offered her first choice. The compensation is as follows.

(a) If an owner gets her second choice, she gets two free one-day ski-lift passes per member of her party as a compensation.

(b) If an owner gets her third choice, she gets four ski-lift passes per member of her party as a compensation.

(c) If an owner gets a week not mentioned in her list of choices, she gets seven ski-lift passes per member plus a cash sum equivalent to 50 passes.

(d) If an owner cannot be offered any week at all, the compensation is a cash sum equivalent to 1000 passes.

The task is to assign to each owner a week in a way that does not violate the constraints such that the total compensation paid by the company is minimized. The problem can be restated in a clearer way as follows.

Inputs: Given the preference lists and the party sizes of all owners.

Objective: Allocate weeks to owners in such a way that it obeys the constraints and minimizes the total compensation.

Constraints:

1. No more than 22 people be resident in the ski-lodge in any week.
2. No more than 4 owners be resident in the ski-lodge in any week.

3 Hyper-Heuristics

Hyper-heuristics are high level strategies that seek a solution method instead of directly seeking a solution [5, 7, 10]. Therefore, the main difference between hyper-heuristics and other search methods, including metaheuristics, is that hyper-heuristics work on the search space of the heuristics while other search methods work on the search space of the solutions. Hyper-heuristics are defined as *"automated methodologies for selecting and generating heuristics to solve hard computational problems"* [6]. The key idea of hyper-heuristics is to combine a set of problem-specific, low-level heuristics in such a way that they compensate for the weakness of the others [10]. This idea is motivated by the fact that heuristic performance is not uniform across search problems. Thus, it is natural to try to automate the design of heuristics by letting a high-level manager decide which heuristics (or combination of them) to apply to a problem (or a problem state) and because of this, hyper-heuristics are also defined as *heuristics to choose heuristics* [7, 9].

4 Selection Perturbative Hyper-Heuristic

A hyper-heuristic is used to solve the ski-lodge problem. The proposed hyper-heuristic falls under the category of *heuristic selection* as defined in [5]. As in the study conducted by [8] although a genetic algorithm is used to produce a sequence of low-level heuristics, this sequence does not present a new heuristic applied to a

candidate solution, i.e. new heuristic is not generated. Instead each element of the
sequence is applied to improve the solution and hence is equivalent to selecting a
heuristic iteratively to improve a solution. The low-level heuristics managed by the
hyper-heuristics are *perturbative*.

The hyper-heuristic is an evolutionary algorithm. Unlike the genetic algorithm of
[1], this evolutionary algorithm hyper-heuristic does not encode solutions in chromo-
somal representations but it is the low-level heuristics that are encoded in chromosomes and thus the proposed evolutionary algorithm hyper-heuristic does not work
directly on the solution space, rather it operates on the search space of the low-level
heuristics. Consequently, a virtual barrier is set to prevent any problem-specific information leaking to the hyper-heuristic. In fact, the proposed hyper-heuristic does not
know that it is actually solving the ski-lodge problem. The genetic algorithm model
used to implement the perturbative hyper-heuristic (PHH) is the steady-state model
where at each generation two parents are chosen through the tournament selection
operator. Then, genetic operators are used to produce the descendants which replace
the parents if they are strictly better than the parents.

4.1 Low-Level Perturbative Heuristics

The implementation of PHH requires defining a set of perturbative low-level heuristics which will be encoded in chromosomes that will be subject to evolution. The
low-level heuristics are listed below.

- H01: Swap two owners' choices in a way that does not violate the constraints.
- H02: Same as H01 but accepts the first swap that does not increase the compensation.
- H03: Same as H01 but accepts the first swap that strictly reduces the compensation.[1]
- H04: Choose an owner at random and accept the best swap (involving the chosen
owner) among all possible swaps.
- H05: Swap the owner with the highest compensation with a randomly chosen
owner if possible.

Any heuristic that tries to alter the current solution by moving a single owner
from her assigned week to any other week will have no effect on difficult problems
which are characterized by having the total number of people close to 352 people
(16[weeks] $\times$ 22 [people in each week] = 352). This should be obvious since when
the initial solution is created for those problems it is often the case where there is
no apartment left empty. Thus, we cannot move an owner from her assigned week
to any other week without violating the constraint of not having more than 4 owners
resident in the ski-lodge in any week.

---

[1]H03 may seem very similar to H02. However there is a reported difference between the performance of $\leq$ and $<$ [1]. Bear in mind that the genetic algorithm of [1] is not a hyper-heuristic.

4.2 Initial Solution

PHH is a selection perturbative hyper-heuristic meaning that it selects heuristics to be applied to an existent, initial solution in an attempt to refine the solution by perturbing it. Thus, PHH requires an initial solution to be constructed for it. A method for creating an initial solution is described in this section.

It is tempting to just use the method of [1] to create an initial solution. However, this would be a fatal mistake because that method may create a partially complete solution and the hyper-heuristic will keep swapping the owners without achieving much because some owners are not assigned any week leading to a very high compensation of 1000 passes. Therefore, an alternative method for building a complete, initial solution is used. The method consists of two phases. The first phase allocates the weeks to owners based on their preferential lists. The second phase finds places for owners who have not been offered any of their choices in the first phase.

Phase I
Process the owners one at a time based on their ID. For each owner, assign her to a week chosen from her preference list. The preference list is considered in order, i.e., an attempt is made to meet the first choice before the second one and the second choice before the third one.

Phase II
If an owner cannot be offered either of her choices in Phase I, label her as a "loser". Try to accommodate the losers in the weeks that have sufficient spaces for them. Remove the label "loser" from any owner who is offered a week. While there are still losers, swap two non-losers *if possible*. Then check if the swap generates a space that is capable of holding one of the losers without violating the constraints. If so, allocate that space to the loser and remove her label.

Although it is straightforward, this method of initializing the solution is capable of generating an assignment where there is no owner left without a week, even for the hardest instances in the dataset being considered.

4.3 General Description of PHH

The algorithm starts by creating an initial population randomly where each individual in the population is a chromosome of variable length whose genes are the low-level heuristics presented in Sect. 4.1. For instance, "H02H03H05" is a chromosome consisting of three genes which are H02, H03 and H05. In this case, the current solution will be modified three times starting by H02 followed by H03 and finally by H05. The fitness of each individual is measured by the total compensation of the assignment generated by applying the chromosome to the current solution. At each generation, the best solution becomes the current solution. Tournament selection is used to select the parents. The standard one-point crossover operator is applied to two parents to produce offspring. The produced offspring are mutated in two genes chosen at random. The offspring replace their parents if they are better than the parents. The solution produced by the best of the two offspring replaces the current solution if it is better. The algorithm keeps evolving the population until the maximum number of generations is reached. A formal description of PHH is provided in Algorithm 1.

Algorithm 1 PHH

Input: Problem instance Π, Number of generations N, Population size M
              Tournament size T, Upper bound on chromosomes' length L
Output: An assignment of weeks to owners
              The cost of the assignment
\# Initialization
Generate at random M chromosomes of variable length bounded by L
and store them in *Pop*
Create an initial solution to Π and store the solution in *current solution*
Determine the fitness of each chromosome in *Pop*
\# Evolve *Pop* over N generations
for i := 1 to N **do**
    \# Parent selection
    Select two parents p_1 and p_2 from *Pop* through the tournament
    selection operator where the size of the tournament is T
    \# Descendants production
    Apply the standard one-point crossover operator to p_1 and p_2 to
    produce two descendants d_1 and d_2
    Mutate d_1 and d_2 in two genes chosen at random
    Determine the fitness of d_1 and d_2
    \# Population refinement
    Replace the parents by the descendants only if the descendants are better
    \# Update the current solution
    The best solution produced by the fittest chromosome in *Pop* becomes
    the *current solution*
end for
return *current solution* and its cost

It should be noted that PHH implements almost the same model used for the genetic algorithm [1] tailored to the ski-lodge problem. The performance of these two algorithms will be compared in Sect. 6. The crude design of PHH is intended to demonstrate that the success of PHH does not rise from detailed knowledge about the problem which is a feature that metaheuristics are criticized for.

5 Experimental Setup

PHH is tuned through trial runs. The automation of the tuning process is not attempted due to the satisfactory performance of PHH. The parameters involved are the number of generations, the size of the population, the tournament size and the

Table 1 The parameters involved in PHH and their values determined through trial runs

The parameter	The value
Number of generations	20,000
Population size	100
Tournament size	5
Heuristic length upper bound	50
Swap limit	50

upper bound on the length of chromosomes. Recall that in PHH, the number of genes to be mutated is fixed to 2 following the genetic algorithm of [1]. However, these two genes are chosen randomly. The heuristics H02 and H03 mentioned in Sect. 4.1 introduce an additional parameter to be tuned which is the limit on the number of swaps to be performed. The suitable values of these parameters are shown in Table 1.

6 Results and Discussion

PHH is able to find feasible solutions for all 10 problem instances. The quality of the solutions produced was also good with competitive compensation payouts. Although the aim of a hyper-heuristic is to provide good results over the problem set rather than producing best results for certain problem instances, for completeness and to get an idea of the quality of solutions the results are compared to that of methods used to solve this problem in previous work, namely, genetic algorithms and simulated annealing. The hyper-heuristic is tested against the genetic algorithm of [1]. In fact, this genetic algorithm qualifies as a memetic algorithm since it applies a local search to descendants produced by the crossover operator. In addition, a repair procedure is integrated into the algorithm to deal with invalid solutions produced by the recombination process. We choose not to consider a repair procedure because it is criticized for three reasons [2]. Firstly, the repairing might result in poor genetic material and thus inhibits the evolution. Secondly, repairing invalid solutions is not always easy. Finally, the repairing can be a computational burden. Instead, we apply the low-level heuristics in a way that respects the constraints of the problem.

PHH and the genetic algorithm are run on the 10 instances and the results of both methods are averaged over 25 runs due to the randomness involved in both methods, see Table 2.

The table lists the minimum, maximum and average compensation paid out over the 25 runs. The minimum compensation obtained for each instance is highlighted in bold. PHH has produced the minimum result for 5 of the problem instances. It is often the case that metaheuristics applied directly to the solution space produce better results when compared to hyper-heuristics, see for instance [3]. Hyper-heuristics, however, offset this drawback by offering a higher level of generality as well as automating the design of solution methods. Interestingly, in our case, the table

Table 2 Comparison between PHH and the genetic algorithm of [1]

Instance	Size	PHH				Genetic algorithm			
		Min	Max	Average	Performance	Min	Max	Average	Performance
Instance-01	344	**641**	714	675.40	7.83	**641**	707	667.48	7.47
Instance-02	337	**404**	440	419.06	6.33	**404**	457	415.88	6.44
Instance-03	338	**448**	486	461.04	6.20	450	502	479.92	7.35
Instance-04	351	**727**	805	761.64	7.90	732	1616	1362.68	13.23
Instance-05	315	316	338	325.02	5.31	**304**	308	305.76	1.97
Instance-06	328	364	388	376.08	5.67	**360**	392	373.84	6.09
Instance-07	347	749	811	771.72	7.25	**730**	842	787.76	8.78
Instance-08	326	487	511	500.84	5.80	**481**	493	484.12	3.62
Instance-09	316	406	428	411.68	4.83	**404**	412	406.00	2.77
Instance-10	351	**643**	705	668.36	7.36	684	1604	1164.20	13.00

indicates that on easier instances, the genetic algorithm has slightly better averaged results whereas on difficult instances (03, 04, 07, 10), the hyper-heuristic produces noticeably better averaged results. It is worth mentioning that we do not compare the performance of the three methods based on the average since we propose a better formula for that purpose and yet we list the averages in the table since they give an idea about the quality of the solutions beside being easily understood.

Table 3 compares the performance of PHH with that of simulated annealing on the problem instances. The minimum compensation obtained is highlighted in bold. The PHH has produced solutions with the minimum compensation for 7 of the problem instances.

Table 3 Comparison between PHH and the simulated annealing of [1]

Instance	Size	PHH				Genetic algorithm			
		Min	Max	Average	Performance	Min	Max	Average	Performance
Instance-01	344	**641**	714	675.40	7.83	691	2644	1816.88	14.60
Instance-02	337	**404**	440	419.06	6.33	410	1436	868.36	13.06
Instance-03	338	**448**	486	461.04	6.20	510	1396	1321.96	13.49
Instance-04	351	**727**	805	761.64	7.90	732	1616	1362.68	13.23
Instance-05	315	316	338	325.02	5.31	**306**	310	307.28	1.65
Instance-06	328	364	388	376.08	5.67	**362**	396	373.28	5.95
Instance-07	347	749	811	771.72	7.25	**747**	2668	1652.68	14.37
Instance-08	326	**487**	511	500.84	5.80	493	515	503.32	5.42
Instance-09	316	**406**	428	411.68	4.83	406	424	413.20	4.86
Instance-10	351	**643**	705	668.36	7.36	807	2633	1933.84	14.54

Since the performance is not only determined by the average, we propose a formula for measuring the performance of different algorithms. In particular, we define the performance of Algorithm A on Instance i as follows.

$$P_i^A = \ln(s_i^A \, |\mu_i^A - o_i^A|),\tag{1}$$

where μ_i^A is the averaged performance of Algorithm A over Instance i and o_i^A is the known optimal value for Instance i. The difference is multiplied by s_i^A which can be any measure of dispersion to capture the variation in the performance of Algorithm A on Instance i. We decide to use the range (the difference between the maximum value and the minimum value) as a measure of the variability in the performance of the three competing algorithms on the instances being considered. The use of the natural logarithm is motivated by the fact that (1) will be used to measure the overall performance across the problem set beside measuring the performance per instances. When comparing the overall performance of different algorithms over problem sets, it is desirable to tune down the effect of extreme performance on specific instances and thus the natural logarithm is introduced.

The performance of PHH and the genetic algorithm per instances is determined according to (1) and presented in Table 2 and that of PHH and simulated annealing is presented in Table 3.

With hyper-heuristics the stress should be on the overall performance across problem sets. Thus, we propose a measure for the overall performance based on (1). Specifically, the performance of Algorithm A over a problem set containing n instances is defined as

$$P^A = \frac{1}{n} \sum_{i=1}^{n} P_i^A.\tag{2}$$

It is clear that low values of P^A reflect better performance. Interestingly, according to (2) the overall performance of PHH over the problem set is 6.45, that of the genetic algorithm is 7.07 and that of the simulated annealing is 10.12 which implies that PHH has better overall performance on the problem set compared to the genetic algorithm and the simulated annealing.

7 Conclusion

In this paper an evolutionary algorithm is implemented on the hyper-heuristic level under the paradigm of perturbative heuristics selection where the algorithm evolves a population of low-level heuristic combinations. The performance of the hyper-heuristic is compared with the performance of a genetic algorithm as well as that of simulated annealing. The experimental results show that the hyper-heuristic and the genetic algorithm have almost equivalent performance on easy instances while the hyper-heuristic outperforms the genetic algorithm on the hard instances.

Furthermore, we propose a formula to measure the overall performance of the hyper-heuristic, the genetic algorithm and the simulated annealing over the problem set. According to the formula, the hyper-heuristic has a better overall performance across the problem set compared to both the genetic algorithm and the simulated annealing.

This study has revealed the potential of hyper-heuristics in solving the ski-lodge problem. Future work will investigate this further by examining the use of other hyper-heuristics in solving this problem and possible hybridization of hyper-heuristics. A selection constructive hyper-heuristics that builds solutions incrementally from scratch instead of perturbing a pre-existing solution will be examined as well as hybridizing both the selection perturbative and selection constructive hyper-heuristics to solve this problem. In addition, the equivalence of the ski-lodge problem to other combinatorial optimization problems such as the one dimensional bin packing problem with additional constraints will be considered. Finally, the scalability of the proposed approaches with ski-lodge problems involving more people, more weeks and larger housing capacity is also an interesting question to investigate.

References

1. The ski-lodge problem, http://www.soc.napier.ac.uk/~peter/ski-lodge/
2. Aickelin, U., Dowsland, K.A.: An indirect genetic algorithm for a nurse-scheduling problem. Comput. Oper. Res. **31**(5), 761–778 (2004)
3. CODeS Research Group at Katholieke Universiteit Leuven in Belgium, in Norway, S.G., the University of Udine in Italy: The First International Nurse Roster Competition, http://www.kuleuven-kulak.be/nrpcompetition (2015). Accessed July 2015
4. Bucci, M.: Optimization with simulated annealing. C/C++ Users J. **19**(11), 10–27 (2001)
5. Burke, E.K., Gendreau, M., Hyde, M., Kendall, G., Ochoa, G., Özcan, E., Qu, R.: Hyper-heuristics: a survey of the state of the art. J. Oper. Res. Soc. **64**(12), 1695–1724 (2013)
6. Burke, E.K., Hyde, M., Kendall, G., Ochoa, G., Özcan, E., Woodward, J.R.: A classification of hyper-heuristic approaches. In: Handbook of Metaheuristics, pp. 449–468. Springer (2010)
7. Burke, E., Kendall, G.: Search Methodologies: Introductory Tutorials in Optimization and Decision Support Techniques. Springer, Berlin (2013)
8. López-Camacho, E., Terashima-Marin, H., Ross, P., Ochoa, G.: A unified hyper-heuristic framework for solving bin packing problems. Expert Syst. Appl. **41**(15), 6876–6889 (2014)
9. Ochoa, G., Qu, R., Burke, E.K.: Analyzing the landscape of a graph based hyper-heuristic for timetabling problems. In: Proceedings of the 11th Annual conference on Genetic and Evolutionary Computation, pp. 341–348. ACM (2009)
10. Ross, P.: Hyper-heuristics. In: Search Methodologies, pp. 529–556. Springer, Berlin (2005)

Real-Time Vehicle Emission Monitoring and Location Tracking Framework

Eyob Shiferaw Abera, Ayalew Belay and Ajith Abraham

Abstract Cars, trucks, and other vehicles typically run on gasoline or diesel, fuels release harmful chemicals in the air. These emissions can create a multitude of problems, including health issues and environmental degradation due to pollution. To reduce these emissions and the problems that they create, it is really important to monitor and control them more frequently. This paper tries to address these issues proposing a real-time vehicle emission monitoring and location tracking framework. The proposed framework uses two main technologies Vehicle On-Board Diagnostic (OBD-II) to monitor vehicle emission information and Assisted Global Positioning System (A-GPS) to get location of the vehicle at real-time. The existing wireless network infrastructure is used in supporting the emission and location data collection on a central server and further processing and presentation of the information is done. To evaluate the operational effectiveness of the developed framework, simulation based experiment is conducted. Real-time vehicle flow on selected roads is conducted using the microscopic simulation SUMO. Within 99 time interval integrated gas emission and location information about individual vehicle at different points of traffic roads is produced. The collected data is pre-processed, categorized and the information is proposed to be displayed on road users' mobile phone as well as on computer connected to the Internet.

Keywords Vehicle emission monitoring · Vehicle tracking · Vehicle condition monitoring · A-GPS positioning · OBD-II

E.S. Abera (✉)
HiLCoE, School of Computer Science and Technology, Addis Ababa, Ethiopia
e-mail: eeyyoobb@gmail.com

A. Belay
Department of Computer Science, Addis Ababa University, Addis Ababa, Ethiopia
e-mail: ayalew.belay@aau.edu.et

A. Abraham
Machine Intelligence Research Labs (MIR Labs), Auburn, WA, USA
e-mail: ajith.abraham@ieee.org

© Springer International Publishing Switzerland 2016
N. Pillay et al. (eds.), *Advances in Nature and Biologically Inspired Computing*,
Advances in Intelligent Systems and Computing 419,
DOI 10.1007/978-3-319-27400-3_19

211

1 Introduction

Transport creates a favorable condition for transaction, exchange, knowledge transfer and economic efficiency. On the other hand economic development and technological advancement enable nations to invest in transport improvements. These relations have not been without difficulties. Congestion, traffic accidents, pollution, and access problems have been the critical challenges of the transport and traffic management sector throughout the world.

Different research activities have been conducted in different parts of the world to address road transport problems. To address congestion problem for example, gathering traffic flow information at every part of the road network was given higher emphasis. In relation to this fixed sensor technologies as well as non-fixed technologies were proposed and developed to gather traffic flow and communicate the state of traffic flow to road users, for instance modern vehicle tracking systems use GPS technologies to locate moving vehicles.

Gas emission is another problem of the current road transport system. There are different types of emissions that can come from a car or other type of vehicle. These emissions are pollutants in the form of fumes, vapors, or even particles that come from the exhaust of a car. They are created when there is incomplete combustion of a car's gasoline. There are several different types of emissions that may be released from vehicles to pollute the air. These emissions include sulfur oxides, nitrogen oxides, carbon dioxide, and hydrocarbons. These pollutants can be dangerous and often have a negative effect on the environment and/or on one's health [1].

Most of the researchers conducted and addressed gas emission and congestion problems, using vehicle tracking technologies, separately. Integrating the vehicle emission and its location along the entire journey of the vehicle will enable users to access real-time status/condition information about the vehicles. Hence, in this paper an extensible vehicle emission monitoring and tracking framework is developed exploiting the existing cellular network infrastructure.

The organization of this paper is as follows. Section 2 discusses the literature review. Section 3 presents the proposed real-time vehicle emission monitoring and location tracking framework. Section 4 depicts the evaluation of the framework and the result. Finally conclusion will be presented on Sect. 5.

2 Literature Review

Vehicle emission monitoring and location tracking has been the interest of many researchers, for example the paper presented by El-Medany et al. [2] describes a real time tracking system that provides accurate localizations of the tracked vehicle with low cost. GM862 cellular quad band module is used for implementation. A monitoring server and a graphical user interface on a website is also developed using Microsoft SQL Server 2003 and ASP.net to view the proper location of a vehicle on a specific map [2]. But the proposed framework is based on Standalone/self-ruling GPS, which depend solely on radio indicators from satellites. Unlike GPS, A-GPS uses cell tower data to enhance unwavering quality and precision during poor satellite signal conditions. In exceptionally poor signal conditions, for example, in urban territories, satellite signs may endure multipath propagation where signals skip off structures, or are weakened by barometrical conditions, dividers, or tree spread. Hu Jian-ming et al. [3] proposed an automobile anti-theft system using GSM and GPS module. Similarly Fleischer, P.B.; Nelson et al. [4] also describes development and deployment of Global Positioning System (GPS), Global System for Mobile Communications (GSM) based Vehicle Tracking and Alert System but both systems developed doesn't provide real-time vehicle emission conditions information.

A system architecture designed by Ravindra et al. with a title of "Sensor architecture allows real-time auto emissions monitoring" [5] proposed framework utilizing Micro Controllers (MCUs), and memory come together in an updated approach to real-time exhaust monitoring for improved pollution control. Since every vehicle data including emission related is retrieved and controlled by Engine Control Unit (ECU) and this information is accessible using OBD-II interface of vehicles, additional tool is not needed.

Though a lot of researches are made for emission monitoring and location tracking, but no research has been done combining the emission monitoring and location tracking together at every point the vehicles moves on a road and the paper is focused in designing and evaluating of a real-time vehicle emission monitoring and location tracking framework.

3 Real-Time Vehicle Emission Monitoring and Location Tracking Framework

3.1 Vehicle Emission Monitoring and Location Tracking Technologies

To gather the emission of every vehicle at every spot of the road network OBD-II technology is used. On-Board Diagnostics generation two (OBD-II) is when vehicle's malfunction indicator lamp (MIL) comes on, vehicle has stored a code that identifies the problem detected. This code is known as a diagnostic trouble code (DTC). Diagnostic trouble codes are alphanumeric codes that are used to identify a problem that is present on any of the systems that are monitored by the on-board diagnostic system. The goal of the On-Board Diagnostics is to alert the driver to the presence of a malfunction of the emission control system, and to identify the location of the problem in order to assist mechanics in properly performing repairs. In addition, the OBD-II system should illuminate the Malfunction Indicator Light (MIL) and store the Trouble Code in the computer memory for all malfunctions that will contribute to increased Hydro Carbon (HC) emissions. OBD-II has to communicate the diagnostic information to the vehicle mechanic via a communication vehicle network using Diagnostic Trouble Codes (DTCs) [6].

Besides, to locate the moving vehicle the UMTS positioning technology, A-GPS will be used. Assisted GPS covers the drawback of GPS for detecting a device indoor. Assisted GPS receiver in the mobile can detect and demodulate the weaker signals which are needed by GPS receivers for accuracy. The air interface traffic is optimized by the A-GPS. Another advantage of A-GPS that the user can hold the data for privacy and the network operator restricts the assistance to service providers. To be more precise, "A-GPS" features are mostly dependent on an Internet network and/or connection to an ISP or CNP, in the case of CP/mobile-phone device linked to a Cellular Network Provider data service) [7]. By integrating this two technologies, both the emission and location data is gathered from every vehicle moving on the sample road network.

3.2 Proposed Framework

Based on the way vehicle emission and location data is gathered and the technology used, we build a real-time vehicle emission and location tracking framework depicted on Fig. 1. The framework is designed in a way that every vehicle has integrated device which will be plugged on vehicles OBD-II interface and used to gather emission and location data from the vehicle. The devices have two

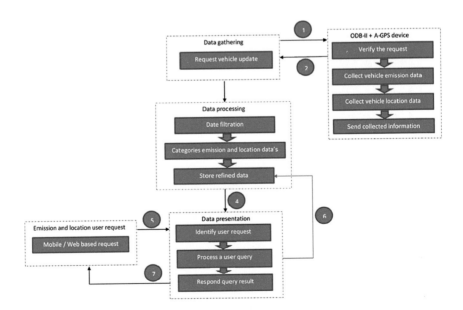

Fig. 1 Detail system architecture

components: OBD-II and A-GPS. These components are used to collect vehicle emission and location data every time a request comes from the server. For privacy reason, the device only responds for the server after performing verification.

The central server gathers collected data from every vehicle on a timely manner. Since we are not using everything from the collected data, data filtration, and then categorized into emission and location information is done. The central server is also responds for every user's query. Users are authenticated in order to process their requests, when the request is from valid user, query is processed from the stored data and the response is sent to the user to be displayed.

For simplicity of explanation, the proposed framework is categorized into four basic components: Information Gathering, Data Processing, Data Storage and Query processing

Information Gathering: The server collects emission and location information from every vehicle on a timely manner using OBD-II and A-GPS which are built into the vehicle. 3G wireless communication is used to transfer the data between vehicles and a server.

Data Processing: Data processing is done on the central server. Activities like data filtration, categorization of emission and location data takes place.

Data Storage: Central server stores received data and make it available for any incoming requests. Storing the data allows to keep histories, for inventory and the data might be also used as an input for some other applications working on data mining and road traffic management activity.

Individual Vehicle Emission and Location Information Processing: The server is responsible to respond for any authorized requests coming to get real-time vehicle emission information and location data.

4   Application and Evaluation of the Framework

To evaluate the feasibility of the proposed hybrid vehicle emission monitoring and tracking scheme depicted on Fig. 1, simulation based experiment is employed. The simulation tool used is SUMO [8] which has been used within several projects for answering a large variety of research questions including the following.

- Vehicle route choice has been investigated, including the development of new methods, the evaluation of eco-aware routing based on pollutant emission, and investigations on network-wide influences of autonomous route choice [9].
- SUMO was used to provide traffic forecasts for authorities of the City of Cologne during the Pope's visit in 2005 and during the Soccer World Cup 2006 [9].
- SUMO was used to support simulated in-vehicle telephony behaviour for evaluating the performance of GSM-based traffic surveillance [9].

To use the tool a sample road network depicted on Fig. 2 is used from openStreetMap [10]. Beside a free flow type of traffic demand is applied.

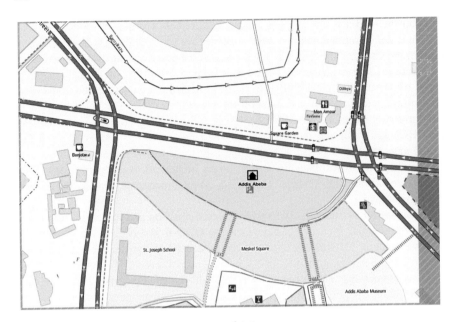

Fig. 2 Sample road network Addis Ababa, Meskel Square area

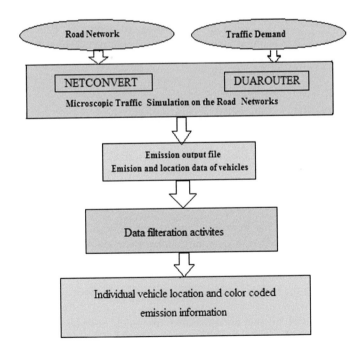

Fig. 3 General simulation architecture

After the road network file is created, it is customized with Java OpenStreetMap [11] in order to remove the road edges which cannot be used by vehicles, such as railway, roadways for motorcycle, bicycle, and pedestrian, etc. In addition, all the edges are set as one-way for simplicity. The next step is to convert the network file using NETCONVERT, required by SUMO simulation. Once the road network is ready, traffic flow is generated on the sample road. Random routes of vehicles traveling on the road network are generated for a 99s time interval using the DUAROUTER. These randomly generated vehicles and their routes are also exported as the SUMO file. The steps we followed and general architecture of the simulation environment is depicted on Fig. 3.

As a result of a 99-second-interval, 35 vehicles have been generated with random trips. The SUMO simulation is conducted for 100-second without incidents. The real-time traffic data, i.e., emission and location is collected together with the output file.

Table 1 Sample simulation output file

Time	Id	CO2	CO	HC	NOx	PMx	x	y
0	Vehicle.0	601.69	6.37	0.35	0.84	0.02	1991.72	2434.98
1	Vehicle.0	1583.16	19.45	0.62	2.99	0.15	1988.27	2434.14
2	Vehicle.0	2430.46	30.86	0.85	4.85	0.25	1986.2	2436.85
3	Vehicle.0	4218.17	66.7	1.48	9.2	0.57	1982.47	2440.7
4	Vehicle.0	4115.73	55.57	1.34	8.63	0.48	1977.37	2445.95
5	Vehicle.0	0	0	0	0	0	1973.89	2454.99
6	Vehicle.0	0	0	0	0	0	1968.3	2460.08
7	Vehicle.0	0	0	0	0	0	1963.35	2463.87
8	Vehicle.0	4213.2	70.88	1.53	9.35	0.6	1959.1	2464.13
9	Vehicle.0	5086.59	83.65	1.78	11.3	0.71	1952.76	2463.43
10	Vehicle.0	6546.39	111.69	2.28	14.8	0.95	1944.54	2462.26
26	Vehicle.0	0	0	0	0	0	1729.79	2498.25
26	Vehicle.4	3870.74	33.84	1.01	7.51	0.32	1896.01	2468.94
26	Vehicle.5	4381.66	53.25	1.33	9.01	0.46	1935.93	2463.45
26	Vehicle.6	3950.95	66.1	1.44	8.73	0.56	1959.56	2464.1
88	Vehicle.29	1953.08	28.38	0.77	3.94	0.22	1988.26	2434.16
89	Vehicle.29	2683.11	37.25	0.96	5.51	0.31	1986.21	2436.83
90	Vehicle.29	3110.7	40.09	1.04	6.35	0.34	1982.59	2440.61
91	Vehicle.29	4205.1	58.99	1.39	8.9	0.5	1977.58	2445.5
92	Vehicle.29	4043.66	48.88	1.24	8.28	0.43	1974.16	2454.22
93	Vehicle.29	0	0	0	0	0	1968.86	2459.97
94	Vehicle.29	0	0	0	0	0	1963.72	2463.85
95	Vehicle.29	4028.11	67.5	1.47	8.91	0.57	1959.59	2464.1
96	Vehicle.29	3654.09	48.61	1.2	7.59	0.42	1953.38	2463.66
97	Vehicle.29	4304.99	56.98	1.37	9	0.49	1945.39	2462.15
98	Vehicle.29	4424.87	53.84	1.34	9.1	0.47	1935.34	2463.53

All output files written by SUMO are in XML-format by default. However, with the python tool xml2csv.py you can convert any of them to a flat-file (CSV) format. By filtering emission and location informations from the output we get the following sample data (Table 1).

Simulation result: After importing the generated output file to central server database, the refined data is ready for use in any client side application. In this paper we have developed a web application to present Emission and location of vehicles for users. The application is developed using MySQL database, apache webserver and PHP as scripting language. Figure 4 depicts the application interface while a user is accessing emission and location information using vehicle id as a search criteria.

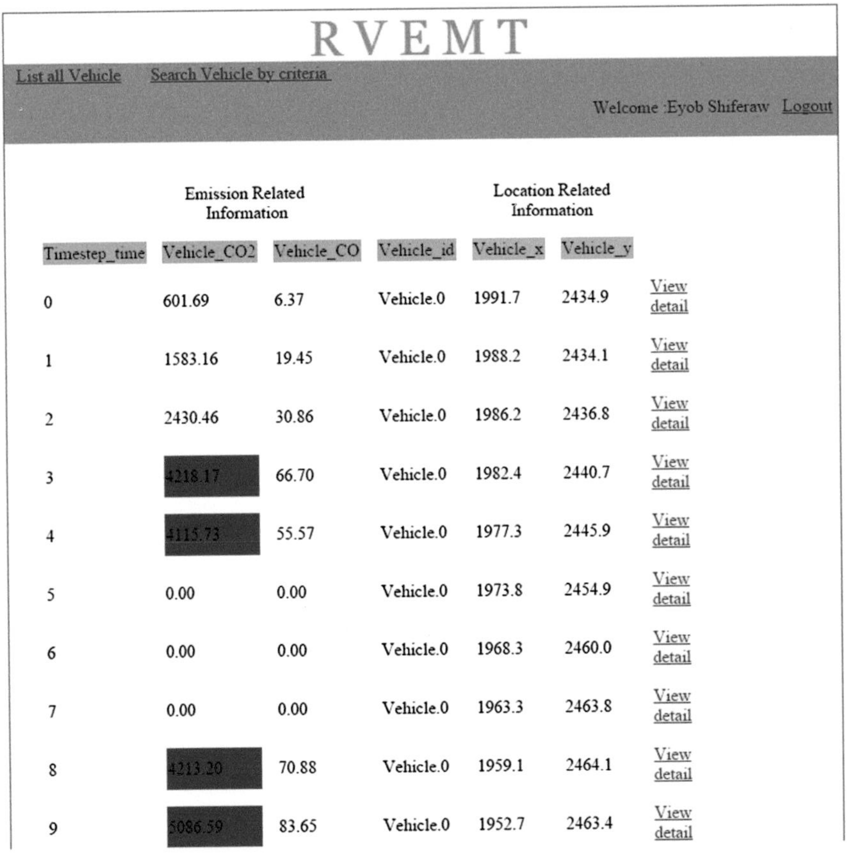

RVEMT

List all Vehicle   Search Vehicle by criteria

Welcome :Eyob Shiferaw Logout

	Emission Related Information			Location Related Information		
Timestep_time	Vehicle_CO2	Vehicle_CO	Vehicle_id	Vehicle_x	Vehicle_y	
0	601.69	6.37	Vehicle.0	1991.7	2434.9	View detail
1	1583.16	19.45	Vehicle.0	1988.2	2434.1	View detail
2	2430.46	30.86	Vehicle.0	1986.2	2436.8	View detail
3	4218.17	66.70	Vehicle.0	1982.4	2440.7	View detail
4	4115.73	55.57	Vehicle.0	1977.3	2445.9	View detail
5	0.00	0.00	Vehicle.0	1973.8	2454.9	View detail
6	0.00	0.00	Vehicle.0	1968.3	2460.0	View detail
7	0.00	0.00	Vehicle.0	1963.3	2463.8	View detail
8	4213.20	70.88	Vehicle.0	1959.1	2464.1	View detail
9	5086.59	83.65	Vehicle.0	1952.7	2463.4	View detail

Fig. 4 Real-time vehicle emission and location information

To demonstrate, how a system can identify vehicles which have exceeded the emission standards, color coded information is displayed to the users with respect to the individual vehicles. View detail link displays detail information about the selected vehicle. Using the collected information from the vehicle moving on the sample road, different types of queries and reports were generated and depicted on Figs. 5, 6 and 7.

In Fig. 5a CO2 emission of vehicle.0 is presented within 10 min. As we can see from the figure from time 5 to 7 the vehicle didn't generate any CO2 emission, or the vehicle engine is stopped. Figure 5b shows the X and Y coordinates of vehicle.0 between 1–10 time intervals.

In Fig. 6a, b CO, HC, NOx, PMx emission and location coordinates of vehicle 4, 5, 6 is presented at the time of 26.

In Fig. 7a, b different Emission and location informations of a vehicle.29 is presented for the time 88–99. From the graph we can say that CO emission was higher at the time of 99, and the vehicle has lower emissions in HC and PMx.

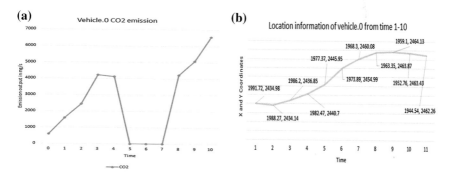

Fig. 5 Emission and location information of vehicle.0 from time 1–10. **a** CO_2 emission on vehicle.0. **b** Location information of vehicle.0

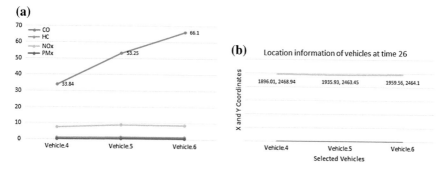

Fig. 6 Different emission measures and location information of selected vehicles at time 26. **a** Different emission measures of selected vehicles. **b** Location information of selected vehicles

(a)

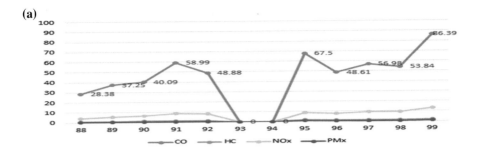

(b)

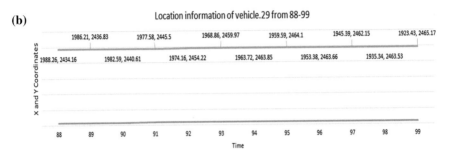

Fig. 7 Emission information of individual vehicle at the time 88–99. **a** Different emission measures of selected vehicles.29 at time 88-99. **b** Location of vehicles.29 at the time of 88–99

5 Conclusion

In this paper an integrated vehicle positioning and emission monitoring framework was presented. By considering standards of OBD-II and by taking advantage of the recently booming A-GPS positioning technology, a real-time vehicle emission and location tracking framework is proposed. Presenting a detail system architecture we have discussed about every components used in the proposed framework. And to evaluate the feasibility of the proposed framework, a simulation based experiments is conducted taking a sample road network of Addis Ababa. Java OpenStreetMap is used in order to remove the road edges which cannot be used by vehicles, such as railway, roadways for motorcycle, bicycle, and pedestrian, etc. In addition, all the edges are set as one-way for simplicity. The next step is to convert the network file using NETCONVERT, required by SUMO simulation. Once the road network is ready, the next step is to generate traffic on the road edges/links. Random routes of vehicles traveling on the road network are generated for a specific time interval using the DUAROUTER. These randomly generated vehicles and their routes are also exported as the SUMO file.

As a result of a 99-second-interval, 35 vehicles have been generated with random trips. Finally the real-time traffic data, i.e., emission and location is collected together with the emission output file. By taking emission output from the simulation we have presented different type of reports. By doing this we are able to monitor emission and location information of individual vehicle at real-time.

References

1. http://www.fleetmatics.com/resources/articles/a-guide-to-vehicle-emissions-and-pollutants. Accessed 4 Feb 2015
2. El-Medany, W., Al-Omary, A., Al-Hakim, R., Al-Irhayim, S., Nusaif, M.: A cost effective real time tracking system prototype using integrated GPS/GPRS module. In: 2010 6th International Conference on Wireless and Mobile Communications (ICWMC), pp. 521, 525, 20–25 Sept 2010
3. Jian-ming, H., Jie, L., Guang-Hui, L., Automobile anti-theft system based on GSM and GPS module. In: 2012 Fifth International Conference on Intelligent Networks and Intelligent Systems (ICINIS), pp. 199, 201, 1–3 Nov 2012
4. Fleischer, P.B., Nelson, A.Y., Sowah, R.A., Bremang, A.: Design and development of GPS/GSM based vehicle tracking and alert system for commercial inter-city buses. In: 2012 IEEE 4th International Conference on Adaptive Science and Technology (ICAST), pp. 1, 6, 25–27 Oct. 2012
5. http://www.edn.com/Pdf/ViewPdf?contentItemId=4371208. Accessed 4 Feb 2015
6. https://en.wikipedia.org/wiki/On-board_diagnostics. Accessed 21 Mar 2015
7. http://en.wikipedia.org/w/index.php?title=Special:Book&bookcmd=download&collection_ id=abfbad961b220f5e2a69897d4b8708df723cb95e&writer=rdf2latex&return_to=Assisted +GPS. Accessed Feb 23 2015
8. "SUMO [Online]: http://sumo.sourceforge.net/
9. dlr.de/ts/sumo. Accessed Feb 4 2015
10. OpenStreetMap. [Online]: http://www.openstreetmap.org/
11. Java OSM Editor. [Online]: http://josm.openstreetmap.de/

Applying Design Science Research to Design and Evaluate Real-Time Road Traffic State Estimation Framework

Ayalew Belay Habtie, Ajith Abraham and Dida Midekso

Abstract This paper presents how design science research can be used to design and evaluate real-time road traffic state estimation framework. An integrated framework of the six process steps of design science research process model and the Matching Analysis-Projection-Synthesis (MAPS) tool was used as a research design to develop the proposed state estimation framework. The utility and efficiency of the framework was evaluated based on the adapted design science research evaluation guideline through simulation and the estimation accuracy indicated that reliable road traffic state estimation can be generated based on the developed framework.

Keywords Design science research · Framework · Artifact · Matching analysis-projection-synthesis (MAPS) · Artificial neural network (ANN) · Simulation of urban mobility (SUMO)

1 Introduction

Road traffic state estimation is a fundamental work in Intelligent Transport System (ITS). It is applicable for dynamic vehicle navigation, intelligent transport management, traffic signal control, vehicle emission monitoring and so on. The two major components of ITS, Advanced Traveler Information System (ATIS) and

A.B. Habtie (✉) · D. Midekso
Department of Computer Science, Addis Ababa University, Addis Ababa, Ethiopia
e-mail: ayalew.belay@aau.edu.et

A. Abraham
Machine Intelligence Research Labs (MIR Labs), Auburn, WA, USA
e-mail: ajith.abraham@ieee.org

A. Abraham · D. Midekso
IT4Innovations, VSB - Technical University of Ostrava, Ostrava, Czech Republic
e-mail: dida.midekso@aau.edu.et

© Springer International Publishing Switzerland 2016
N. Pillay et al. (eds.), *Advances in Nature and Biologically Inspired Computing*,
Advances in Intelligent Systems and Computing 419,
DOI 10.1007/978-3-319-27400-3_20

223

Advanced Traffic Management System (ATMS) particularly need accurate current road traffic sate estimation and short term prediction of the future for smooth traffic flow [1].

In recent times, with the increasing interest in Advanced Traveler Information Systems (ATIS) and Advanced Traffic Management Systems (ATMS), providing travelers with accurate and timely travel time information has gained paramount importance. Operators, for example, can use travel time information (current and/or predicted) to improve control on their networks as part of ATMS. Drivers also can choose their optimal route, either pre-trip or en-route, provided that traffic information from an ATIS is available together with the drivers' individual preferences. For transport companies the knowledge of travel time can help to improve their service quality. Moreover, they can choose their routes dynamically according to the current and predicted state of traffic and thus increase their efficiency and reliability.

A large number of research studies and literature reviews are concerned with the estimation, prediction and application of travel time in various areas of road traffic monitoring and management activities. One of the major issue in travel time estimation and prediction is the selection of appropriate methodological approach [2]. Existing road traffic state estimation techniques can be grouped into model based approach which are used for offline traffic state estimation and data driven approach for online state estimation [3]. Habtie et al. [4] proposed hybrid method of combining data driven approach and model based approach to estimate road traffic flow at real-time on urban road networks.

Developing road traffic flow estimation framework offers different opportunities to practitioners and researchers. For practitioners, the framework provide practical and applied knowledge by providing high-level solutions to traffic flow prediction problems that can be converted into practices. For researchers, the developed framework offers a method to synthesize and capture knowledge in road traffic management as well as highlight areas for future research.

Within design science research, researchers develop artifacts to solve a problem and evaluate their utility [5, 6]. The framework to be developed can be considered as an artifact [5] of design science research in which the artifact that is created (i.e., state estimation framework) is helping a practitioner to solve road traffic problem. Besides, design science research method is applicable in solving real-world problem [7] and can proceed in the absence of consolidated theories [8]. By applying design science research, it is possible to conduct rigorous evaluation on the framework, refine and enhance the design and its performance. Therefore the purpose of this paper is to propose and demonstrate a design science research process model used in designing and evaluating real-time road traffic state estimation framework.

The rest of this paper is organized as follows. Section 2 presents description of design science research approach and presents the research design employed to design and develops the state estimation framework. Explanation about each of the activities done on the different phases of the research design is also discussed. The proposed framework together with its component descriptions is presented in the design and development phase of the proposed research approach. The evaluation on the performance of the framework together with the result and the dissemination of the contribution of the research is discussed in evaluation and communication part of the design science research method. Conclusion of the research is presented in Sect. 3.

2 Design Science Research

Design activities exist in many disciplines such as engineering, architecture, education and fine arts, and therefore, research on design has long history. But design science research approach is a problem solving paradigm which pursues to create innovative constructs, ideas, methods, models through analysis, design, development and evaluation processes [9].

According to Hevner et al. [5] "Design science...creates and evaluates IT artifacts intended to solve identified problems." It involves a rigorous process to design artifacts to solve observed problems, to make research contributions, to evaluate the designs, and to communicate the results to appropriate audiences. Such artifacts may include constructs, models, methods, and instantiations [5]. Besides design science research method is applicable in solving real-world problem [7] and can proceed in the absence of consolidated theories [8].

Considering the above characteristics of design science research method, employing this research method in designing and evaluation of the proposed real-time road traffic state estimation framework is appropriate. Hence the objective of this research is to use design science research method in designing real-time road traffic state estimation framework and evaluate its efficiency and utility based on design science research evaluation guidelines.

There are different process models, descriptions, diagrams of design science research process given by different researchers including Hevner et al. [5], Purao and Sandeep [10], Peffers et al. [11], Vaishnavi and Kuechler [12]. The difference in the process models is due to the fact that the activities to be carryout at similar phases of a design science research are different depending to the design science research setting [13]. To design and evaluate the study artifact, the design science research process model developed by Peffers et al. [11] is adapted due to the following reasons.

- The model has problem identification and motivation phase that can support for the synthesis of selected prior literature to the topic of study.
- Although all design science research models focus on the contribution of new knowledge, the design process model given by Peffers et al. [11] has a separate phase, communication phase, used to communicate the problem and its importance, the artifact, its utility and novelty, the rigor of its design, and its effectiveness to researchers and other audiences.

- For research publications, the structure of this design process model can be used to structure the paper [11], just as the nominal structure of an empirical research process (problem definition, literature review, hypothesis development, data collection, analysis, results, discussion, and conclusion).
- The design process model fits with the research setting of this study. For example, the artifact design and development is within a single phase and during evaluation, demonstration of the developed artifact using simulation to observe its performance is carried out.

Peffers et al. [11] prescribe six processes for design science research steps: identify problem, define objectives of a solution, design and development, demonstration, evaluation, and communication. But detail guideline on the design steps is not properly presented [14]. Offermann et al. [15], based on the work in Peffers et al. [14] presented a three-process framework entailing problem identification, solution design and evaluation and considered Matching Analysis Projection Synthesis (MAPS) [16] as a design steps though no detail process was given.

Hence, to design and evaluate real-time road traffic state estimation framework, the design science research process model offered by Peffers et al. [14] is adapted. As a design guide for the design steps in the design science research process Matching Analysis-Projection-Synthesis (MAPS) tool [16] is adapted. MAPS is a tool for systematic planning of design research and innovation projects and it is equivalent to the concepts of 'the true' how it is today, 'the ideal' how it could be and 'the real' how it is tomorrow [16]. The next section explains the integration of the research methods in the research design.

2.1 Research Design

The research design aims at facilitating the collection of data to develop real-time road traffic state estimation framework. The strategy in this work follows what Creswell [17] calls "explanatory sequential mixed methods design". The overall research design was designed following the design science research guidelines as defined by Peffers et al. [14] in designing the research process, and Chow and Jonas [16] as a design guidelines for the design steps.

The study began with detail literature review on past and current research studies conducted so as to identify the need for better road traffic flow estimation accuracy together with improved performance of the procedures. The literature review undertaken also provided experimentation tools, procedures, framework evaluation guidelines and parameters. Although the review of literature was the initial activity in the research, the literature in the topic was continuously reviewed along the rest of the work in order to keep up to date with related works.

The other research method employed in this research is content analysis. Content Analysis according to Leedy [18] is a detail and systematic examination of the content of a particular body of material for the purpose of identifying patterns, themes, or biases within that material. And in the context of this research, content analysis was used to contribute to a better understanding on the needs for better performance on road traffic flow estimation procedures and tools by identifying the different steps needed to execute estimation of real-time road traffic state information including accuracy.

Experimental research method is also used in this research. According to Goddard et al. [19], the purpose of experimental research method is to collect data that can be used to validate theories, models used in the process of the research. In this study an experimental research method is employed in evaluating the accuracy of the proposed hybridized positioning technology and validating the performance of road traffic state estimation framework. For evaluation and validation, data is gathered based on simulation experiment and also from filed test conducted using J2ME location API (JSR-179) [20] software installed on A-GPS enabled mobile phone moving in all journey of a moving vehicle. These research methods were integrated with selected design science research process model. The overall research steps are summarized in Fig. 1 are discussed below.

Objective-Centered Solution: Over the years, different researches have been conducted and different types of traffic state like travel time and traffic flow estimation models based on different traffic data collection technologies have been developed. But no single road traffic state estimation model with traffic data collection technique covering the whole road network, with better accuracy and acceptable deployment and maintenance cost had yet been developed that presented itself to be universally accepted as the best. The existing literature showed road traffic state estimation utilizing a better road traffic data collection method is open

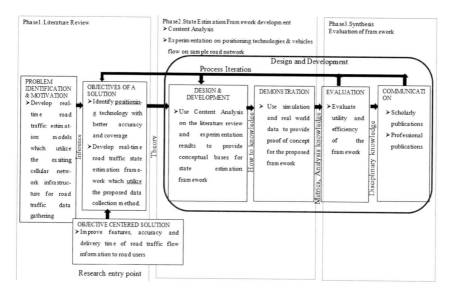

Fig. 1 Research design framework

for research. Developing road traffic state estimation framework improving the traffic data collection coverage, accuracy and traffic flow information delivery time to road users is the motivation behind this research.

Problem Identification and Motivation: With the rapid increase of urban development and vehicles number on the road, traffic flow information is very important for road users as well as in deploying Intelligent Transport System applications.

Road traffic data collection technologies plays vital role in traffic state estimation activity. The current state-of-the-practice data collection technologies are fixed sensor technologies which suffer from deployment and maintenance cost [21]. Floating vehicle technologies or dedicated vehicle probes are also used in road traffic data collection but these traffic data collection technologies provide limited sample size [22] and the data collected can't be representative for all vehicles on the entire road network [23]. In our literature review [4], we propose to use the existing cellular network infrastructure to gather and process road traffic data so that coverage, cost and accuracy problems can be addressed.

Next to road traffic data collection, traffic flow estimation for the different road links of the entire road network is done using the processed data. Existing road traffic state estimation can be either model based or data driven approach. To overcome limitation of with the advantage of the other we proposed to employ hybrid method combining model based and data driven approach [4]. This hybrid approach will use artificial neural network (data driven approach) as traffic flow estimator and the microscopic simulator SUMO (model based technique) in representing the traffic flow demand on the sample road networks.

Objective of the Solution: The objective of this research is to identify road traffic data collection technology utilizing the cellular network with better road network coverage capability and accuracy and to develop traffic flow estimation framework that can help to deliver traffic flow information to road users and other stakeholders at real-time.

Design and Development: The literature review on the different road traffic data collection technologies and traffic flow estimation models is the base in the development of the state estimation framework. Using content analysis on the literature review, the hybrid positioning technology combining Assisted Global Positioning system (A-GPS) and Uplink Time difference Of Arrival (UTDOA) is identified. To fuse the measurements of the positioning technologies, measurement fusion algorithm is found to be superior to state vector fusion algorithm in tracking moving vehicle on the road [24]. And in our experiment, the hybrid A-GPS and UTDOA based road traffic data collection technology is found to be reliable in localizing a moving vehicle with better accuracy than the individual positioning technologies [25] and also meet the E-911 positioning technology requirements [26].

Real-Time Road Traffic State Estimation Framework: We build our real-time road traffic state estimation framework, depicted on Fig. 2, based on design science research process model and using the cellular network traffic data collection system,

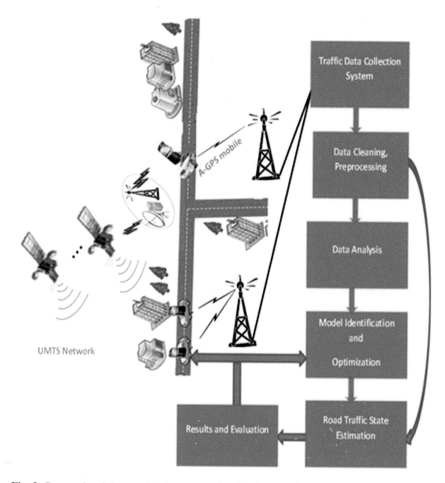

Fig. 2 Proposed real time road traffic state estimation framework [27]

i.e. hybrid A-GPS and U-TDOA positioning method discussed above. As it is indicated in Fig. 2, there are five basic components on the framework.

The components are traffic data collection system which represents the data sources and the hybrid A-GPS and UTDOA is used as data collection system, data pre-processing component where data cleaning, mobility classification, fusion of data and map matching activities are done, data analysis component represents the different statistical activities done in sampling and analysis purpose, model identification and optimization component comprising hybrid of model based and data driven approach as a traffic flow predicator, road traffic state estimation component where the traffic flow estimation technique/procedure is represented and estimation results dissemination and evaluation component which transmits road traffic flow information to road users on different devices connected to the Internet.

Demonstration: The accuracy and vehicle localization capability of the proposed hybrid A-GPS and UTDOA positioning technology for road traffic data collection is demonstrated based on simulation as well as real-world data. And to evaluate the performance of state estimation framework, the microscopic simulator Simulation of Urban Mobility (SUMO) is used in representing road traffic demand flow.

Evaluation: The evaluation of the artifact (framework) is performed based on experimental evaluation through simulation. To model the arterial road traffic, a microscopic simulation package "Simulation of Urban Mobility" (SUMO) [28] is employed. To generate road network and vehicle route, the SUMO packages NETCONVERT and DFROUTER are used. In order to mimic the real traffic situation on sample road, free flow traffic demand is used with maximum speed 30 km/h. (8 m/s), which is the speed limit on the real situation. From SUMO simulation aggregated speed information for each road link named "aggregate edge states" and floating car data (FCD) export file are generated and used for further analysis. The information on aggregate edge states include road edge IDs, time intervals, mean speeds, etc. The aggregate speeds are used to determine ground truth of traffic flow. Whereas the FCD output file records the location coordinates of every moving vehicle on the sample road at every timestamp, vehicle speed, edge IDs and this data file is used as simulation for in-vehicle mobile based traffic information system.

From the 1 h SUMO simulation, the simulation output file, FCD output, generates large amount of simulated mobile probes at real time. At every second, location updates (in terms of x, y or longitude, latitude) for the mobile probes is recorded. There are in total 437 probes and the time they spent range from 10 to 202 s. The FCD output file contains detail information of each vehicle/mobile and grows extremely large. Hence converting this location data into more compressed one is necessary [29]. Accordingly, to degrade location data, one can set up a specified percentage of simulated vehicles/mobiles to be traffic probe. Previous studies suggest that for arterial roads reliable speed estimation, a minimum penetration rate of 7 %, i.e. at least 10 probe vehicles traversing a road section (every road link) successfully is required [30–32] although factors like road type, link length and sample frequency affects the minimum sample size.

After the sampling process, the extracted data were used for training the neural network, which was proposed as a data driven state estimation model in the framework [4], and estimating the link travel speed. Total of 165 probe data were simulated and using random approach of dividing the available dataset for ANN development [33], 110 probe data (two-third of the data) were used for training process and the other 55 probe data (one-third of the data) were used for performance evaluation. During the training process different hidden neurons like 10, 15, 20, 25 were chosen. During testing the performance in terms of Mean square error (MSE) for the case of 10, 15, 20 and 25 neurons is compared and 15 hidden neurons were used to build the network. During training Levenerg-Marquardt algorithm (Trainlm) [34] was chosen so that the over fitting phenomenon can be avoided. To evaluate how the ANN model performs, the performance indicators

Root Mean Square Error (RMSE) and Mean Absolute Percentage Error (MAPE) are used.

Evaluation Results: The trained Artificial Neural Network (ANN) model is used to estimate link travel speed with a simulation data input. A correlation between the estimated link travel speed (speed of vehicle between two nodes) based on ANN model and the true link travel speed which is computed from the simulation data is performed. And as it is depicted on Fig. 3, the estimated link travel speed has very high correlation with the true link travel speed ($R^2 > 99$ %). the linear regression between the estimated and true (simulated) link speed that predicts the best performance among these values has an equation $y = 0.997 * x + 0$ indicating the trained ANN model performs reasonably well, where x represents true speed and y estimated link travel speed.

The performance of the estimation method in terms of RMSE and MAPE is 0.029325 and 0.127 % respectively with average speed of 7.191 m/s. The trained ANN model was also applied to estimate travel speed based on real A-GPS data. A car with A-GPS based mobile phone travelled on the sample road network and location updates in terms of longitude, latitude, timestamp, speed and accuracy is recorded at every 3 s for about 45 min. A sample with 10 s time interval is taken and at every road link, 15 location points i.e. total of 195 A-GPS based vehicle location points are taken. The estimation result is shown on Fig. 4. Each point represents travel speed on each tripe i.e., at every road link. From the regression formula in the figure, it can be seen that the trained ANN model performs very good. The RMSE and MAPE are 0.101034 and 1.1877 % respectively which show the possible application of the ANN model to real link travel speed measurements.

Communication: The contribution of this effort is disseminated in peer reviewed scholarly publications. The literature review on the different road traffic data collection technologies is published in a journal [4]. The hybrid A-GPS and UTDOA based road traffic data collection scheme is appeared in springer [25]. The framework proposed using hybrid A-GPS, UTDOA data collection method is published in [27].

Fig. 3 Correlation between estimated link travel speed and true link travel speed

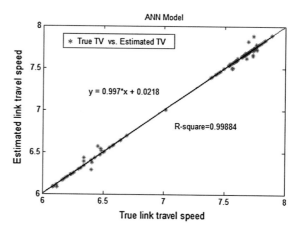

Fig. 4 Correlation between estimated link travel speed and true link travel speed using real A-GPS data

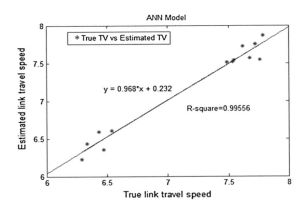

3   Conclusions

This paper presented a design science research process model used to develop real-time road traffic state estimation framework. The adapted design science process model blended the six processes of design science research and Matching Analysis-Projection-Synthesis (MAPS) tool as a guide to design the proposed framework. The evaluation done through simulation on the designed framework proved that reliable road traffic flow information can be generated using the proposed state estimation framework.

References

1. Kumar, S.V., Vanajakshi, L., Subramanian, S.C.: Traffic State Estimation and Prediction under Heterogeneous Traffic Conditions (2011)
2. Hu, T.-Y., Ho, W.-M.: Travel Time Prediction for Urban Networks: The Comparisons of Simulation-based and Time-Series Models (2010)
3. Aydos, J., Hengst, B., Uther, W., Blair, A., Zhang, J.: Stochastic Real-Time Urban Traffic State Estimation: Searching for the Most Likely Hypothesis with Limited and Heterogeneous Sensor Data (2012)
4. Habtie, A.B., Abraham, A., Dida, M.: In-vehicle mobile phone-based road traffic flow estimation: a review. J. Netw. Innovative Comput. **2**, 331–358 (2013)
5. von Alan, R.H., March, S.T., Park, J., Ram, S.: Design science in information systems research. MIS Q. **28**, 75–105 (2004)
6. Vaishnavi, V., Kuechler, W.: Design research in information systems (2004)
7. Rodríguez, P., Kuvaja, P., Oivo, M.: Lessons learned on applying design science for bridging the collaboration gap between industry and academia in empirical software engineering. In: Proceedings of the 2nd International Workshop on Conducting Empirical Studies in Industry, pp. 9–14 (2014)
8. Kuechler, B., Vaishnavi, V.: On theory development in design science research: anatomy of a research project. Eur. J. Inf. Syst. **17**, 489–504 (2008)
9. Simon, H.: The Sciences of the Artificals. MIT press, Cambridge (1969)

10. Purao, S.: Design research in the technology of information systems: truth or dare. GSU Department of CIS Working Paper, 2002
11. Peffers, K., Tuunanen, T., Rothenberger, M.A., Chatterjee, S.: A design science research methodology for information systems research. J. Manage. Inf. Syst. **24**, 45–77 (2007)
12. Vaishnavi, V., Kuechler, W.: Design Research in Information Systems (2004)
13. Vaishnavi, V., Kuechler, W.: Design Science Research in Information Systems, 20 Jan 2004
14. Gacenga, F., Cater-Steel, A., Toleman, M., Tan, W.-G.: A proposal and evaluation of a design method in design science research. Electron. J. Bus. Res. Meth. **10**, 89–100 (2012)
15. Offermann, P., Levina, O., Schönherr, M., Bub, U.: Outline of a design science research process. In: Proceedings of the 4th International Conference on Design Science Research in Information Systems and Technology (2009)
16. Chow, R., Jonas, W.: Beyond Dualisms in Methodology: An Integrative Design Research Medium. MAPS "and some Reflections". Undisciplined, pp. 1–18 (2008)
17. Creswell, J.W.: Research Design: Qualitative, Quantitative, and Mixed Methods Approaches. Sage publications (2013)
18. Leedy, P.D., Ormrod, J.E.: Practical research. Plan. Des. **8** (2005)
19. Goddard, W., Melville, S.: Research methodology: an introduction. Juta and Company Ltd (2004)
20. JCP: Java Specification Request (JSR) 179: Location API for J2METM (2011)
21. Rewadkar, D., Dixit, T.: Review of Different Methods Used for Large-Scale Urban Road Networks Traffic State Estimation
22. Calabrese, F., Colonna, M., Lovisolo, P., Parata, D., Ratti, C.: Real-time urban monitoring using cell phones: a case study in Rome. IEEE Trans. Intell. Transp. Syst. **12**, 141–151 (2011)
23. Pueboobpaphan, R., Nakatsuji, T.: Real-time traffic state estimation on urban road network: the application of unscented Kalman filter. In: Proceedings of the Ninth International Conference on Applications of Advanced Technology in Transportation, pp. 542–547 (2006)
24. Habtie, A.B., Abraham A., Dida, M.: Comparing Measurement and State Vector Data Fusion Algorithms for Mobile Phone Tracking Using A-GPS and U-TDOA Measurements, Springer International Publishing Switzerland HAIS 2015, LNAI 9121, pp. 1–13 (2015)
25. Habtie, A.B., Abraham A., Dida, M.: Hybrid U-TDOA and A-GPS for Vehicle Positioning and Tracking, Springer International Publishing Switzerland, HAIS 2015, LNAI 9121, pp. 1–13, 2015
26. FCC: FCC Acts to Promote Competition and Public Safety in Enhanced Wireless 911 Services (1999)
27. Habtie, A.B., Abraham A., Dida, M.: Road Traffic State Estimation Framework Based on Hybrid Assisted Global Positioning System and Uplink Time Difference of Arrival Data Collection Methods. AFRICON. IEEE, pp. 308 (2015)

Application of Biologically Inspired Methods to Improve Adaptive Ensemble Learning

Gabriela Grmanová, Viera Rozinajová, Anna Bou Ezzedine, Mária Lucká, Peter Lacko, Marek Lóderer, Petra Vrablecová and Peter Laurinec

Abstract Ensemble learning is one of the machine learning approaches, which can be described as the process of combining diverse models to solve a particular computational intelligence problem. We can find the analogy to this approach in human behavior (e.g. consulting more experts before taking an important decision). Ensemble learning is advantageously used for improving the performance of classification or prediction models. The whole process strongly depends on the process of determining the weights of base methods. In this paper we investigate different weighting schemes of predictive base models including biologically inspired genetic algorithm (GA) and particle swarm optimization (PSO) in the domain of electricity consumption. We were particularly interested in their ability to improve the performance of ensemble learning in the presence of different types of concept drift that naturally occur in electricity load measurements. The PSO proves to have the best ability to adapt to the sudden changes.

G. Grmanová (✉) · V. Rozinajová · A.B. Ezzedine · M. Lucká · P. Lacko · M. Lóderer ·
P. Vrablecová · P. Laurinec
Faculty of Informatics and Information Technologies, Slovak University
of Technology in Bratislava, Ilkovičova, 2, 842 16 Bratislava, Slovak Republic
e-mail: gabriela.grmanova@stuba.sk

V. Rozinajová
e-mail: viera.rozinajova@stuba.sk

A.B. Ezzedine
e-mail: anna.bou.ezzeddine@stuba.sk

P. Lacko
e-mail: peter.lacko@stuba.sk

M. Lóderer
e-mail: marek_loderer@stuba.sk

P. Vrablecová
e-mail: petra.vrablecova@stuba.sk

P. Laurinec
e-mail: peter.laurinec@stuba.sk

© Springer International Publishing Switzerland 2016
N. Pillay et al. (eds.), *Advances in Nature and Biologically Inspired Computing*,
Advances in Intelligent Systems and Computing 419,
DOI 10.1007/978-3-319-27400-3_21

235

Keywords Ensemble learning · Power load prediction · Particle swarm optimization · Genetic algorithm

1 Introduction

Reliable predictions are valuable in many areas of our life. In the case of time series forecasting, classical approaches are broadly used—mostly regression and time series analysis. However, these methods have some limitations—mainly in the case that they have to adapt to sudden or gradual changes of data characteristics, i.e. *concept drift*. Therefore new approaches have been introduced—i.e. incremental and ensemble methods. Incremental learning algorithms alone are usually not able to cover the problem of changes in target variables sufficiently. Ensemble models [24], which use a set of diverse base models, achieve better results in this case. In our work we have focused on heterogeneous ensemble learning systems, where the rationale id to combine the advantages of different base algorithms. The essential part of the ensemble learning is the method of combining the base models. The weights of the base models are computed by different statistical or AI based methods. We have concentrated on finding the most suitable method for setting up these weights by investigating three types of weight computation—i.e. our method of weighting based on median error (MB), the genetic algorithm (GA) and particle swarm optimization (PSO). To achieve more targeted experimental cases we have identified some typical concept drift patterns and explored which method fits the best for the particular shape of the drift. The achieved results let us assume that we have found the most appropriate weighting method for specific patterns. These findings were subsequently verified on real data from the domain of electric energy —namely on the data coming from smart metering.

This paper is organized as follows: Sect. 2 contains a summary of the related works part. In Sect. 3 we describe our proposed approach (the incremental heterogeneous ensemble model for time series prediction and its improvement in terms of new ways of weighting base models). Experimental evaluation and results are presented in Sect. 4 and the conclusion is in Sect. 5.

2 Related Work

To compute time series predictions, classical approaches are mainly applied. Recently incremental and ensemble approaches were also introduced to create methods that are able to adapt to changes.

Classical approaches to time series prediction are represented by regression and time series analysis. Regression approaches, such as a stepwise regression model, neural network and decision tree [27], model the dependencies of target variable on independent variables. For electricity load prediction, the independent variables are the day of the week, hour of the day, temperature, etc. Because of the strong seasonal periodicities in electricity load data, time series models are often used to make predictions. The most applied models originate from Box-Jenkins methodology [5], such as AR, MA, ARMA, ARIMA and derived models.

However, the classical approaches are only able to model seasonal dependencies. They are not able to adapt to different types of concept drift (changes in the target variable). Thus, they are not suitable to make predictions of electricity loads evolving in time, so incremental learning algorithms are needed.

Incremental algorithms process new data in chunks of appropriate size. They can possibly process the chunks by classical algorithms. The examples of the incremental learning algorithms are incremental support vector regression [31], extreme learning machines [8] and incremental ARIMA [21].

Usually, the incremental learning algorithms alone cannot treat concept drifts sufficiently well and therefore ensemble models are used to achieve better predictions (see [20] for introduction to ensemble learning). The ensemble model uses a set of diverse base models, where each base model provides an estimate of a target variable—a real number for a regression task. The estimates are combined together to make a single ensemble estimate. The combination of base estimates is usually made by taking a weighted sum of base estimates. The idea behind it is that the combination of several models has the potential to provide more accurate estimates than single (even incremental) models, namely in the presence of concept drifts [32].

The accuracy of the ensemble depends on the accuracy of base models and on their diversity [16]. The diversity of base models may be accomplished either by homogeneous or by heterogeneous ensemble learning approaches [32]. In the homogeneous approach, the ensemble is comprised by models of the same type learned on different subsets of available data [12, 23]. The heterogeneous learning process applies different types of base models [24, 34]. From the literature, several combinations of heterogeneous and homogeneous learning are known [35].

Ensemble learning was also used to predict values of time series [25, 29]. All of these approaches use ensembles of regression models for generating time series predictions. They do not take explicitly any seasonal dependence into account and do not use time series analysis methods to make predictions.

The essential part of the ensemble learning process is the approach used to combine the predictions of base models. Weights are computed by different methods, such as the linear regression model, gradient descent or by evolutionary or biologically inspired algorithms, e.g. particle swarm optimization or cuckoo search [19, 30].

We identify typical types of concept drifts from time series of electricity load data and investigate the ability of different weighting schemes in base model prediction combinations to decrease the error for those types of concept drifts.

3   The Incremental Heterogeneous Ensemble Model
   for Time Series Prediction

We propose the *incremental heterogeneous ensemble model for time series prediction*. The main goal is to investigate different approaches for computing weights that are used to combine predictions of base models—MB, GA and PSO. The ensemble approach was chosen for its ability to adapt to changes in the distribution of a target variable and its potential to be more accurate than a single method. As base models we use several chosen regression and time series models to capture seasonal dependencies.

3.1   Incorporating Different Types of Models

Our ensemble model incorporates several models to capture different seasonal dependencies. The models differ in *algorithm, size of data chunk* and *training period* (see Fig. 1). We chose the heterogeneous approach that assumes different algorithms for particular base models that bring diversity into the ensemble. The size of each data chunk is chosen in order to capture particular seasonal variation, e.g. data from the last 4 days for daily seasonal dependence. The training period represents the period after which a particular base model is re-trained. The model that is trained on a data chunk that contains data from the last 4 days, can be re-trained the next day (using a 1-day training period) on a data chunk that overlaps with the previous one.

The ensemble model is used to make one-day ahead predictions. Let h be the number of observations that are available daily, i.e. h is also the number of the next predictions to be computed daily by the ensemble model. After the observations for the current day are available, the prediction errors are computed. Based on computed errors, the weights are updated by one of the weighting schemes.

Let $\hat{\mathbf{Y}}^t$ be the matrix of predictions of m base models for the next h observations at day t (see Eq. 1) and $\mathbf{w}^t = \begin{pmatrix} w_1^t & \cdots & w_m^t \end{pmatrix}^T$ is a vector of weights for m base

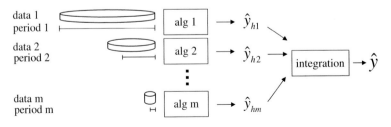

Fig. 1 Schematic of ensemble learning

models at day t before observations of day t are available. Weights and particular predictions are combined to make an ensemble prediction $\hat{y}^t = (\hat{y}_1^t \quad \cdots \quad \hat{y}_m^t)^T$.

$$\hat{\mathbf{Y}}^t = \begin{pmatrix} \hat{y}_{11}^t & \cdots & \hat{y}_{1m}^t \\ \vdots & \ddots & \vdots \\ \hat{y}_{h1}^t & \cdots & \hat{y}_{hm}^t \end{pmatrix} = (\hat{\mathbf{y}}_1^t \quad \cdots \quad \hat{\mathbf{y}}_m^t) \qquad (1)$$

The ensemble prediction for kth ($k = 1, \ldots, h$) observation is calculated by Eq. 2.

$$\hat{y}_k^t = \frac{\sum_{j=1}^m \hat{y}_{kj}^t \bar{w}_j^t}{\sum_{j=1}^m \bar{w}_j^t} \qquad (2)$$

From the matrix $\hat{\mathbf{Y}}^t$ and vector of h current observations $\mathbf{y}^t = (y_1^t \quad \cdots \quad y_h^t)^T$, the vector of errors $\mathbf{e}^t = (e_1^t \quad \cdots \quad e_h^t)^T$ for m models is computed. The error of each model is given by the median absolute error $e_j^t = \text{median}\left(\left|\hat{\mathbf{y}}_j^t - \mathbf{y}^t\right|\right)$. A vector of errors $\mathbf{e}^t$ is used to update the weights vector of base models in the ensemble by one of the selected weighting schemes. The integration method is robust since it is not sensitive to outliers.

3.2 Base Models

Table 1 contains the base models we used in our ensemble model. Each base model is able to capture seasonal dependencies. According to prediction approach, our selected base models can be divided into these groups:

- *regression-based models* (multiple linear regression, robust linear regression),
- *AI based models* (support vector regression, recursive regression trees, random forests, artificial neural network),
- *time series analysis models* (seasonal and trend decomposition using Loess and combinations, moving average, moving median, random walk).

3.3 Base Model Weighting

We investigate three types of weighting computation—MB, GA and PSO.

Weighting based on median error (MB). We proposed a gradual weighting method that changes each weight proportionally to the ratio of the median error and error of the particular based method. Initially weights w_j^1 are set to 1, for each

Table 1 The 13 base models that were included in the ensemble model incorporated both short-term (first 11 models) and long-term (last 2 models) seasonal dependencies

RW	Random walk
MLR	Multiple linear regression
RoLR	Robust linear regression [2]
SVR	Support vector regression [33]
RPART	Recursive partitioning and regression trees [1]
RF	Random forests [6]
STL + EXP	Seasonal and trend decomposition using Loess (STL) [7] in combination with simple exponential smoothing
STL + ARIMA	STL in combination with Box-Jenkins method (ARIMA) [5]
STL + HW + ANN + MLR	STL in combination with Holt-Winters exponential smoothing (HW), artificial neural networks (ANN) and multiple linear regression (MLR)
STL + HW + SVR	STL in combination with HW and support vector regression (SVR) [4]
ANN	Artificial neural network
MA	Moving average model [22]
MMed	Moving median model

$j = 1, \ldots, m$. At time t, the weight for jth method is recalculated by the following equation:

$$w_j^{t+1} = w_j^t \frac{\text{median}(\mathbf{e}^t)}{e_j^t} \tag{3}$$

and rescaled to interval $\langle 1, 10 \rangle$. The positive lower boundary ensures that no base method is excluded from the ensemble. The advantages of the presented type of weighting is its robustness and the ability to recover the impact of base methods. The weighting is robust since it uses a median of errors. In contrast to the average, median is not sensitive to large fluctuations and abnormal prediction errors.

After the application of this weighting method on test data we were not satisfied with its performance on segments with sudden or gradual changes (see Fig. 2 in Sect. 4.1). We found the prediction errors to be substantial in these cases. Therefore we decided to investigate the behavior of biologically inspired weighting methods and their suitability for the various types of concept drifts we identified in the data.

Biologically inspired methods. To solve complex problems, scientists have often been inspired by the behavior of animals which are observed in nature. At first view the behavior of individuals may seem very primitive, but as a whole, they can simply and very effectively solve complex problems.

Particle swarm optimization. PSO [15] is an artificial intelligence (AI) technique that can be used to find approximate solutions to extremely difficult or impossible numeric maximization and minimization problems. PSO is a population based stochastic optimization technique. It is one of the most widely used Evolutionary

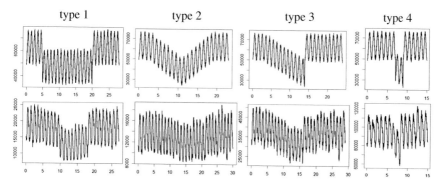

Fig. 2 Examples of concept drift patterns. Synthetic (*top*) and real (*bottom*) data

Computation Techniques; it takes inspiration from the natural swarm behavior of birds and fishes. PSO has roots in two main component methodologies—artificial life in general and swarming theory in particular [13]. It is based on the collective intelligence of a swarm of particles. Each particle explores a part of the search space looking for the optimal position and adjusts its position according to two factors: the first is its own experience and the second is the collective experience of the whole swarm.

Genetic algorithm. GA [10] is a stochastic optimization method. Its basic principle is inspired by real biological evolution process. The fundamental thesis of Darwin evolution theory states that only the best adapted individuals of a population will survive and reproduce. After reproduction of individuals with high fitness (assessment of the individual adaptation in environment) new individual is generated, such a descendant will reach high fitness with high probability. The goal of the evolution is the emergence of an individual with maximum possible fitness in a given environment. When minimizing errors, higher fitness corresponds to lower error.

GA works with a set of individuals who make up the population. Individual features are encoded using vectors (chromosomes in biology). Each individual is rated with his fitness—value which expresses his success in a given environment. New individuals are created and selected by different selection, crossover and mutation methods [17].

4 Experimental Evaluation

We performed a series of experiments to evaluate the performance of our ensemble model that incorporates the thirteen base models. Particularly, we examined the behavior of three weighting methods and their ability to react to the concept drifts. Our goal was to find out which weighting method is the most suitable (i.e. shows the smallest prediction error) for data with the occurrence of sudden or gradual changes. We identified four concept drift patterns and further investigated what type of weighting is appropriate for the particular patterns. In the next sections we describe the data we used, the experiments and the results.

4.1 Data

For the experiments we used the Slovak smart metering data. The quarter-hourly measurements of power load for 20 regions of Slovakia were available. We observed concept drifts in data mainly on the holidays or summer leave. We identified the four common patterns of drifts (see Fig. 2) and prepared two datasets from the real data for each type of drift. Additionally, we created two synthetic datasets for each pattern so we could evaluate to what extent is the accuracy of a weighting method affected by the noise of the real data. We could compare then whether the method performs equally good on the real data as on the synthetic data. To create the synthetic data we used the sinus function to simulate the daily power load shape. We added the noise from $\mathcal{N}(0, 1000^2)$ for the first datasets and $\mathcal{N}(0, 3000^2)$ for the second ones.

4.2 Measures

We utilized the *mean absolute percentage error* (MAPE) to evaluate the prediction accuracy. MAPE is based on relative (percentage) errors. This enables us to compare errors for time series with different absolute values. It is computed by Eq. 4.

$$MAPE = \frac{1}{n} \sum_{t=1}^{n} \frac{|\hat{y}_t - y_t|}{y_t} \times 100 \qquad (4)$$

4.3 Experiments

To design the experiments, the best data chunk sizes for particular models were found experimentally. The most precise predictions for regression methods (MLR, RoLR, SVR, RPART, RF) were for 4 days. The other short-term models (RW, STL + EXP, STL + ARIMA, STL + HW + ANN + MLR, STL + HW + SVR, ANN) coping with daily seasonality performed the best with a data chunk size equal to 10 days. MA and MMed need 1-year data chunks. We observed that models which incorporate trend (e.g., STL-based models), perform better with a greater size of data chunk.

Training periods for methods working with short-term seasonal dependency were 1 day and for the long-term methods were 1 year. Since we had less than 2 years' data available, they were trained only once and were not further retrained. There were available data for training (both for the previous 10 days and previous 1 year observations) only for the period July 1, 2014—February 15, 2015. Only non-holiday Tue-Fri days were assumed, because they contain similar daily consumption patterns. The base models were incrementally trained based on particular sizes of data chunks and their periods on synthetic and real data, while subsequently adding new data and ignoring old data. At each day, ensemble weights are recomputed by the chosen weighting method.

Biologically inspired methods (PSO and GA) which we have used have several characteristics in common. Both minimize the median absolute error that stands for the fitness function (see Sect. 3.1). They use populations of individuals (known as particles in PSO). The dimension of those individuals is given by the number of base methods m. New weights are optimized based on results from k last days (k-window). At the beginning, when there are measurements from only 1 day available, the parameter k is set to 1. As there are more days available, k increases to its maximal value. The maximal value was experimentally set to 4. We implemented a dynamic change of window length that sets k to value 1, when e^t is larger than given threshold T. The dynamic change of window length is applied only if it decreases objective function. T was set experimentally. The implemented values were 3, 4 and 5. Weights computed by GA and PSO are rescaled such that the smallest value is 1 and the largest value is 10.

For GA we set the population size to 80 and the number of generations to 200. The next population is composed of the best 50 % of individuals. The next half of the population is generated by crossover. All but the best individual are mutated. Mutation adds noise from $N(0, 0.05^2)$ to a randomly selected weight.

The population size of the PSO method was set to 20. The algorithm ends when the objective function is less than 0.5 or if the maximal number of iterations (100) is reached.

Three different weighting schemes were used to investigate their ability to decrease error for different types of concept drift patterns on synthetic and real data. The experiments were provided in the R environment. We used methods implemented in *stats, forecast* [11], *rpart* [26], *kernlab* [14], *randomForest* [18] and *pso* [3] packages.

4.4 Results

Table 2 contains MAPE prediction errors for four selected types of electric load change. Values in bold represent the minimal prediction error for each test case. Every experiment was performed 3 times with different weighting schemes. The best result from each experiment was selected.

Table 2 Average daily MAPE errors of ensemble predictions with various weighting methods

Type	Synthetic datasets			Real datasets		
	MB	PSO	GA	MB	PSO	GA
1	3.608057	**3.400082**	3.438460	5.879536	**5.115981**	5.425064
	6.337597	6.263065	**6.221111**	4.167873	**3.909871**	4.015005
2	3.087152	**2.794140**	2.949461	3.784411	**3.603070**	3.771126
	5.766299	**5.630975**	5.778790	4.517974	**4.353529**	4.452641
3	3.366592	**3.251221**	3.699905	4.358098	**3.907345**	4.013147
	6.210916	**6.210729**	6.493604	3.548538	**3.100759**	3.129184
4	9.311810	**8.766447**	9.978095	5.846086	**5.479890**	5.733959
	10.742680	**10.270954**	10,977823	4.101759	4.084168	**4.075274**

Results for synthetic and real datasets of each type of change. Each row is the result for one dataset of respective concept drift types

Average daily results (see Table 2) show that biologically inspired methods achieved better prediction results than median-based methods in most cases. The average prediction error of PSO and GA was 5.31 % smaller than the prediction error of the MB method on the synthetic dataset and 5.80 % on the real dataset.

Tests on real datasets showed that PSO achieved 7.16 % better prediction results than MB and 2.84 % better prediction results than GA. On average, PSO achieved a 5.61 % smaller prediction error than MB and a 4.08 % smaller prediction error than GA on both datasets combined. The PSO outperformed the other two methods on synthetic and real datasets. The GA method gained better results than the MB method, but not significantly.

According to the results, PSO seems to be the best weighting method to decrease error for all types of concept drift patterns, especially for type 3 and 4. The average relative prediction error of PSO was smaller than GA and MB by 4.20 % on type 1, 3.94 % on type 2, 5.40 % on type 3 and 5.85 % on type 4. In two cases GA obtained better results than PSO, but the difference between the results was not significant.

5   Conclusion

The presented paper describes our results in the area of exploring new ways of predictive base models weighting in ensemble learning. Our goal was to minimize the error of the proposed prediction model. The main concern was to examine the behavior of three weighting methods in their ability to react to concept drifts and to find out the most suitable method for data where sudden or gradual changes occur. We identified four concept drift patterns and investigated which type of weighting fits best to each particular pattern. During the evaluation we have considered various types of concept drift and explored how the precision of the predictive model is affected by individual methods (i.e. weighting based on median error—MB, particle swarm optimization—PSO and genetic algorithm—GA). The experiments were performed on synthetic and real datasets. The best results were achieved by weighting computed by the PSO method (in real and synthetic datasets in all cases except one). We would like to remark that in the case of electric load prediction even slight improvement of prediction is of great significance. The accuracy of electricity load prediction is particularly important, as even minor prediction errors emerging simultaneously in a large number of sampling sites is considerable. Our experiments, similar to the works [28, 9], proved that PSO can be very successfully employed to solve various complex problems.

Acknowledgement This work was partially supported by the Research and Development Operational Programme for the project "International Centre of Excellence for Research of Intelligent and Secure Information-Communication Technologies and Systems", ITMS 26,240,120,039, co-funded by the ERDF and the Scientific Grant Agency of The Slovak Republic, grant No. VG 1/0752/14 and VG 1/1221/12.

References

1. Archer, K.J.: rpartOrdinal: an R package for deriving a classification tree for predicting an ordinal response. J. Stat. Softw. **34**(7), 1–17 (2010)
2. Bellio, R., Ventura, L.: An Introduction to Robust Estimation with R Functions, http://www.dst.unive.it/rsr/BelVenTutorial.pdf (2005)
3. Bendtsen, C.: pso: Particle Swarm Optimization. R package version 1.0.3. http://cran.r-project.org/package=pso (2012)
4. Boser, B.E. et al.: A training algorithm for optimal margin classifiers. In: Proceedings 5th Annual ACM Workshop on Computational Learning Theory. pp. 144–152 (1992)
5. Box, G.E.P., et al.: Time Series Analysis: Forecasting and Control. Wiley, New Jersey (2008)
6. Breiman, L. et al.: Classification and Regression Trees. Chapman and Hall (1984)
7. Cleveland, R.B., et al.: STL: A Seasonal-Trend Decomposition Procedure Based on Loess. J. Off. Stat. **6**(1), 3–73 (1990)
8. Guo, L., et al.: An incremental extreme learning machine for online sequential learning problems. Neurocomputing **128**, 50–58 (2014)
9. Hadavandi, E. et al.: Developing a time series model based on particle swarm optimization for gold price forecasting. In: 2010 Third International Conference on Business Intelligence and Financial Engineering (BIFE) (2010)
10. Holland, J.: Adaptation in Natural and Artificial Systems. MIT Press, Cambridge (1992)
11. Hyndman, R.J. et al.: Package "forecast." http://cran.r-project.org/web/packages/forecast/forecast.pdf (2015)
12. Ikonomovska, E., et al.: Learning model trees from evolving data streams. Data Min. Knowl. Discov. **23**(1), 128–168 (2011)
13. Kennedy, J. et al.: Swarm Intelligence. Morgan Kaufmann Publishers (2001)
14. Karatzoglou, A., et al.: kernlab— An S4 Package for Kernel Methods in R. J. Stat. Softw. **11**(9), 1–20 (2004)
15. Kennedy, J., Eberhart, R.: Particle swarm optimization. In: Proceedings of the ICNN'95—International Conference on Neural Networks, vol. 4 (1995)
16. Kuncheva, L.I., Whitaker, C.J.: Measures of diversity in classifier ensembles and their relationship with the ensemble accuracy. Mach. Learn. **51**(2), 181–207 (2003)
17. Kvasnička, V. et al.: Evolučné algoritmy. STU Bratislava (2000)
18. Liaw, A., Wiener, M.: Classification and regression by random forest. R News. **2**(3), 18–22 (2002)

19. Mendes-Moreira, J., et al.: Ensemble approaches for regression: a survey. ACM Comput. Surv. **45**(1), 1–40 (2012)
20. Minku, L.L.: Online ensemble learning in the presence of concept drift. University of Birmingham (2011)
21. Moreira-Matias, L. et al.: On predicting the taxi-passenger demand: A real-time approach. Lecture Notes in Computer Science (including Subser. Lect. Notes Artif. Intell. Lect. Notes Bioinformatics), pp. 54–65 (2013)
22. Nau, R.: Moving average and exponential smoothing models. http://people.duke.edu/~rnau/411avg.htm, (2015)
23. Oza, N.C.: Online bagging and boosting. In: 2005 IEEE International Conference on Systems, Man and Cybernetics (2005)
24. Reid, S.: A Review of Heterogeneous Ensemble Methods. http://www.colorado.edu/physics/pion/srr/research/heterogeneous-ensemble-review-reid.pdf (2007)
25. Shen, W. et al.: An ensemble model for day-ahead electricity demand time series forecasting. In: Proceedings of the Fourth International Conference on Future Energy Systems—e-Energy '13, p. 51 (2013)
26. Therneau, T. et al.: rpart: Recursive Partitioning and Regression Trees. R package version 4.1–8. http://cran.r-project.org/package=rpart (2014)
27. Tso, G.K.F., Yau, K.K.W.: Predicting electricity energy consumption: a comparison of regression analysis, decision tree and neural networks. Energy **32**(9), 1761–1768 (2007)
28. Wang, Y. et al.: Crystal structure prediction via particle-swarm optimization. Phys. Rev. B—Condens. Matter Mater. Phys. **82**, 9 (2010)
29. Wichard, J.D., Ogorzałek, M.: Time series prediction with ensemble models. In: IEEE International Conference on Neural Networks—Conference Proceedings, pp. 1625–1630 (2004)
30. Xiao, L. et al.: A combined model based on data pre-analysis and weight coefficients optimization for electrical load forecasting. Energy (2015)
31. Xie, W. et al.: incremental learning with support vector data description. In: 22nd International Conference on Pattern Recognition, pp. 3904–3909 (2014)
32. Zang, W. et al.: Comparative study between incremental and ensemble learning on data streams: case study. J. Big Data. **1**, 1, 5 (2014)
33. Zhang, F., et al.: A conjunction method of wavelet transform-particle swarm optimization-support vector machine for streamflow forecasting. J. Appl. Math. **2014**, 1–10 (2014)
34. Zhang, P. et al.: Categorizing and mining concept drifting data streams. Int. Conf. Knowl. Discov. Data Min. **8** (2008)
35. Zhang, P., et al.: Robust ensemble learning for mining noisy data streams. Decis. Support Syst. **50**(2), 469–479 (2011)

A Variable Neighbourhood Search for the Workforce Scheduling and Routing Problem

Rodrigo Lankaites Pinheiro, Dario Landa-Silva and Jason Atkin

Abstract The workforce scheduling and routing problem (WSRP) is a combinatorial optimisation problem where a set of workers must perform visits to geographically scattered locations. We present a Variable Neighbourhood Search (VNS) metaheuristic algorithm to tackle this problem, incorporating two novel heuristics tailored to the problem-domain. The first heuristic restricts the search space using a priority list of candidate workers and the second heuristic seeks to reduce the violation of specific soft constraints. We also present two greedy constructive heuristics to give the VNS a good starting point. We show that the use of domain-knowledge in the design of the algorithm can provide substantial improvements in the quality of solutions. The proposed VNS provides the first benchmark results for the set of real-world WSRP scenarios considered.

Keywords Workforce scheduling and routing problems · Home healthcare scheduling · Variable neighbourhood search · constructive heuristics

1 Introduction

In workforce scheduling and routing problems (WSRP) a mobile workforce must perform tasks in scattered geographical locations. Solving this problem requires defining a schedule and a route plan for each worker such that all tasks (where possible) are covered. These problems combine features from both scheduling and routing problems, making them very challenging optimisation problems [1–3]. Examples of

R.L. Pinheiro (✉) · D. Landa-Silva · J. Atkin
School of Computer Science, ASAP Research Group, The University
of Nottingham, Nottingham, UK
e-mail: psxrp2@nottingham.ac.uk

D. Landa-Silva
e-mail: pszjds@nottingham.ac.uk

J. Atkin
e-mail: pszja@nottingham.ac.uk

© Springer International Publishing Switzerland 2016
N. Pillay et al. (eds.), *Advances in Nature and Biologically Inspired Computing*,
Advances in Intelligent Systems and Computing 419,
DOI 10.1007/978-3-319-27400-3_22

247

practical applications of WSRP are home healthcare scheduling [4, 5], technician scheduling [6, 7] and security personnel scheduling [8]. Here we consider the problem of scheduling nurses and care workers to visit and provide care services to patients in their homes. Data from four distinct home healthcare companies is used, provided by our industrial partner, who provide an enterprise resource planning software for home healthcare companies. This software includes a scheduling tool where a decision maker must manually set the schedules and routes for each worker; our work aims to automate this process.

The four real-world scenarios used here have been tackled before. Exact methods for these problems were investigated in [3], and the large size and complexity of the mixed integer programming model required a decomposition method before applying an exact solver. A study of genetic operators for the same problem scenarios was later performed in [9]. These two previous studies focused on understanding the problem scenarios and the behaviour of specific solution techniques rather than providing benchmark results.

Here we present a Variable Neighbourhood Search (VNS) algorithm to tackle the aforementioned scenarios and provide benchmark results. VNS algorithms have been successfully applied to both scheduling and routing problems, e.g. nurse scheduling [10, 11], job shop scheduling [12], vehicle routing problem with time windows [13] and the multi-depot version of the vehicle routing problem with time windows [14], a problem inherently similar to the WSRP.

The proposed VNS employs two domain-specific local search neighbourhoods. The first one sorts the workers by priority, identifying the best workers for each task and restricting the search space to the highest priority workers. The second one attempts to eliminate time and area violations from a solution. As mentioned above, results for the WSRP scenarios considered here are not currently available for many techniques. We therefore compare the variants of our VNS against each other, and also propose two constructive greedy heuristics to generate initial solutions quickly, showing that the proposed VNS outperforms simpler versions with less domain-knowledge integrated.

The contributions of this paper are twofold. The first is the proposed VNS algorithm that produces benchmark best known solutions for these real-world WSRP scenarios. The second is an improved understanding of the WSRP, obtained through assessing both the performance of the proposed algorithms and the impact of the tailored techniques which incorporate the domain knowledge. The remainder of this paper is structured as follows: Sect. 2 presents an overview of the WSRP. Section 3 describes the proposed algorithmic approach. Section 4 details the experiments and presents the results, while Sect. 5 concludes the paper.

2 The Workforce Scheduling and Routing Problem

For reasons of space, the WSRP is only briefly explained in this section. More details and a survey can be found in [15], while a full mathematical formulation of the problem can be found in [3]. The WSRP involves both scheduling and routing. A set

Assignment	a_1	a_2	a_3	a_4	...	a_s
Visit	v_1	v_1	v_2	v_3	...	v_n
Worker	w_5	w_1	w_3	w_9	...	w_5

Fig. 1 Example of solution representation. Note that two distinct workers are assigned to visit v_1

of m workers $\{w_1, w_2, \ldots, w_m\}$, must perform tasks at a set of n visits $\{v_1, v_2, \ldots, v_n\}$, which are located at various geographical locations. Each worker possesses a set of skills, time and area availabilities, and working area preferences. Visits have both required and preferred skills, and possibly have specific preferred workers. A worker-visit match requires matching the required skills, the worker's contract must allow him/her to perform that visit, and the specific allocation may incur additional costs.

A solution for a WSRP instance is a set of s assignments or pairs (v_j, w_i), specifying that worker w_i ($i \in \{1 \ldots m\}$) is assigned to perform visit v_j ($j \in \{1 \ldots n\}$). Note that $s \geq n$ because some visits might require more than one worker. An example of a solution is shown in Fig. 1, where workers 5 and 1 are assigned to visit 1, worker 5 is also assigned to visit n, and workers 3 and 9 are assigned to visits 2 and 3 respectively.

There are requirements that should be met if possible when assigning workers to visits. Such visit requirements include preferred skills (patient preferences), preferred workers (service provider preferences) and preferred working areas (staff preferences). Also, the workers availability in terms of time and geographical areas should be observed. A solution requires all visits to be served hence it is not always possible to meet all visit requirements and workers availability. To evaluate the quality of a solution, the tier-based minimisation objective function shown in Eq. (1) is utilised, which is employed by our industrial partner and also commonly used in the literature [2, 16].

$$f(S) = \lambda_1(d + c) + \lambda_2(3s - \rho_s - \rho_w - \rho_a) + \lambda_3(\psi_a + \psi_t) + \lambda_4\omega + \lambda_5\phi \quad (1)$$

The objective function has five main components each multiplied by a coefficient ($\lambda_1, \ldots \lambda_5$) to enforce tier-based objectives, i.e. the component multiplied by λ_5 is more important than the one multiplied by λ_4, and so on. The first component represents operational cost in terms of total travel distance (d) and staff cost (c). The second component represents visit requirements for preferred skills (ρ_s), workers (ρ_w) and areas (ρ_a). The value $3s$ is used because the values of ρ_s, ρ_w and ρ_a can be between 0 and 1 for each of the s visits. The third component represents the number of violations of workers availability in terms of area (ψ_a) and time (ψ_t). The fourth and fifth components represent the number of unassigned visits (ω) and the number of time conflicts (ϕ) respectively; a time conflict occurs when a worker is assigned to visits overlapping in time.

3 Proposed Algorithms

A VNS algorithm is proposed for solving the instances of the WSRP which were provided by our industrial partner and a number of algorithm variants are considered. VNS is an improvement metaheuristic proposed in [17]. It starts from an initial solution and performs successive local searches using multiple neighbourhoods to improve the solution. In order to escape local optima, VNS randomly disturbs the current solution (possibly making it worse) at the end of each iteration.

The VNS variant used in this paper has two stages, which are repeated: a *shaking phase* and a *local search phase*. In the *shaking phase*, one of seven shaking neighbourhoods is randomly selected and applied. If no change is made to the solution, another shaking neighbourhood is selected. This process is repeated until a change is made to the solution. The changed solution is then passed to the *local search phase*, which uses two neighbourhood search operators to hopefully generate better neighbouring solutions. One iteration of the local search phase consists of applying the two neighbourhood searches in some random order. If an improved solution is obtained from the current iteration, then another local search iteration (execution of both neighbourhood searches) takes place. When no improvement has been achieved in an iteration, the algorithm goes back to the shaking phase. We evaluate different configurations of the local search in our experiments.

3.1 Shaking Neighbourhood Structures

Solutions generated with the seven shaking neighbourhoods described below may be infeasible (e.g. one worker assigned to two simultaneous visits), but will still be kept.

Random Flip. Randomly picks a visit and changes the assigned worker to a random different worker that is skilled to perform that visit.

Area Availability Flip. Randomly picks a visit where the area availability constraint is violated and attempts to fix the violation by picking any other worker that is skilled to perform that visit and is available to work in that area.

Time Availability Flip. Randomly picks a visit where the time availability constraint is violated and attempts to fix the violation by picking any other worker that is skilled to perform that visit and is available to work at the visit time.

Preferred Worker Flip. Randomly picks a visit where the assigned worker is not the most preferred and replaces the worker by the most preferred worker for the visit.

Preferred Skills Flip. As *Preferred Worker Flip*, but uses the preferred skills value.

Preferred Area Flip. As *Preferred Worker Flip*, but uses the preferred areas value.

Priority-Based Flip. Uses a priority list (defined in Sect. 3.3). Selects a random visit for which the currently assigned worker is not the top one in the priority list and assigns the top priority worker instead.

3.2 Neighbourhood Searches

We define the following three neighbourhood searches to be evaluated as the operators used in the local search phase of the proposed VNS algorithm.

Randomised Hill Climbing (RHC). This is a very simple hill climbing local search which iteratively processes all unassigned visits in random order, greedily selecting the best worker in terms of the overall cost $f(S)$ and assigning that worker to the visit.

Priority-Based Search (PBS). This is detailed in Sect. 3.3 and exploits problem domain knowledge to make an estimation of the overall cost $f(S)$. Such estimation considers the objective function components associated to λ_1 (except travel distance), λ_2 and λ_3.

Availability Violations Search (AVS). It was observed that minimising violations of time and area workers availability is very difficult to achieve for the majority of the scenarios and that high costs resulted from these violations. This local search aims to resolve this and is explained in detail in Sect. 3.4.

3.3 Priority-Based Search (PBS)

This search is applied to prioritise workers for each visit. The concept is shown in Fig. 2 and involves the following steps:

1. An $m \times n$ cost matrix C is defined, containing the estimated costs of assigning each worker w_i to each visit v_j. As explained above, an estimation of $f(S)$ is used for each assignment. More specifically, the cost c_{ij} for each assignment in C is the weighted sum of the *staff costs* (greedily selecting the cheapest contract); preferred *skills*, *workers* and *areas*; and workers availability for *area* and *time*. For each assignment where w_i does not have the required skills or a valid contract to perform the visit v_j we set $c_{ij} = \infty$.
2. A priority list P^j is built for each visit v_j, sorting the workers into ascending order of c_{ij}. All visits are recorded as 'unmarked'.
3. Pick a random unmarked visit, select it for use in steps 4 to 6 and mark it.
4. Use P^j to determine if there is a worker with lower costs for the current visit v_j.
5. If one or more such workers exist, pick the first in the list and check if assigning that worker will generate a time conflict.

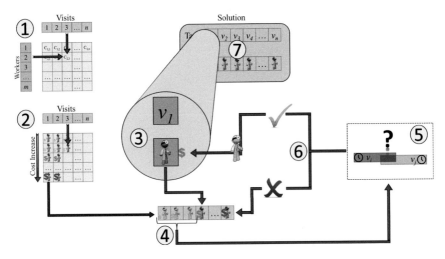

Fig. 2 Diagram for the Priority-Based Search (PBS)

6. If the assignment can be done (no time conflict), then assign the worker and evaluate the solution. If this reduces the cost, the assignment is accepted, otherwise the assignment is reverted. If the assignment generates a time conflict (hence reverted), repeat from step 5, selecting the next valid worker on the list.
7. Repeat from step 3 until all visits have been marked.

3.4 Availability Violations Search (AVS)

The concept of this search is presented in Fig. 3 and involves the following steps:

1. For each visit, identify the candidate worker with the minimum number of time and area availability violations for that visit. If multiple workers meet that criterion, choose the one with the lowest value in the cost matrix C used by the PBS.
2. Pick an unmarked visit in which the number of area and time availability violations is larger than the selected candidate for that visit. Mark the current visit.
3. Replace the current assigned worker with the candidate worker.
4. Identify whether time conflicts were created in step 3.
5. Where time conflicts occur, each conflicting visit is unassigned then perform steps 4 to 6 of the PBS heuristic to find a new worker for the visit. This will eliminate all time conflicts.
6. Evaluate the solution. If the costs improved (compared with those prior to step 3), accept the changes, otherwise revert the changes.
7. Repeat from step 2 while there are still visits for which candidates were identified in step 1.

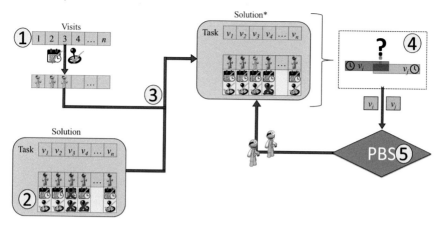

Fig. 3 Diagram for the Availability Violations Seeker Search

3.5 Constructive Heuristics

VNS algorithms are well known for producing faster and better results when given improved initial solutions. As suggested in [1], constructive heuristics were considered for finding initial feasible solutions.

Greedy Heuristic. This simple greedy heuristic achieved competitive results on its own (as shown in Sect. 4). The algorithm starts with an empty solution and considers each visit in a random order, choosing the best of the remaining workers for that visit, i.e. the worker that results on the best value of $f(S)$.

Flat Costs Heuristic (FCH). This heuristic performs a greedy search using the estimation of the overall cost $f(S)$ based on the matrix C as explained in the PBS method. The heuristic iterates through the visits in a random order and identifies the best (lowest cost in C) of the compatible remaining workers (skills and times match) for the visit, then assigning that worker to the visit.

4 Experiments and Results

We use four real-world scenarios provided by an industrial partner. Each dataset, A, B, C and D, is composed of 7 instances for a total of 28 problem instances. These scenarios come from different home healthcare companies, hence having different requirements and features. Set A has small instances (number of visits and workers) while set D has the largest instances.

All experiments were performed on Intel quad-core i7 machines with 16GB DDR2 RAM memory and each algorithm was executed eight times, computing the average solution. For scenarios A, B and C the runtime limit was set to 15 minutes, for scenarios D it was one hour. The following experiments were performed to evaluate the VNS algorithm.

4.1 Evaluation of Individual Components

The performance of the PBS and AVS local searches was evaluated. Four configurations of the algorithm were considered, all starting from the same feasible solution which was obtained by the constructive heuristics:

- **Full-VNS** is the full proposed algorithm including both PBS and AVS local searches and all seven shaking neighbourhoods.
- **PBS-VNS** is the Full-VNS without AVS (using only PBS as the local search).
- **AVS-VNS** is the Full-VNS without PBS (using only AVS as the local search).
- **HC-VNS** uses only six shaking neighbourhoods (excluding the Priority-based Flip) and uses only RHC as the local search (not PBS or AVS).

Results are shown in Fig. 4, where the y axis presents the gap between the average solution found by each algorithm for each of the problem instances in each dataset, compared to the best solution found by all runs of all algorithms for that instance. For some instances (sets A and B and instances C2, C4, C5 and C7), this is also the optimal solution, obtained by a mathematical solver. All three algorithm variants perform well for the smaller sets (A and B). However, for sets C and D PBS-VNS and AVS-VNS alternate, showing that on some scenarios one local search has an edge over the other. Notably, the AVS provided better solutions than the PBS for scenarios where the number of time and area availability violations is high, whereas PBS provided better solutions for the remaining scenarios. Importantly, the Full-VNS produced better results overall, especially for sets C and D.

Fig. 4 Relative gap comparison between the Full-VNS, the PBS-VNS and the AVS-VNS

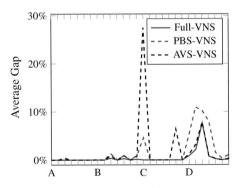

Fig. 5 Relative gap comparison between the Full-VNS and the HC-VNS

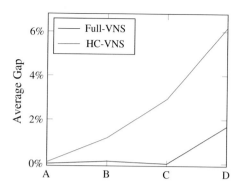

4.2 Overall Performance

The overall performance of the Full-VNS algorithm was evaluated and compared to HC-VNS. Results are shown in Fig. 5. We can see that HC-VNS produced good solutions (the gap is always below 7 %). However, its performance worsens as the size of the problem grows. This is due to hill climbing being slow for this problem. A single iteration of the HC-VNS algorithm for the largest D instance took over ten minutes while the Full-VNS could iterate in less than one minute. The proposed VNS produced better solutions for all instances, and did so in a faster computation time than HC-VNS. Additionally, the proposed VNS was able to reach the optimal solution for all instances for which the optimal solution is known. None of the other algorithm variants were able to do this. Full detailed results are shown in Table 1, where the best average results obtained by the VNS variants for each dataset are shown in bold.

4.3 Constructive Heuristics

The performance of the constructive heuristics was also evaluated. Figure 6 shows the average gaps obtained by the constructive algorithms alone (without the VNS). Both techniques show similar performance, with the Greedy outperforming the FCH on sets B and C and the FCH outperforming the Greedy on set D, however, the difference is always small. Figure 7 presents a comparison between the runtimes for the Greedy and the FCH heuristics. Both are fast for sets A, B and C, providing solutions in less than a second. The greedy algorithm is much slower for set D (up to 107 seconds compared to less than one second for FCH).

Table 1 Average results for each algorithm

		Greedy	FCH	HC-VNS	PBS-VNS	AVS-VNS	Full-VNS	Best Known[a]
A	1	3.812	3.812	**3.487**	**3.487**	**3.487**	**3.487**	3.487
	2	3.355	3.519	**2.491**	**2.491**	**2.491**	**2.491**	2.491
	3	4.908	4.999	3.007	3.009	3.000	**2.995**	2.995
	4	1.568	1.562	1.422	**1.421**	**1.421**	**1.421**	1.421
	5	3.103	2.959	**2.420**	**2.420**	**2.420**	**2.420**	2.420
	6	3.632	3.916	**3.549**	**3.549**	**3.549**	**3.549**	3.549
	7	4.345	4.102	**3.714**	**3.714**	**3.714**	**3.714**	3.714
B	1	1.861	1.897	1.705	**1.703**	1.704	**1.703**	1.703
	2	1.812	1.825	**1.755**	**1.755**	**1.755**	**1.755**	1.755
	3	2.011	2.013	1.779	1.740	1.733	1.724	1.718
	4	2.131	2.130	2.077	**2.074**	**2.074**	**2.074**	2.074
	5	2.027	2.080	1.859	1.840	1.840	1.828	1.823
	6	1.835	1.864	1.638	**1.620**	**1.620**	**1.620**	1.620
	7	1.966	2.010	1.816	1.807	1.803	**1.792**	1.790

		Greedy	FCH	HC-VNS	PBS-VNS	AVS-VNS	Full-VNS	Best Known[a]
C	1	148.65	149.09	141.59	119.69	157.47	**114.21**	114.21
	2	**3.05**	3.05	**3.05**	**3.05**	**3.05**	**3.05**	3.05
	3	141.79	135.51	104.37	103.54	**103.52**	**103.52**	103.52
	4	11.48	11.68	11.15	**11.15**	**11.15**	**11.15**	11.15
	5	12.81	12.89	**12.34**	**12.34**	**12.34**	**12.34**	12.34
	6	176.19	181.58	141.13	140.47	150.07	**140.44**	140.44
	7	**4.30**	4.30	**4.30**	**4.30**	**4.30**	**4.30**	4.30
D	1	178.17	177.98	180.62	178.30	170.92	170.33	168.84
	2	179.37	178.63	183.62	180.12	165.66	163.85	160.36
	3	185.18	185.18	183.98	183.23	178.53	178.19	164.56
	4	178.44	178.75	183.77	180.33	167.27	167.11	165.89
	5	164.69	164.51	163.36	163.22	161.20	161.11	160.74
	6	179.50	179.93	178.46	178.02	177.46	**177.43**	177.42
	7	180.17	180.48	180.32	179.37	177.94	177.86	177.42

[a]Best solutions found among all runs of all algorithms

Fig. 6 Performance
comparison of the FCH
heuristic against the Greedy
heuristic

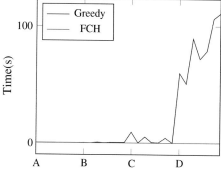

Fig. 7 Time comparison
between the FCH heuristic
and the Greedy heuristic

4.4 Discussion

The sizes of the instances in set D provoke unacceptably high runtimes for a simple
greedy algorithm, potentially hindering the performance of search methodologies
that rely on the systematic exploration of neighbourhoods (such as the HC-VNS).
The design of the Priority-Based Search (PBS) proved to be much more efficient,
allowing a much faster exploration of the neighbourhoods.

The constructive heuristics produce reasonable results compared to the other techniques, with the average gap to the best known solution being roughly 9.5 %. This is
due to the structure of the problem, which favours the assignment of the best worker
to each visit unless there are time conflicts. The preferred worker requirement is
related to the continuity of care, i.e. the same worker providing care to the same
patient. So, a worker that is preferred for a visit is usually the worker who most often
performs the visits for that patient. Thus, a worker with a high preference value is
often also available for that area and time, and is likely to live nearby. In fact, since the
FCH provided results only 9 % worse on average than the proposed VNS, this constructive heuristic may be sufficient to provide good enough solutions very quickly,
illustrating the efficacy of the developed approach for combining the objectives considered here.

Both PBS-VNS and AVS-VNS outperformed the HC-VNS. Since PBS outperformed RHC in all scenarios (even the small ones where runtime was less of an issue), the restriction of workers by the priority list does not appear to have eliminated the best candidates for assignments. Moreover, the AVS heuristic, which fixes the time and area availabilities using the first worker available in the priority list, could also reach the optimal and provide some of the best results. Comparing the results for the PBS and the AVS shows differences between sets. The latter technique performs 3.5 % better for set D. This shows that the occurrence of area and time violations are common in these datasets and that a focus on fixing these violations is beneficial. On the other hand, for set C the priority-based approach performs 4 % better.

Finally, when comparing all algorithm variants, it can be observed that increased domain knowledge gives improved results. HC-VNS includes some domain-specific shaking methods, giving reasonably good solutions. AVS-VNS and PBS-VNS use more domain-specific knowledge, giving even better results. The full proposed VNS makes the most use of domain-knowledge and provides the best results overall.

5 Conclusion

Heuristic algorithms to tackle difficult instances of workforce scheduling and routing problems (WSRP) were presented. First we introduced a VNS that employs two domain specific local search procedures. The first local search procedure attempts to reduce the search space while focusing on the interesting assignments. The second local search procedure tries to reduce the number of area and time availability violations. We also introduced two greedy heuristics—one straightforward and another that uses an estimation of costs to provide faster results.

We assessed the proposed algorithms on a set of real-world problem instances. We showed that it may be difficult for the VNS to find a feasible solution in some scenarios, hence the use of constructive heuristics is a useful strategy. We also showed that each local search procedure can perform well on its own, however the combined effect of both local searches provides improved results. Additionally, we observed that the greedy algorithms show good performance on this problem, considering the trade-off *quality versus computation time*. Finally, we discussed how the algorithms exhibited distinct performances on different sets of the WSRP, which allowed us to understand more of the nature of the problem and its difficulties. It became clear that adding problem domain knowledge to the solution algorithms improves their performance.

Our future work will further investigate mechanisms to make a fast yet accurate enough estimation of costs when exploring local moves and neighbour solutions. The results in this paper show that much computational effort can be avoided by using this technique and that the quality of the solutions is not affected much. We could apply this concept to other local search procedures for the VNS or even exact methods [3].

References

1. Castillo-Salazar, J.A., Landa-Silva, D., Qu, R.: Computational study for workforce scheduling and routing problems. In: ICORES 2014—Proceedings of the 3rd International Conference on Operations Research and Enterprise Systems, pp. 434–444 (2014)
2. Castillo-Salazar, J.A., Landa-Silva, D., Qu, R.: Workforce scheduling and routing problems: literature survey and computational study. Ann. Oper. Res. (2014)
3. Laesanklang, W., Landa-Silva, D., Salazar, J.A.C.: Mixed integer programming with decomposition to solve a workforce scheduling and routing problem. In: Proceedings of the International Conference on Operations Research and Enterprise Systems, pp. 283–293 (2015)
4. Cheng, E., Rich, J.L.: A home health care routing and scheduling problem (1998)
5. Akjiratikarl, C., Yenradee, P., Drake, P.R.: Pso-based algorithm for home care worker scheduling in the UK. Comput. Ind. Eng. 53, 559–583 (2007)
6. Xu, J., Chiu, S.: Effective heuristic procedures for a field technician scheduling problem. J. Heuristics 7(5), 495–509 (2001)
7. Cordeau, J.F., Laporte, G., Pasin, F., Ropke, S.: Scheduling technicians and tasks in a telecommunications company. J. Sched. 13(4), 393–409 (2010)
8. Misir, M., Smet, P., Verbeeck, K., Vanden Berghe, G.: Security personnel routing and rostering: a hyper-heuristic approach. In: Proceedings of the 3rd International Conference on Applied Operational Research, vol. 3 (2011)
9. Algethami, H., Landa-Silva, D.: A study of genetic operators for the workforce scheduling and routing problem. In: Proceedings of the XI Metaheuristics International Conference (MIC 2015) (2015)
10. Burke, E., De Causmaecker, P., Petrovic, S., Berghe, G.: Variable neighborhood search for nurse rostering problems. In: Metaheuristics: Computer Decision-Making. Applied Optimization. vol. 86, pp. 153–172. Springer (2004)
11. Constantino, A.A., Tozzo, E., Pinheiro, R.L., Landa-Silva, D., Romão, W.: A variable neighbourhood search for nurse scheduling with balanced preference satisfaction. In: 17th International Conference on Enterprise Information Systems (ICEIS 2015), Barcelona, Spain, Scitepress, Scitepress (2015)
12. Roshanaei, V., Naderi, B., Jolai, F., Khalili, M.: A variable neighborhood search for job shop scheduling with set-up times to minimize makespan. Future Gen. Comput. Syst. 25(6), 654–661 (2009)
13. Bräysy, O.: A reactive variable neighborhood search for the vehicle routing problem with time windows. INFORMS J. Comput. 15, 347–368 (2003)
14. Polacek, M., Hartl, R.F., Doerner, K., Reimann, M.: A variable neighborhood search for the multi depot vehicle routing problem with time windows. J. Heuristics 10(6), 613–627 (2004)
15. Castillo-Salazar, J.A., Landa-Silva, D., Qu, R.: A survey on workforce scheduling and routing problems. In: Proceedings of the 9th International Conference on the Practice and Theory of Automated Timetabling (PATAT 2012), pp. 283–302. Son, Norway, August 2012
16. Rasmussen, M.S., Justesen, T., Dohn, A., Larsen, J.: The home care crew scheduling problem: preference-based visit clustering and temporal dependencies. Eur. J. Oper. Res. 219(3), 598–610 (2012)
17. Mladenović, N., Hansen, P.: Variable neighborhood search. Comput. Oper. Res. 24(11), 1097–1100 (1997)

Modelling Image Processing with Discrete First-Order Swarms

Leif Bergerhoff and Joachim Weickert

Abstract So far most applications of swarm behaviour in image analysis use swarms as models for *optimisation* tasks. In our paper, we follow a different philosophy and propose to exploit them as valuable tools for *modelling* image processing problems. To this end, we consider models of swarming that are individual-based and of first order. We show that a suitable adaptation of the potential forces allows us to model three classical image processing tasks: grey scale quantisation, contrast enhancement, and line detection. These proof-of-concept applications demonstrate that modelling image analysis tasks with swarms can be simple, intuitive, and highly flexible.

Keywords Swarms · Image processing · Dynamical systems · Collective behaviour

1 Introduction

Understanding and simulating swarm behaviour continues to be an exciting interdisciplinary research area for more than six decades [2]. Numerous researchers have investigated models of swarming in such different fields as biology [4], computer science [16], mathematics [5], physics [22], and even philosophy [20]. For recent reviews and comparisons, we refer the reader to [7, 23] and the references therein. Generally, there exist two different model classes:

1. continuum / population-based / Eulerian / macroscopic models,
2. discrete / individual-based / Lagrangian / microscopic models.

L. Bergerhoff (✉) · J. Weickert
Faculty of Mathematics and Computer Science, Mathematical Image Analysis Group,
Campus E1.7, Saarland University, 66041 Saarbrücken, Germany
e-mail: bergerhoff@mia.uni-saarland.de

J. Weickert
e-mail: weickert@mia.uni-saarland.de

© Springer International Publishing Switzerland 2016
N. Pillay et al. (eds.), *Advances in Nature and Biologically Inspired Computing*,
Advances in Intelligent Systems and Computing 419,
DOI 10.1007/978-3-319-27400-3_23

Continuum models describe the evolution of a swarm's population density in space and time. These models provide a large scale description of general swarm attributes, but they cannot distinguish between individual swarm members.

Such a distinction requires discrete models that address each swarm member individually by characterising its position, velocity and other properties. Discrete models define simple rules that affect each individual. These rules are either based on the sociological behaviour of animals or—considering an artificial setup—motivated by a given task, which needs to be fulfilled. They control the attraction, repulsion, and orientation behaviour of the swarm members. Pairwise potentials model the effects between individuals; see e.g. [8] and the references therein for some commonly used potential functions. Integrating these rules into equations of motion describes the temporal evolution of the individuals. The literature distinguishes between first-order and second-order models [9]: First-order models use a set of equations that describe the velocities of each individual, while second-order models involve their acceleration.

Due to its heuristic character, it is common practice to apply discrete models to approximate solutions for difficult optimisation problems. Two well-known representatives are ant colony optimization (ACO) [3] and particle swarm optimization (PSO) [11]. By reducing problems to a pure *optimisation* task, ACO, PSO, and related models have already been applied numerous times in digital image analysis. However, apart from the idea of optimisation, discrete swarm methods have been used only rarely in order to *model* problems in the domain of image analysis. Notable exceptions deal with image halftoning [17], colour correction [19], segmentation [14], contour detection [12], boundary identification and tracking [15, 21], and the detection of fibre pathways [1]. Most of these modelling applications are fairly new and show convincing performance. However, all of these authors have focussed on a specific application. It seems that they have not been interested in exploiting the genericity behind the models of swarming.

Goals of our Paper. Motivated by these recent encouraging results, the goal of our article is to present novel applications of discrete first-order models of swarming in image analysis. To this end, we define behavioural rules for three fairly different image processing problems: grey scale quantisation, contrast enhancement, and line detection with the Hough transform. In all scenarios we use essentially the same model and modify only some of its features. This emphasises the versatility and genericity of models of swarming.

Paper Structure. Section 2 reviews the modelling of swarm behaviour in a discrete setup. We present our different behavioural rules and the corresponding potentials, and we discuss some model characteristics and a time discretisation. Section 3 adapts this modelling framework to three different applications in image processing and shows experimental results. Section 4 summarises our contributions and gives an outlook to future work.

2 Discrete Modelling of Swarm Behaviour

Basic Notations and Definitions. We consider a set $S = \{A_i \mid i = 1, \ldots, N\}$, called *swarm*, which is composed of N *agents* A_i. In the following, we use the terms agent, *particle*, and *individual* interchangeably. By $x_i \in \mathbb{R}^d$ we denote the position of an individual A_i, and $v_i \in \mathbb{R}^d$ describes its velocity. Both the particle position and its velocity are functions over time $t \in [0, \infty)$:

$$x_i = x_i(t), \qquad v_i = v_i(t). \tag{1}$$

If the agents are intended to have a limited field of perception, many discrete models such as [16] make use of a disk-shaped neighbourhood with radius δ. For an agent A_i, it is given by

$$\mathcal{N}_{i,\delta}(t) = \left\{ A_j \in S \mid j \neq i, \ |x_i - x_j| \leq \delta \right\} \tag{2}$$

where $|.|$ denotes the Euclidean norm. If the neighbourhoods $\mathcal{N}_{i,\delta}(t)$ contain all swarm mates A_j for all times t, a model is said to be *global*. Otherwise, it is called *local*.

Potential Energies and Forces. To describe a desired collective behaviour, discrete models of swarming define update rules for the positions and velocities of their individuals. These rules include effects based on attractive, repulsive, and orientating behaviour among agents [4, 5, 16, 22], as well as on the environment [11], or a combination of both [8, 17]. In our paper, we restrict ourselves to the treatment of attraction and repulsion among the agents.

The influence of the swarm mates on an agent A_i is described by a pairwise function $W: \mathbb{R}^d \to \mathbb{R}$ that denotes the *potential energy*. The total potential energy of the swarm is given by

$$E_{pot}(S) = \frac{1}{2} \sum_{A_i \in S} \sum_{A_j \in S \setminus \{A_i\}} W(x_i - x_j). \tag{3}$$

Our potential functions model either attraction,

$$W_a(x_i - x_j) = \frac{1}{2} \cdot |x_i - x_j|^2, \tag{4}$$

or repulsion,

$$W_r(x_i - x_j) = \frac{c^2}{2} \cdot \exp\left(-\frac{|x_i - x_j|^2}{c^2}\right) \qquad (c \neq 0), \tag{5}$$

where c serves as a spatial scale of repulsion. If we compute the partial derivative w.r.t. x_i, we arrive at the *potential forces* $-\nabla_{x_i} W : \mathbb{R}^d \to \mathbb{R}^d$ that act on the agent A_i:

$$-\nabla_{x_i} W_a(x_i - x_j) = -(x_i - x_j),\tag{6}$$

$$-\nabla_{x_i} W_r(x_i - x_j) = \exp\left(-\frac{|x_i - x_j|^2}{c^2}\right) \cdot (x_i - x_j).\tag{7}$$

For further details on potential functions and their use in discrete models of swarming, we refer the reader to [8].

First-Order Models. First-order models are based on the (physically simplified [10]) assumption that the particle velocity v_i can be expressed in terms of the potential forces $-\nabla_{x_i} W$:

$$\frac{dx_i(t)}{dt} = v_i(t) = -\sum_{A_j \in S \setminus \{A_i\}} \nabla_{x_i} W(x_i(t) - x_j(t)) \qquad \forall A_i \in S,\tag{8}$$

where we assume that we know the initial state at time $t = 0$.

Time Discretisation. Since we cannot expect to find an analytical solution to the dynamical system (8), we have to approximate it numerically on the computer. This requires to discretise it in time.

Let $\tau > 0$ denote some time step size, and let $t_k := k\tau$. Moreover, we abbreviate $x_i(t_k)$ by x_i^k. The simplest time discretisation of Eq. 8 approximates the time derivative by its forward difference:

$$\frac{dx_i^k}{dt} \approx \frac{x_i^{k+1} - x_i^k}{\tau}.\tag{9}$$

This turns (8) into the following explicit update scheme:

$$x_i^{k+1} = x_i^k - \tau \cdot \sum_{A_j \in S \setminus \{A_i\}} \nabla_{x_i^k} W\left(x_i^k - x_j^k\right) \qquad (k = 0, 1, \ldots)\tag{10}$$

with some appropriate initialisation x_i^0 for all $A_i \in S$.

If we restrict the interactions of agent A_i to its δ-neighbourhood $\mathcal{N}_{i,\delta}(t_k)$ from (2), we can replace (10) by the local update rule

$$x_i^{k+1} = x_i^k - \tau \cdot \sum_{A_j \in \mathcal{N}_{i,\delta}^k} \nabla_{x_i^k} W(x_i^k - x_j^k).\tag{11}$$

It is well-known from the theory of numerical methods for differential equations that such explicit schemes may require a fairly small time step size τ in order to be stable [13], in particular if the right hand side fluctuates strongly w.r.t. its argument.[1]

The experiments below make use of the explicit schemes (10) and (11) with the potential forces $\nabla_{x_i^k} W$ given by (6) or (7).

3 Application to Image Processing Problems

Let us now apply our discrete first-order model of swarming to three different image processing problems: grey scale quantisation, contrast enhancement, and line detection. This requires to interpret the specific image processing problem in terms of swarming agents, and to specify our potential forces in an appropriate way.

In our application scenarios, we consider a *digital greyscale image f* that is discrete in its domain and its codomain. The domain contains n_x equally spaced pixels in x-direction and n_y pixels in y-direction. The grey value range is given by the set $\{0, \ldots, 255\}$, which results from a bytewise encoding:

$$ f : \{1, \ldots, n_x\} \times \{1, \ldots, n_y\} \ \rightarrow \ \{0, \ldots, 255\}. \tag{12} $$

For such a two-dimensional greyscale image f, its *histogram* $h[f](k)$ counts how often each grey value $k \in \{0, \ldots, 255\}$ is attained. Thus, $h[f]$ is a one-dimensional function from $\{0, \ldots, 255\}$ to $\mathbb{N}_0$.

3.1 Grey Scale Quantisation

The discretisation of the codomain of an image is called *quantisation*. Obviously, the number of different greyscales in an image determines how expensive it is to store them: While 256 different values require a full byte, 8 values can be encoded already with 3 bits. Since humans cannot distinguish many greyscales, one can compress image data without severe visual degradations by reducing the number of quantisation levels.

To design a model of swarming for obtaining a coarser quantisation of some digital greyscale image f, we proceed as follows. We consider its histogram $h[f]$ and identify some histogram value $h[f](n) = c_n$ with c_n agents sharing the same position $x_i = n$. Thus, we have a one-dimensional model of swarming. Note that multiple agents that share the same position have to undergo the same joint motion.

---

[1]If this becomes too time-consuming, one can also consider more efficient, so-called implicit schemes [13]. However, they require to solve linear or nonlinear systems of equations.

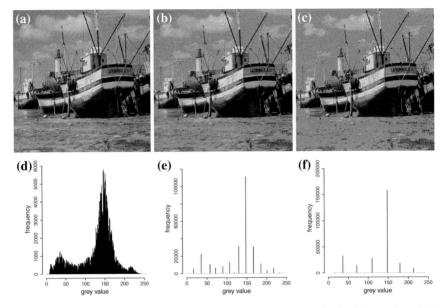

Fig. 1 Swarm-based image quantisation. *Top*, from *left* to *right* **a** Original image from [18], 512×512 pixels, $q = 255$ *greyscales*. **b** Swarm-based quantisation with $\delta = 8$, yielding $q = 16$ *greyscales*. **c** $\delta = 16$, $q = 8$ *greyscales*. *Bottom*, from *left* to *right* **d–f** Corresponding histograms

This reduces the computational complexity in a substantial way: The computational effort becomes proportional to the number of greyscales instead of the number of pixels.

In order to cluster multiple quantisation levels into a single level, we use the linear attraction force from (6). As we will see below, it makes sense to localise the interaction to a δ-neighbourhood, which requires the update scheme (11).

In our quantisation experiments we have chosen $\tau = 10^{-5}$. This leads to a stable steady state solution after at most $4 \cdot 10^4$ iterations. For a 512×512 image, this can be accomplished in far less than one minute on a single core of a standard PC. Figure 1 illustrates the effect of our model of swarming for different δ values. We observe that increasing δ reduces the number q of quantisation levels. Interestingly there seems to be an almost inverse relation, such that $2\delta q$ is roughly equal to the length of the original greyscale interval (255 in our case). This suggests that our model of swarming clusters the grey scales into q bins of approximately[2] the same size 2δ. Note that the interval length 2δ is the diameter of the neighbourhood $\mathcal{N}_{i,\delta}$. Hence, the model of swarming can be interpreted and handled in a very intuitive way.

---

[2]It is clear from the structure of our approach and the experiments that the quantisation levels depend on the actual image histogram and are not necessarily equidistant.

3.2 Contrast Enhancement

The contrast of an image is characterised by the modulus of the difference between the greyvalues of neighbouring pixels. For recognising interesting image structures, their contrast should be sufficiently high. This may require some preprocessing that enhances the image contrast.

Let us now adapt our model of swarming to this application. To this end, it is sufficient to find a mapping of the greyvalues that yields a better contrast. As before, we consider the histogram $h[f]$ of the image f, and we assign c_n agents to a grey value n if $h[f](n) = c_n$. However, since we want to increase the global contrast this time, we use the global explicit scheme (10), and we equip it with the repulsion forces from (7). Moreover, we employ reflecting boundary conditions to prevent that agents leave the admissible greyscale range $[0, 255]$. For $t \to \infty$, the swarm converges to a steady state distribution, where the grade of contrast enhancement grows with the repulsion parameter c. Once the histogram is enhanced, one simply replaces the grey values of the image by their transformed values.

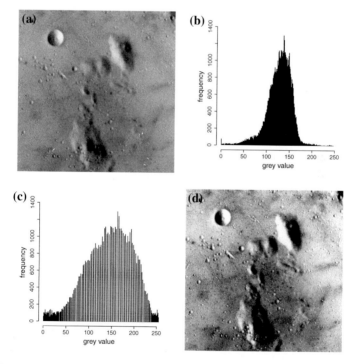

Fig. 2 Swarm-based contrast enhancement. **a** *Top left* Moon surface image from [18], 256×256 pixels. **b** *Top right* Its histogram. **c** *Bottom left* After swarm-based histogram enhancement with $c = 1$, and $2 \cdot 10^6$ iterations with step size $\tau = 10^{-3}$. **d** *Bottom right* Enhanced image using the *grey values* from the transformed histogram

Figure 2 illustrates this procedure, where the evolution reaches a steady state. We observe a clear visual contrast improvement of the test image.This is also confirmed quantitatively by its standard deviation, which has increased from 27.74 to 56.86.

3.3 Line Detection

Our third application scenario for models of swarming is concerned with another important image processing problem, the detection of lines. Our goal is to improve a classical method which is based on the so-called Hough transform [6].

The basic idea behind line detection with the Hough transform is as follows. For some greyscale image f, one searches for locations that may lie on significant lines by computing the gradient magnitude $|\nabla f|$. For a digital image, this requires finite difference approximations. A location is significant if its gradient magnitude exceeds a certain threshold T_g. In a next step, the line candidate pixels vote for all lines that pass through them. All lines through a pixel (x, y) satisfy the normal representation

$$\rho = x \cdot \cos\theta + y \cdot \sin\theta, \tag{13}$$

where θ denotes the angle between the line normal and the x-axis, and ρ is the distance to the origin. Thus, a candidate point is mapped to a trigonometric curve $\rho(\theta)$ in the Hough space (θ, ρ). If n candidate points lie on a line with parameters $(\tilde{\theta}, \tilde{\rho})$, then their corresponding n trigonometric curves in Hough space intersect in $(\tilde{\theta}, \tilde{\rho})$. Therefore, one can find lines in the input image f by searching for clustering points in its Hough space: One discretises the Hough space (θ, ρ), and each trigonometric curve votes for all cells that it crosses. The cells with the most votes characterise the most significant lines in the original image. Typically one finds these clustering points by applying a threshold T_a on the votes in Hough space.

While this sounds nice in theory, in practice it is not easy to find appropriate thresholds that avoid false negatives and false positives. Also the bin size of the discrete Hough space is problematic: If the discretisation is too fine, it is unlikely that many votes will fall in the same cell. If it is too coarse, the line parameters are prone to imprecisions.

As a remedy, we propose the following procedure. First we consider a relatively fine discretsation in Hough space and threshold the votes. Afterwards we process the surviving votes with a swarm-based clustering. To this end, we set up n agents at every position (ρ, θ) that received n votes. Note that in contrast to our clustering for quantisation—which took place in the one-dimensional histogram space—this is a two-dimensional clustering. In analogy to the quantisation setting, we use the linear attraction force (6) within the localised update scheme (11), and compute its steady state.

Figure 3 shows how this works in a real-world setting. We observe that the classical Hough transform suffers from the fact that lines in the image cluster in several

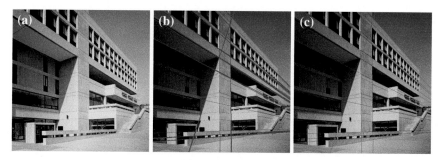

Fig. 3 Swarm-based line detection. **a** *Left* Test image, 512×512 pixels. **b** *Middle* 71 lines detected with the Hough transform. $(T_g = 19, T_a = 244)$. **c** *Right* 13 lines detected with the Hough transform with swarm-based postprocessing $(T_g = 19, T_a = 244, \delta = 5, \tau = 10^{-4}, 300$ iterations$)$

adjacent cells in Hough space. As a consequence, we obtain a bundle of almost parallel lines instead of a single line. Our swarm-based clustering in Hough space is well-suited to solve this problem, since votes from the neighbours move towards the local centroids. In this way they sharpen the clusters and avoid multiple almost parallel lines.

4 Conclusions

Our paper shows that discrete first-order models of swarming have a high potential in image processing that goes far beyond classical applications as tools for difficult *optimisation* tasks: By means of three proof-of-concept applications we have demonstrated their usefulness as powerful *modelling* methods. The fact that these applications serve fairly different goals underlines the genericity of the swarm-based paradigm: It is a highly versatile framework that can be adapted in an intuitive way to a broad spectrum of problems.

In our ongoing work, we intend to study more efficient numerical algorithms, evaluate different ways to incorporate neighbourhood information, equip our models with more problem-specific features, and compare them to non-swarm based approaches. Last but not least, we will also study other models of swarming and apply them to further problems in the broad area of visual computing.

Acknowledgments Our research activities have been supported financially by the *Deutsche Forschungsgemeinschaft (DFG)* through a *Gottfried Wilhelm Leibniz Prize*. This is gratefully acknowledged.

References

1. Aranda, R., Rivera, M., Ramirez-Manzanares, A.: A flocking based method for brain tractography. Med. Image Anal. **18**(3), 515–530 (2014)
2. Clark, P.J., Evans, F.C.: Distance to nearest neighbor as a measure of spatial relationships in populations. Ecology **35**(4), 445–453 (1954)
3. Colorni, A., Dorigo, M., Maniezzo, V.: Distributed optimization by ant colonies. In: Proceedings of the First European Conference On Artificial Life. pp. 134–142. Paris, France (1991)
4. Couzin, I.D., Krause, J., James, R., Ruxton, G.D., Franks, N.R.: Collective memory and spatial sorting in animal groups. J. Theor. Biol. **218**(1), 1–11 (2002)
5. Cucker, F., Smale, S.: Emergent behavior in flocks. IEEE Trans. Autom. Control **52**(5), 852–862 (2007)
6. Duda, R.O., Hart, P.E.: Use of the Hough transformation to detect lines and curves in pictures. Commun. ACM **15**(1), 11–15 (1972)
7. Fine, B.T., Shell, D.A.: Unifying microscopic flocking motion models for virtual, robotic, and biological flock members. Auton. Robot. **35**(2–3), 195–219 (2013)
8. Gazi, V.: On Lagrangian dynamics based modeling of swarm behavior. Phys. D Nonlinear Phenom. **260**, 159–175 (2013)
9. Gazi, V., Fidan, B.: Coordination and control of multi-agent dynamic systems: models and approaches. In: Sahin, E., Spears, W.M., Winfield, A.F.T. (eds.) Swarm Robotics. Lecture Notes in Computer Science, vol. 4433, pp. 71–102. Springer, Berlin (2007)
10. Gazi, V., Passino, K.M.: Stability analysis of social foraging swarms. IEEE Trans. Syst. Man Cybern. Part B Cybern. **34**(1), 539–557 (2004)
11. Kennedy, J., Eberhart, R.: Particle swarm optimization. In: Proceedings of the IEEE International Conference on Neural Networks. vol. 4, pp. 1942–1948. Perth, WA (1995)
12. Kirchmaier, U., Hawe, S., Diepold, K.: A swarm intelligence inspired algorithm for contour detection in images. Appl. Soft Comput. **13**(6), 3118–3129 (2013)
13. LeVeque, R.J.: Finite Difference Methods for Ordinary and Partial Differential Equations. SIAM, Philadelphia (2007)
14. Liu, J., Tang, Y.Y.: Adaptive image segmentation with distributed behavior-based agents. IEEE Trans. Pattern Anal. Mach. Intell. **21**(6), 544–551 (1999)
15. Marthaler, D., Bertozzi, A.L.: Tracking environmental level sets with autonomous vehicles. Recent Developments in Cooperative Control and Optimization. Cooperative Systems, vol. 3, pp. 317–332. Springer, New York (2004)
16. Reynolds, C.W.: Flocks, herds and schools: a distributed behavioral model. ACM SIGGRAPH Comput. Graph. **21**(4), 25–34 (1987)
17. Schmaltz, C., Gwosdek, P., Weickert, J.: Electrostatic halftoning. Comput. Graph. Forum **29**(8), 2313–2327 (2010)
18. Signal and Image Processing Institute of the University of Southern California: The USC-SIPI image database (2015). Last visited August 16, 2015, http://sipi.usc.edu/database/
19. Simone, G., Audino, G., Farup, I., Albregtsen, F., Rizzi, A.: Termite retinex: a new implementation based on a colony of intelligent agents. J. Electron. Imaging **23**(1), 013006 (2014)
20. Sumpter, D.J.T.: The principles of collective animal behaviour. Philos. Trans. R. Soc. B: Biol. Sci. **361**(1465), 5–22 (2005)
21. Triandaf, I., Schwartz, I.B.: A collective motion algorithm for tracking time-dependent boundaries. Math. Comput. Simul. **70**(4), 187–202 (2005)
22. Vicsek, T., Czirók, A., Ben-Jacob, E., Cohen, I., Shochet, O.: Novel type of phase transition in a system of self-driven particles. Phys. Rev. Lett. **75**, 1226–1229 (1995)
23. Vicsek, T., Zafeiris, A.: Collective motion. Phys. Rep. **517**(3–4), 71–140 (2012)

Training Pattern Classifiers with Physiological Cepstral Features to Recognise Human Emotion

Abdultaofeek Abayomi, Oludayo O. Olugbara, Emmanuel Adetiba and Delene Heukelman

Abstract The choice of a suitable set of features based on physiological signals to be utilised in enhancing the recognition of human emotion remains a burning issue in affective computing research. In this study, using the MAHNOB-HCI corpus, we extracted cepstral features from the physiological signals of galvanic skin response, electrocardiogram, electroencephalogram, skin temperature and respiration amplitude to train two state of the art pattern classifiers to recognise seven classes of human emotions. The important task of emotion recognition is largely considered a classification problem and on this basis, we carried out experiments in which the extracted physiological cepstral features were transmitted to Gaussian Radial Basis Function (RBF) neural network and Support Vector Machines (SVM) pattern classifiers for human emotion recognition. The RBF neural network pattern classifier gave the recognition accuracy of 99.5 %, while the SVM pattern classifier posted 75.0 % recognition accuracy. These results indicate the suitability of using cepstral features extracted from fused modality physiological signals with the Gaussian RBF neural network pattern classifier for efficient recognition of human emotion in affective computing systems.

Keywords Classifier · Emotion · Radial · Recognition · Signal · Support · Vector

A. Abayomi (✉) · O.O. Olugbara · E. Adetiba · D. Heukelman
ICT and Society Research Group, Durban University of Technology, Durban, South Africa
e-mail: 21451441@dut4life.ac.za

O.O. Olugbara
e-mail: oludayoo@dut.ac.za

E. Adetiba
e-mail: emmanuela1@dut.ac.za

D. Heukelman
e-mail: deleneh@dut.ac.za

© Springer International Publishing Switzerland 2016
N. Pillay et al. (eds.), *Advances in Nature and Biologically Inspired Computing*,
Advances in Intelligent Systems and Computing 419,
DOI 10.1007/978-3-319-27400-3_24

271

1   Introduction

Emotion is a complex state of the mind, which connotes a positive or a negative reaction to stimuli by humans. The basic human emotions reported by Ekman [1] are fear, anger, sadness, disgust, surprise and joy. These six basic classes of emotions exist in every culture worldwide and could be easily recognized from facial expressions. The recognition of human emotions plays essential roles in diverse application areas, such as human communication, tutoring, mother-infant interaction, psychiatric disorder, safe driving, robotics, crime control, stress detection, telemedicine, call centre conversation and customer services [2–6]. Some of the external stimuli that have been used to elicit emotions in a human subject are pictures, audiovisuals, odour, self-elicitation and music [7–9]. However, for automatic emotion recognition, pattern classifiers such as Support Vector Machine (SVM), Hidden Markov Model (HMM), Naïve Bayes (NB), Decision Trees, k-Nearest Neighbour, Radial Basis Functions (RBF) neural network and Multilayer Perceptron (MLP) neural network have been utilised in various studies [10–14]. To train these pattern classifiers, different features have been extracted from physiological signals using methods such as Fourier, cosine, wavelet and Hilbert Huang transforms [6, 13, 15, 16]. Such features include statistical information in time and frequency domains, sub-band spectra and entropy [6, 8, 10, 16]. However, despite the efforts so far invested by researchers, an efficient automatic recognition of human emotions using physiological signals is still faced with a number of challenges. These challenges include the choice of appropriate emotional stimuli, suitability of pattern classifiers and appropriate choice of features to be extracted from physiological signals [6, 7, 15].

In the study at hand, we have utilised cepstral analysis to extract physiological cepstral features from the MAHNOB-HCI emotional database [10]. Cepstral analysis is a concept of signal processing that has been extensively used with good results in other research domains such as speaker recognition, biomedical engineering, radar analysis and mechanical vibration [17–22]. Gaussian Radial Basis Function (RBF) neural network and Support Vector Machine (SVM) pattern classifiers were configured and trained with the extracted physiological cepstral features to recognise human emotions. The recognition results of these two state of the art pattern classifiers were subsequently compared in order to make an informed decision regarding a suitable combination of pattern classifier and physiological cepstral features that gives improved recognition of human emotion.

2   Materials and Methods

2.1   Dataset

The MAHNOB-HCI emotional corpus [10] was used in this study for human emotion recognition. This corpus consists of data collected from 27 subjects

(16Females, 11 Males) who were stimulated with video clips to elicit a target emotion while concurrently collecting their physiological data of galvanic skin response, electrocardiogram, electroencephalogram, skin temperature and respiration amplitude. In building the corpus, 20 trials under varied target emotions were administered per subject. Each data collection session lasted about 2 min and the physiological responses were down sampled to 256 Hz rate. However, only the physiological data whose self-report emotion of a subject tallies with the induced target emotion were considered as valid. Out of the available 532 samples in the corpus, 437 samples were analysed in this study. 403 samples were used as training data and 34 samples for testing. These physiological data samples were labelled and grouped under seven emotion classes of amusement, anger, disgust, fear, joyhappiness, neutral and sadness for an enhanced supervised emotion recognition task.

2.2 Features Extraction

Feature extraction entails converting raw physiological signals into a sequence of feature vectors of numbers that characterises information about the signals. Power cepstrum are originally utilised to represent human voices as well as musical sounds [23, 24]. A cepstrum is named from the spectrum by reversing its first four letters and it is obtained from the Inverse Fourier Transform (IFT) of the logarithm of the signal [20, 23, 24]. To efficiently extract cepstral features in this study, using the deconvolution approach as shown in Eq. (1), the input physiological signal "x" is first converted into a spectrogram using appropriate functions in MATLAB R2012a according to the following expression:

$$x[t] = h[t] * e[t] \tag{1}$$

where "x" represents the input feature vector of physiological signals, "h" is the spectral envelope and vector "e" contains the spectral details. The spectrogram represents the distribution of spectrum components within a given time and it is generated by framing the physiological signal into smaller parts while applying spectral transformation algorithms such as the wavelet transform and the Fourier transform to the framed signals [20, 23, 24], thus transforming the Eq. (1) to obtain:

$$X[k] = H[k] \, E[k] \tag{2}$$

Taking the magnitudes of both sides, we have the following expression:

$$||X[k]|| = ||H[k]|| \, ||E[k]|| \tag{3}$$

The different channels in the spectrogram need to be regrouped through the frequency component mapping using the logarithmic scale:

$$\log||X[k]|| = \log||H[k]|| + \log||E[k]|| \tag{4}$$

The mapped values are nonlinearly transformed as power is converted to decibels (dB) and to obtain a feature vector representation of each frame, the mapped spectrogram is transformed using the discrete cosine transform to decorrelate the values from different channels by transforming the Eq. (3) to obtain:

$$x[k] = h[k] + e[k] \tag{5}$$

This method computes a single feature vector that captures the overall distributions of the cepstral features for an instance of a physiological signal [24]. The component h[k] in Eq. (5) represents the spectral envelope of the signal from which the cepstral features are obtained by filtering the low frequency region of the spectrum x[k]. Learning from the training data, the classifiers mapped the derived cepstral features to recognise human emotions.

2.3 Pattern Classifier

Pattern classifiers are machine learning algorithms that are used to classify the extracted features from the input signal to their appropriate output classes by learning from the training data. In this study, we compared Gaussian RBF neural network and SVM classifiers in order to select a suitable pattern classifier that can better utilise physiological ceptral features for recognising human emotions. The RBF neural network was chosen because of its short training time, excellent approximation capability, it is robust to input noise, it guarantees high accuracy and simple to design [25–28]. The SVM uses a flexible representation of class boundaries, reduces overfitting, offers a single global minimum, gives a good generalization capability, solves a variety of pattern recognition problems with a little tuning, provides memory economy of memory and different kernel functions [33, 36].

2.3.1 Radial Basis Function Neural Network

Radial Basis Function (RBF) is a feed forward artificial neural network for solving problems of pattern recognition and function approximation [25, 26]. The concepts of RBF are ingrained in earlier pattern recognition techniques such as clustering, spline interpolation, mixture of models and functional approximation [26, 27]. A typical RBF neural network consists of an input layer, one hidden layer and an output layer of artificial neurons [25–28]. Each neuron in the hidden layer implements a radial basis activation function, while the network output is a linear combination of radial basis functions of the input and neuron parameters. The use of RBF neural network involves the specification of the number of input neurons, the activation function to be implemented in the hidden neurons and the number of output neurons. The inputs to our RBF neural network represent the physiological cepstral features of each emotion class and the output neurons correspond to the seven emotion classes to be recognised. The Gaussian basis function is used in this study as the activation function because of its wide usage in many pattern recognition applications that have utilised RBF [25–31].

2.3.2 Support Vector Machine

Support Vector Machine (SVM) [32] is a machine learning method, which was originally used as a binary classifier. It has since been extended to the case of multiple class problems by decomposing the problems into several two-class subproblems and addressing each of the subproblems using SVM [33–35]. Being a discriminative classifier, SVM is very efficient in memory economy and it offers different kernels for its decision making [33, 36]. In training the SVM, a hyperplane that best separates the training feature vectors of the two classes in terms of the support vectors is estimated in the feature space. Pattern recognition using SVM involves determining which side of the hyperplane a feature vector falls while an estimate of the confidence of the decision can use the distance of the feature vector from the hyperplane. For a non-linear SVM, the feature vectors are mapped onto a different space using a nonlinear transform and the distance between a feature vector from the hyperplane in the transformed space.

3 Experimental Results and Discussion

The purpose of the experiments reported in this study is to determine the suitability of the physiological cepstral features for the task of human emotion recognition. The experimental setup was conducted on the fusion of peripheral physiological signals of skin conductance, electrocardiogram, respiration amplitude and skin temperature as well as EEG signals obtained from the MAHNOB-HCI emotional database [10]. The input feature vector was based on the first 12 cepstral coefficients because these are coefficients found to be significant in the literature [37–39]. These 12 coefficients represent the spectral envelope of human emotions captured by the physiological signal.

The Gaussian RBF neural network pattern classifier was trained using a total of 403 samples of the 12 element physiological cepstral features using Matlab R2012a. The default maximum number of neurons in the hidden layer of Gaussian RBF neural network implemented in this study is 403 and the number of output neurons is 7, which represent the total number of samples and number of emotion classes respectively. The goal (mean square error) and spread parameters of the network were also kept at the default values of 0 and 1 respectively to enable an unbiased training procedure. Table 1 depicts the confusion matrix for computing

Table 1 The confusion matrix of the Gaussian RBF neural network pattern classifier showing misclassified instances

a	b	c	d	e	f	g	Classified as
60	0	0	0	0	0	0	a = amusement
0	**58**	0	0	0	0	0	b = anger
0	0	**57**	0	1	0	0	c = disgust
0	0	0	**54**	0	0	0	d = fear
1	0	0	0	**59**	0	0	e = joyhappiness
0	0	0	0	0	**58**	0	f = neutral
0	0	0	0	0	0	**55**	g = sadness

the recognition accuracy result. The recognition accuracy of 99.5 % and a mean square error of 0.0007 were achieved with the Gaussian RBF pattern classifier.

The number of instances in each emotion class that is correctly classified appears on the confusion matrix diagonal (Table 1). This result indicates that a Gaussian RBF pattern classifier is able to classify amusement, anger, fear, neutral and sadness emotions with 100 % precision because all instances were correctly classified. The precision results of joyhappiness and disgust emotions are 98.33 and 98.27 % respectively. This is because an instance of joyhappiness was confused with amusement while an instance of disgust was also confused with joyhappiness. This could be as a result of an overlap in emotion response between joyhappiness and amusement that is traceable to challenges of a discrete approach for emotion labeling and emotion feelings about a subject. However, we tuned the spread parameter of the Gaussian RBF neural network to determine whether the recognition result could be affected, but we observed no effect.

This particular experiment was further extended by utilising the extracted physiological cepstral features to train the SVM pattern classifier for the purpose of establishing comparability of the two pattern classifiers. The Gaussian kernel whose size γ was varied between [0.01, 10] was used in the SVM pattern classifier. Similar set of parameters was used in the research that utilised the same emotional database [10]. The γ is one of the parameters that could be varied in the training of the SVM pattern classifier because it has significant effects on model performance [44–47]. Consequently, 10 experimental trials for each kernel size in the range [0.01, 10] were performed. The best recognition accuracy of 60.3 % and mean square error of 0.0444 were achieved with a kernel size of 1. The cross validation of 20-fold over 50,000 iterations prevented overfitting or underfitting of the SVM pattern classifier. The result obtained in this experiment is shown in Table 2. It can be observed in Table 2 that varying the value of γ increases the recognition accuracy of the SVM pattern classifier in agreement with the literature [44]. This increase in recognition accuracy was from the lowest value $\gamma = 0.01$ to $\gamma = 1$ where the highest recognition accuracy was achieved and thereafter the SVM pattern classifier suffers a decrease with subsequent γ values.

Table 2 Effects of varied kernel sizes on the recognition accuracy and mean square error of the SVM pattern classifier

Serial no.	Kernel size = γ	Recognition accuracy	Mean square error (MSE)
1	0.01	55.1	0.0589
2	0.05	53.8	0.0585
3	0.1	55.1	0.0589
4	0.5	57.7	0.0575
5	1.0	60.3	0.0444
6	1.5	50.0	0.0462
7	2.0	39.7	0.0457
8	2.5	34.6	0.0510
9	3.0	33.3	0.0448
10	3.5	35.9	0.0467

The kernel of size $\gamma = 1$ was used to determine whether we could get a better result using Gaussian, linear, quadratic, perceptron and polynomial kernels. The number of cross validation folds used to train the SVM pattern classifier has a significant impact on its recognition accuracy [45–47]. In this study, the number of folds was varied to experimentally determine their effects on recognition accuracy. This result is shown in Table 3. Since large numbers of folds imply less bias, large values such as 5, 10 and 20 have been used in the literature, therefore following the rule-of-thumb, these values were used in this study. The result shown in Table 3 indicates that the Gaussian kernel retains the best recognition accuracy when compared to the other kernels investigated in this study. The best recognition accuracy of 75.0 % was achieved by the SVM pattern classifier with fivefold cross validation, but with higher mean square error of 0.1833 when compared to 10 and 20-fold cross validation. This result is superior and well preferred to the 60.3 % recognition accuracy achieved by the SVM pattern classifier with 20-fold cross validation (Table 2). The recognition accuracies of the other kernel functions are poor as shown in Table 3. The recognition accuracies of the Gaussian RBF neural

Table 3 Effects of varied cross validation folds and kernel function on the recognition accuracy and mean square error of SVM pattern classifier

Kernel type	Number of folds for cross validation					
	Fivefold		Tenfold		20-folds	
	Recognition accuracy (%)	Mean square error	Recognition accuracy (%)	Mean square error	Recognition accuracy (%)	Mean square error
Gaussian	75.0	0.1833	64.9	0.0895	60.3	0.0444
Linear	18.8	0.2463	13.5	0.1095	17.9	0.0517
Quadratic	12.5	0.2568	18.9	0.1096	17.9	0.0523
Perceptron	18.8	0.2568	18.9	0.1150	12.8	0.0540
Polynomial	25.0	0.2568	27.0	0.1091	32.1	0.0523

network pattern classifier (99.5 %, MSE 0.0007) and the SVM pattern classifier (75.0 %, MSE = 0.1833) indicate the excellent recognition capability of the Gaussian RBF neural network pattern classifier over the SVM pattern classifier for the recognition problem considered in this study. This result compares favourably with the results of SVM pattern classifier that were achieved in research reported in [40–43].

4 Conclusion

We have presented the training of Gaussian RBF neural network and SVM pattern classifiers with physiological cepstral features to automatically recognise human emotion. The physiological cepstral features were extracted from the MAHNOB-HCI corpus of physiological signals of emotionally induced subjects. The suitability and effectiveness of using physiological cepstral features with the Gaussian RBF neural network pattern classifier for human emotion recognition were experimentally established. Gaussian RBF neural network pattern classifier is able to use physiological ceptral features to classify human emotion into seven classes, which are amusement, anger, disgust, fear, joy/happiness, neutral and sadness. The recognition accuracy of 99.5 % recorded by the Gaussian RBF neural network pattern classifier is very high when compared to the recognition accuracy of 75.0 % given by the SVM pattern classifier for the same task of human emotion recognition. The final result of this study shows that Gaussian RBF neural network pattern classifier has huge potential to perceptibly give high recognition accuracy in affective computing applications for human emotion recognition. In the future, we intend to incorporate other emotional databases and perform intensive experiments with other features and classifiers.

References

1. Ekman, P., Rosenberg, E.L.: What the Face Reveals: Basic and Applied Studies of Spontaneous Expression Using the Facial Action Coding System, 2nd edn. Oxford University Press, Oxford (2005)
2. Fried, E.: The impact of nonverbal communication of facial affect on children's learning. Ph.D. dissertation, Rutgers University (1976)
3. Larson, J., Rodriguez, C.: Road Rage to Road-Wise, 1st edn. Tom Doherty Associates Inc., New York (1999)
4. Cohn, J.F., Tronick, E.Z.: Mother-infant interaction: the sequence of dyadic states at three, six, and nine months. Dev. Psychol. **24**(3), 386–392 (1988)
5. Won-Joong Y., Kyu-Sik P.: A study of Emotion Recognition and its Applications. Modeling Decisions for Artificial Intelligence. Lecture notes in Computer Science, vol. 4617, pp. 455–462. Springer-Verlage, Berlin Heidelberg (2007)
6. Ramakrishnan, S.: Recognition of emotion from speech: a review. In: Ramakrishnan, S. (ed.) Speech Enhancement, Modeling and Recognition-Algorithms and Applications. Intec, (2012). ISBN:978-953-51-0291-5

7. Noppadon, J., Setha, P., Pasin, I.: Real-time EEG based happiness detection system. Sci. World J., Article id 618649. Hindawi Publishing Corporation (2013)
8. Jonghwa, K., Andre, E.: Emotion recognition based on physiological changes in music listening. IEEE Trans. Pattern Anal. Mach. Intell. 30(12), 2067–2083 (2008)
9. Nasoz, F., Alvarez, K., Lisetti, C., Finkelstein, N.: Emotion recognition from physiological signals using wireless sensors for presence technologies. Cogn. Technol. Work 6(1), 4–14 (2004)
10. Soleymani, M., Jeroen, L., Pun, T., Pantic, M.: A multimodal database for affect recognition and implicit tagging. IEEE Trans. Affect. Comput. 3(1), 42–55 (2012)
11. Amit, K., Aruna, C.: Emotion Recognition: A Pattern Analysis Approach. Wiley (2014). ISBN:1118130669, 9781118130667
12. Zong, C., Chetouani, M.: Hilbert-Huang transform based physiological signals analysis for emotion recognition. In: Proceedings of IEEE International Symposium on Signal Processing and Information Technology (ISSPIT), pp. 334–339 (2009)
13. Lanjewar, R.B., Chaudhari, D.S.: Speech emotion recognition: a review. Int. J. Innovative Technol. Exploring Eng (IJITEE) 2(4), 68–71 (2013). ISSN:2278–3075
14. James, A.R., Jo-Anne, B., Jose-Miguel, F.: Facial and vocal expressions of emotion. Annu. Rev. Psychol. 54, 329–349 (2003)
15. Heng, Y.P., Lili, N.A., Alfian, A.H., Puteri, S.S.: A study of physiological signals-based emotion recognition systems. Int. J. Comput. Technol. 11(1), 2189–2196 (2013)
16. Jerritta, S., Murugappan, M., Nagarajan, R., Wan, K.: Physiological signals based human emotion Recognition: A review. In: Proceedings of IEEE 7th International Colloquium Signal Processing and Its Applications (CSPA), pp. 410–415 (2011)
17. Soma, B., Shanthi, T., Madhuri, G.: Emotion recognition using combination of MFCC and LPCC with supply vector machine. IOSR J. Comput. Eng. (IOSR-JCE) 17(4), 01–08 (2015). e-ISSN:2278-0661, p-ISSN: 2278-8727
18. Kumar, P., Lahudkar, S.L.: Automatic speaker recognition using LPCC and MFCC. Int. J. Recent Innovation Trends Comput Commun. 3(4), pp. 2106–2109 (2015). ISSN:2321-8169
19. Silvia, M.F., Marius, D.Z.: Emotion recognition in romanian language using LPC features. In: The 4th IEEE International Conference on E-Health and Bioengineering—EHB 2013, Grigore T. Popa University of Medicine and Pharmacy, Iaşi, Romania, Nov 21–23 2013
20. Taabish, G., Anand, S., Sandeep, S.: Comparative analysis of LPCC, MFCC and BFCC for the recognition of Hindi words using artificial neural networks. Int. J. Comput. Appl. (0975–8887) 101(12), 22–27 (2014)
21. Yixiong, P., Peipei, S., Liping, S.: Speech emotion recognition using support vector machine. Int. J. Smart Home. 6(2), 101–108 (2012)
22. Peipei, S., Zhou, C., Xiong, C.: Automatic speech emotion recognition using support vector machine. In: International Conference on Electronic and Mechanical Engineering and Information Technology, pp. 621–625 (2011)
23. Ravelo-Garcial, A.G., Navarro-Mesal, J.L., Hemadez-Perezl, E., Martin-Gonzalezl, S., Quintana-Moralesl, P., Guerra-Morenol, I., et al.: Cepstrum feature selection for the classification of sleep Apnea-Hypopnea syndrome based on heart rate variability. Comput. Cardiol. 40, 959–962 (2013)
24. Zhouyu, F., Guojun, L., Kai, M.T., Dengsheng, Z.: Optimizing cepstral features for audio classification. In: Proceedings of the Twenty-Third International Joint Conference on Artificial Intelligence, pp. 1330–1336 (2013)
25. McCormick, C.: Computer vision and machine learning projects and tutorials. Radial Basis Function Network (RBFN) Tutorial (2013)
26. Bors, A.G.: Introduction of the Radial Basis Functions (RBF) Networks. The York Research database. Online Symposium of Electronics Engineers (2001)
27. Ugur, H.: Artificial Neural Networks. Ch. 9, EE543 Lecture Notes (2004)
28. Xianhai, Guo: Study of emotion recognition based on electrocardiogram and RBF neural network. Procedia Eng. 15, 2408–2412 (2011)

29. Venkatesan, P., Anitha, S.: Application of a radial basis function neural network for diagnosis of diabetes mellitus. Curr. Sci. **91**(9) (2006)
30. Saastamoinen, A., Pietilä, T., Värri, A., Lehtokangas, M., Saarinen, J.: Waveform detection with RBF network—application to automated EEG analysis. Neurocomputing **20**, 1–13 (1998)
31. Yojna, A., Abhishek, S., Abhay, B.: A study of applications of RBF network. Int. J. Comput. Appl. (0975–8887) **94**(2) (2014)
32. Corina, C., Vladimir, V.: Support-vector networks. Machine Learning, vol. 20, pp. 273–297. Kluwer Academic Publishers, Boston. Manufactured in The Netherlands (1995)
33. Pazhanirajan, S., Dhanalakshmi, P.: EEG signal classification using linear predictive cepstral coefficient features. Int. J. Comput. Appl. (0975–8887) **73**(1) (2013)
34. Tokunbo, O., Roberto, T., Madihally, S.N.: Speech and Audio Processing for Coding, Enhancement and Recognition. Springer, New York (2014)
35. Abe, B.T., Olugbara, O.O., Marwala, T.: Experimental comparison of support vector machines with random forests for hyperspectral image land cover classification. J. Earth Syst. Sci. **123** (4), 779–790 (2014)
36. Ahmad, F., Roy, K., O'Connor, B., Shelton, J., Dozier, G., Dworkin, I.: Fly wing biometrics using modified local binary pattern, SVMs and random forest. Int. J. Mach. Learn. Comput. **4**(3), 279–285 (2014)
37. Qi, P.L.: Speaker Authentication, pp. 174–175. Springer Science and Business Media (2011)
38. Lindasalwa, M., Mumtaj, B., Elamvazuthi, I.: Voice recognition algorithms using mel frequency cepstral coefficient (MFCC) and dynamic time warping (DTW) techniques. J. Comput. **2**(3), pp. 138–143 (2010). ISSN:2151-9617
39. Haifeng, S., Jun, G., Gang, L., Qunxia, L.: Non-stationary environment compensation using sequential EM algorithm for robust speech recognition. In: Jorge, A. et al. (eds.) pp. 264–273. Springer-Verlag, Berlin Heinderg (2005)
40. Honig, F., Wagner, J., Batliner, A., Noth, E.: Classification of user states with physiological signals: On-line generic features versus specialized feature sets. In: 17th European Signal Processing Conference 2009, Glasgow, Scotland, pp. 2357–2361 (2009)
41. Maaoui, C., Pruski, A.: Emotion recognition through physiological signals for human-machine communication in cutting edge robotics 2010. In: Kordic, V. (ed.) (2010)
42. Gu, Y., Tan, S.-L., Wong, K.-J., Ho, M.-H. R., Qu, L.: A biometric signature based system for improved emotion recognition using physiological responses from multiple subjects In 8th IEEE International Conference on Industrial Informatics (INDIN), 2010 Osaka, 2010
43. Wan-Hui, W., Yu-Hui, Q., Guang-Yuan, L.: Electrocardiography recording, feature extraction and classification for emotion recognition. In: WRI World Congress on Computer Science and Information Engineering Los Angeles, CA (2009)
44. Renukadevi, N.T., Thangaraj P.: Performance evaluation of SVM–RBF Kernel for medical image. Classifi. Global J. Comput. Sci. Technol. Graph. Vis, **13**(4), Version 1.0, (2013)
45. Anguita, D., Ghelardoni, L., Ghio, A., Oneto, L., Ridella, S.: The 'K' in K-fold cross validation. In: Proceedings, European Symposium on Artificial Neural Networks, Computational Intelligence and Machine Learning. Bruges (Belgium), pp. 441–446 (2012)
46. Hsu, C.W., Chang, C.C., Lin, C.J.: A Practical Guide to Support Vector Classification. Technical report (2000)
47. Hastie, T., Tibshirani, R., Friedman, J., Franklin, J.: The elements of statistical learning: data mining, inference and prediction. Math. Intell. **27**(2), 83–85 (2005)

Identification of Pathogenic Viruses Using Genomic Cepstral Coefficients with Radial Basis Function Neural Network

Emmanuel Adetiba, Oludayo O. Olugbara and Tunmike B. Taiwo

Abstract Human populations are constantly inundated with viruses, some of which are responsible for various deadly diseases. Molecular biology approaches have been employed extensively to identify pathogenic viruses despite the limitations of the approaches. Nevertheless, recent advances in the next generation sequencing technologies have led to a surge in viral genome sequence databases with potentials for Bioinformatics based virus identification. In this study, we have utilised the Gaussian radial basis function neural network to identify pathogenic viruses. To validate the neural network model, samples of sequences of four different pathogenic viruses were extracted from the ViPR corpus. Electron-ion interaction pseudopotential scheme was used to encode the extracted sample sequences while cepstral analysis technique was applied to the encoded sequences to obtain a new set of genomic features, here called Genomic Cepstral Coefficients (GCCs). Experiments were performed to determine the potency of the GCCs to discriminate between different pathogenic viruses. Results show that GCCs are highly discriminating and gave good results when applied to identify some selected pathogenic viruses.

Keywords Cepstral · Dengue · Ebolavirus · Electron · Enterovirus · Genomics · Hepatitis · Radial · Neural · Network

E. Adetiba (✉) · O.O. Olugbara · T.B. Taiwo
ICT and Society Research Group, Durban University of Technology, 1334,
Durban 4000, South Africa
e-mail: emmanuela1@dut.ac.za

O.O. Olugbara
e-mail: oludayoo@dut.ac.za

T.B. Taiwo
e-mail: tunmike.bukola@yahoo.com

© Springer International Publishing Switzerland 2016
N. Pillay et al. (eds.), *Advances in Nature and Biologically Inspired Computing*,
Advances in Intelligent Systems and Computing 419,
DOI 10.1007/978-3-319-27400-3_25

1 Introduction

Application of advanced technologies in molecular biology has greatly accelerated the ease with which known disease pathogens were identified within the last few years. Novel pathogens were discovered with ease using molecular approaches such as immune screening of cDNA libraries and polymerase chain reactions. Examples of pathogenic viruses that were discovered as a result of these efforts are the "hepatitis C" and "sin nombre" [1]. Despite this huge success, the effective identification of several viral pathogens has been elusive. On this account, the development of new techniques for identification of pathogens has become of interest. Modern DNA sequencing technologies hold promise because of the avalanche of genomic sequences of viral from laboratory and environmental surveillance studies that are made publicly available online. The availability of huge datasets has made automatic identification of species of the DNA sequences, an open challenge in Bioinformatics and Genomic Signal Processing (GSP) [2, 3].

In GSP, which is a somewhat new area in Bioinformatics, digital signal processing techniques are employed to analyse genomic data and the biological knowledge gained can be translated to a system based application [3]. Several studies reported in the literature have addressed the identification of species from nucleotide sequences using the digital signal processing techniques alongside the machine learning techniques. A classification model based on data mining and Artificial Neural Network (ANN) was developed for the identification of species from DNA sequences [3]. The authors mined nucleotide patterns from selected DNA sequences and used Multilinear Principal Component Analysis (MPCA) to reduce the dimensions of the mined patterns. The technique was validated on two different species and they reported good classification accuracies. In [4], a classification model was developed to classify eight different eukaryotes species. The authors utilized Frequency Chaos Game Representation (FCGR) to encode genomic sequences and they utilised a neural network technique to obtain the classification accuracy of 92.3 %. A model based on the Naïve Bayesian technique was used to classify archeal and bacterial genomes [5]. The authors used the dinucleotide composition of the genome sequences to report a classification accuracy of 85 %. In [6], the classification of proteins of three different species was reported. The authors used a Markov model to obtain classification accuracies 83.51, 82.12 and 66.63 % of the proteins of microbes, eukaryotes and archaea respectively.

However, the success of any genome identification (or classification) system is hugely dependent on critical factors such as the availability of valid datasets, feature extraction method that truly reflects the attributes of the genetic sequences, the classification algorithm and the objective evaluation of the identification system [7]. The immediate motivation for the study at hand, is the need to obtain a set of discriminating genomic features to improve the identification of species in genomic sequences [7]. The genomic sequences of four pathogenic viruses were extracted from the Virus Pathogen Database and Analysis Resource (ViPR) corpus. The extracted genomic sequences were numerically encoded using a low computational

Electron-Ion Interaction Potential (EIIP) scheme. Thereafter, a set of Genomic Ceptral Coefficients (GCCs) was computed from the encoded sequences and transmitted to the Gaussian Radial Basis Function (RBF) neural network to learn the genome sequences. Section 2 of this paper contains materials and methods, Sect. 3 contain the results and discussion and the paper concludes in Sect. 4.

2 Materials and Methods

In this section, we describe the viral dataset, the EIIP nucleotide mapping scheme, the cepstral analysis technique on which the GCCs is based, the RBF neural network classifier and the experiments performed. All the computational techniques described were implemented in MATLAB R2012a.

2.1 Dataset

We extracted genome sequences of featured viruses from the Virus Pathogen Database and Analysis Resource (ViPR) corpus for our experimentations. The ViPR corpus provides free access to records of gene sequences and proteins of various viral pathogens so as to facilitate research and development of diagnostics, vaccines and therapeutics. The extracted viruses are Ebolavirus, Dengue virus, Hepatitis C and Enterovirus D68. These viruses are currently classified as featured viruses on the ViPR databases because they are responsible for diseases that are presently attracting serious attention from health sectors, scientists and governments worldwide [22]. Complete genome sequences of seven strains of the Bundibuggo specie of Ebolavirus were extracted while ten complete genomes of each of the remaining three viruses were extracted from the ViPR for our experiments. We extracted a small dataset (37 instances of virus sequences) in order to examine the efficacy of the GCCs. The sequence length of each of these viruses is shown in Table 1. As illustrated in the Table, the length of each of the genomic sequences varies from one virus to the other and from one strain of the virus to another strain of the same virus. For instance, the sequence length of the Ebolavirus varies from 18939 to 18941 while that of Enterovirus D68 varies from 420 to 809.

Table 1 The range of the genome sequence length of the selected viruses

Virus	Range of the length of genome sequence
Ebolavirus	18939–18941
Enterovirus D68	420–809
Dengue virus	10176–15287
Hepatitis C virus	9220–9587

2.2 Electron-Ion Interaction Potential (EIIP)

Nucleotide sequence is a string of four distinct characters representing four nucleotides, which are A (Adenine), C (Cytosine), G (Guanine) and T (Thymine) [2, 8]. To apply DSP techniques to process these characters, it is necessary to first convert them into numeric sequences. This problem was solved by Voss [9] using four binary indicator sequences, which are $u_A[n]$, $u_T[n]$, $u_C[n]$ and $u_G[n]$. These indicator sequences take the value of one or zero depending on whether or not the corresponding character exists at n location. These four binary indicator sequences are said to contain some redundancy and they can be transformed into three non-redundant sequences as reported in [10, 11]. Moreover, the authors in [12] proposed a nucleotide encoding scheme by replacing the four indicator sequences with just one sequence and they named the scheme "EIIP indicator sequence". Doing so, they calculated the energy of delocalized electrons in amino acid and nucleotides as the Electron-Ion Interaction Pseudopotential (EIIP). Substituting the EIIP values for A, C, G and T in a nucleotide string $x[n]$, we obtain the "EIIP indicator sequence" that represents the distribution of the energies of free electrons along the sequence [13]. The EIIP values for the four nucleotides are shown in Table 2, where A = 0.1260, G = 0.0806, C = 0.1340 and T = 0.1335. Using the EIIP scheme, the computational overhead of binary indicator sequence is significantly reduced by 75 % [12, 13].

The corresponding Discrete Fourier Transform (DFT) of EIIP encoded sequence $u_i[n]$ where i = A, G, C or T is:

$$U_i[k] = \sum_{n=0}^{N-1} u_i[n] e^{\frac{-j2\pi kn}{N}}, \quad k = 0, 1, 2, \ldots N - 1 \tag{1}$$

and the power spectrum is defined as:

$$P[k] = |U_A[k]|^2 + |U_G[k]|^2 + |U_C[k]|^2 + |U_T[k]|^2 \tag{2}$$

2.3 Cepstral Analysis Technique

A cepstrum is defined as the inverse DFT of the logarithmic magnitude of the DFT of a signal as illustrated with a block diagram in Fig. 1. In other words, a cepstrum

Table 2 Electron-Ion Interaction Pseudopotential for Nucleotides

Nucleotide	EIIP value
A	0.1260
G	0.0806
C	0.1340
T	0.1335

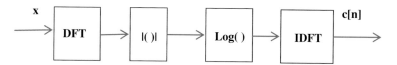

Fig. 1 Block diagram of the computational components of a signal cepstrum

can be considered as a spectrum of logarithmic spectrum that comprises of logarithmic amplitude scale, but linear frequency scale [14]. Cepstrum analysis is used as a tool for detection of periodicity in a spectrum because the harmonic structure of the spectrum is emphasized by the logarithmic amplitude scale. Areas where cepstral coefficients are applied include radar analysis, speech processing, marine exploration, diastolic heart sound analysis and electroencephalogram pattern classification [14–16]. This current study utilizes cepstral analysis to obtain Genomic Cepstral Coefficients (GCCs) for identification of pathogenic viruses from their genome sequences. This effort strongly aligns with the practice in the field of Genomic Signal Processing (GSP) in which DSP techniques are applied to solve biological problems based on nucleotide sequences [9–12].

In this study, the real and complex cepstral [15] are considered to obtain real and complex genomic cepstral features.

The real cepstrum of a signal $x[n]$ is calculated by computing the natural logarithm of the magnitude of its Fourier transform and taking the inverse Fourier transform of the result given as:

$$c_x[n] = \frac{1}{2\pi} \int_{-\pi}^{\pi} \log|X(e^{jw})| e^{jwn} dw \tag{3}$$

The complex cepstrum of the signal is also computed by calculating the complex natural logarithm of the Fourier transform and taking the inverse Fourier transform of the result using:

$$\hat{x}[n] = \frac{1}{2\pi} \int_{-\pi}^{\pi} [\log|X(e^{jw})| e^{jw} + j \arg(X(e^{jw}))] e^{jwn} dw \tag{4}$$

The phase of the signal is represented as $arg()$ in Eq. (4). Both the real and the complex cepstral analysis produce cepstral coefficients and truncating the coefficients at different linear frequency scale allows the preservation of different amount of spectral details. It has been reported in the literature that the first 12 to 15 coefficients of a cepstrum are a compact representation of the spectral envelope [16].

Radial Basis Function Neural Network

An ANN is a parallel biologically inspired computational system that mimics the configuration of the human nervous system. The human brain, which is the central organ in the nervous system is made up of 10^{14}–10^{15} interconnections of neurons and learning involves adjustments to the synaptic connections between these neurons [17]. Artificial neuron was motivated principally from the structure and functions of the human brain and in order to learn problem that cannot be handled by one neuron, an aggregate of several neurons called ANN are engaged. The two main types of ANN architectures are the feed-forward and recurrent networks. In feed-forward architecture, signal flows from input to output stringently in a forward path while there are feedback paths in recurrent networks.

Radial Basis Function (RBF) and Multilayer Perceptron (MLP) neural network are feed-forward networks that have been used extensively in the Bioinformatics and artificial intelligence research communities for pattern classification [18–20]. However, we selected the RBF neural network in this study as a biologically inspired computational platform to identify the GCCs extracted from selected pathogenic viruses. This is because the RBF neural network is reported to be very robust to input noise, always guarantee high accuracy and have lower design rigour than MLP neural network [21]. The input layer of the RBF neural network receives input signal x and transmits it to the hidden layers where the signal is processed and further transmitted in a forward direction to the output layer to generate the output signal y. To process information received in the hidden layer, an RBF network utilizes several kernel functions such as Gaussian, cubic, thin plate spline, Cauchy and inverse multiquadric. However, one of the most commonly used kernel is the Gaussian function, which is used in this study. This kernel function is represented as:

$$f(x) = \exp\left(-\frac{x^2}{\sigma^2}\right) \tag{6}$$

Where σ is the width or scaling parameter that characterises the input space under the influence of the basis function. The output y, from the RBF neural network is defined as:

$$y_k = \sum_{j=1}^{N} \sigma_{j,k} f\left(\left\|c_j - x\right\|_2\right) \tag{7}$$

Where c_j is the centroid of the jth basis function, N is the number of neurons in the hidden layer and $\|x\|$ is the radial distance of its argument, which is usually taken as Euclidean distance.

The RBF neural network used in this study was configured to appropriately identify virus sequences in our experimental dataset of GCCs. The number of neurons in the input layer vary from 12 to 15 because this range of elements was tested for the GCCs. The number of neurons in the hidden layer by default is equal to the number of instances in the training dataset which is equal to 37 in this study. The dataset is partitioned to 70 % training, 15 % validation and 15 % testing. Meanwhile, the output layer contains 4 neurons because there are four virus classes in our experimentation dataset with each virus representing a class. The other configurations of the RBF neural network in this study are MSE goal of 0 and spread of 0.1. To perform an evaluation of the results computed by the RBF neural network, we utilized four different widely used metrics, which are the accuracy, Mean Square Error (MSE), sensitivity and specificity [8, 23].

Experiments

In this study, two experiments were performed to determine the potency of complex and real GCCs to discriminate between different pathogenic viruses. The purpose of the first experiment is to use the complex GCCs to identify pathogenic viruses, while the purpose of the second experiment is to use real GCCs to identify pathogenic viruses. In the experiments, the number of GCCs used as features varied from 12 to 15 in order to determine their effects on the identification accuracy.

In the first experiment, the nucleotide sequences of all the virus sequences were first encoded using the EIIP scheme. Thereafter, we obtained the vectors of complex GCCs, which were transmitted to train the configured RBF neural network to learn virus identification task. In the second experiment, we used the same nucleotide encoding scheme (EIIP) to encode the genome sequences of pathogenic viruses. The real GCCs were thereafter computed to train the RBF neural network to learn the task of virus identification.

3 Experimental Results and Discussion

The spectral plots of the complex GCCs for all the viruses obtained from the first experiment are shown in Fig. 2. These plots clearly show that the complex cepstrum of each virus is unique. For instance, the spectral shape of Dengue virus has a downward peak at the end of the spectrum. The Ebolavirus spectral shape shows both upward and downward peaks at the end of the spectrum. The Enterovirus D68 has a dense spectrum with peaks at the beginning and terminates with a downward peak. The Hepatitis C virus has intermittent dense spectral details across the length of the spectrum that terminates in an upward spike.

In this first experiment, although the number of GCCs varied from 12, 13, 14 to 15, when training the RBF neural network to learn GCCs, we obtained the same results for all the performance metrics. The identification results yield an accuracy of 83.3 %, MSE of 0.0497, sensitivity of 0.9063 and specificity of 0.9520 as shown in Table 3. These results imply that retaining any number of elements of the complex GCCs from 12 to 15 does not affect the performance of the pathogenic virus identification system.

Figure 3 shows the plots of the real GCCs for all the virus sequences in the second experiment. These plots show that the real GCCs of each virus is unique. It can be observed from the plots that Dengue virus has many tiny spikes across the length of the spectrum, Ebola virus has a flat spectrum with a few tiny spikes, Enterovirus has a zig-zag spectrum while Hepatitis C virus has a dense spectrum with a spectral envelope of low amplitude.

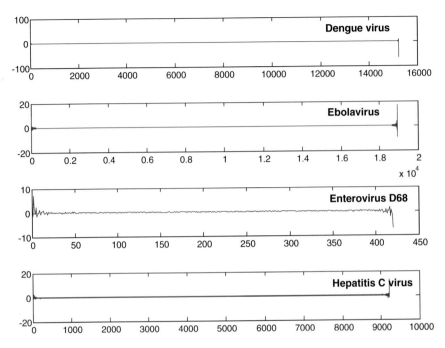

Fig. 2 The complex Genomic Cepstral Coefficients of the four viruses

Table 3 Results of the first experiment

No of elements in GCCs	Accuracy	MSE	Sensitivity	Specificity
12	0.8330	0.0497	0.9063	0.9520
13	0.8330	0.0497	0.9063	0.9520
14	0.8330	0.0497	0.9063	0.9520
15	0.8330	0.0497	0.9063	0.9520

The results obtained in this second experiment after training the RBF neural network to learn the real GCCs are tabulated in Table 4. These results show that the same performance values were obtained for real GCCs with 12 and 13 elements, whereas the results improved for real GCCs with 14 elements. There is also an improvement in the results with respect to sensitivity and specificity when the numbers of elements in the real GCCs were increased from 14 to 15. The best performance accuracy of 97.3 %, MSE of 0.0309, sensitivity of 0.9919 and specificity of 0.9919 were obtained with 15 element real GCCs.

Comparatively, Tables 3 and 4 show the results of real GCCs to be better than the results obtained with complex GCCs, which led to the realization of our experimental objective. The result of this study is significant because each pathogenic virus with sequence length ranges of 18939–18941, 420–809, 10176–15287,

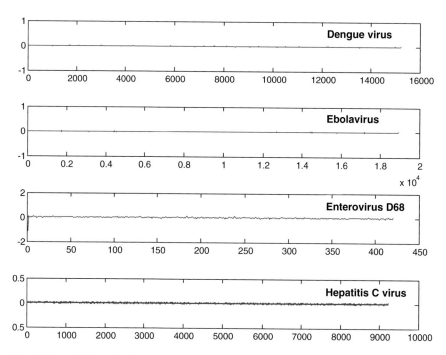

Fig. 3 The real Genomic Cepstral Coefficients of the four viruses

Table 4 Results of the second experiment

No of elements in GCCs	Accuracy	MSE	Sensitivity	Specificity
12	0.9460	0.0541	0.9393	0.9824
13	0.9460	0.0541	0.9393	0.9824
14	0.9730	0.0309	0.9773	0.9773
15	0.9730	0.0309	0.9919	0.9919

and 9220–9587 can be represented using 15 element real GCCs with good discriminating power. Despite an extensive literature search for a similar study that used the same dataset and neural network for pathogenic virus identification, the closest work we found was Karthika et al. [4]. The authors used FCGR to encode the genomic sequences of eight eukaryote species and neural network as the classifier to obtain an accuracy of 92.3 %. This previous study in [4] corroborates our position of the need for discriminating features from genomic sequences for species identification. Hence, the high level of accuracy, sensitivity, specificity and low MSE achieved in this study is a promising endeavour. This result forms the basis for proposing a genomic computational model that incorporates the EIIP scheme, 15 element real GCCs and RBF neural network for accurate identification of pathogenic viruses from genome sequences.

4 Conclusion

In this study, we have been able to apply cepstral analysis technique to obtain genomic cepstral coefficients from genome sequences of pathogenic viruses. These features were incorporated into the proposed computational model for accurate identification of pathogenic viruses from genome sequences. The efficacy of the model has been validated using appropriate data and evaluation metrics. The implementation of the model holds a lot of promises for genomic based diagnosis of microbial diseases in humans, DNA Barcoding, bio-diversity study and wildlife forensic. In the future, we hope to improve the robustness of the proposed model by incorporating more pathogenic viruses and validating the model on a large data set. In addition, we hope to experiment with other classification algorithms that can enhance the performance of the model when more pathogenic virus sequences are elicited.

References

1. Wang, D., Urisman, A., Liu, Y.T., Springer, M., Ksiazek, T.G., Erdman, D.D., DeRisi, J.L.: Viral discovery and sequence recovery using DNA microarrays. PLoS Biol. 1(2), 257–260 (2003)
2. Mabrouk, M.S.: A study of the potential of EIIP mapping method in exon prediction using the frequency domain techniques. Am. J. Biomed. Eng. 2(2), 17–22 (2012)
3. Sathish Kumar, S., Duraipandian, N.: An effective identification of species from DNA sequence: a classification technique by integrating DM and ANN. Int. J. Adv. Comput. Sci. Appl. 3(8), 104–114 (2012)
4. Karthika, V., Nair, V.V., Gopinath, D.P.: Classification of organisms using frequency-chaos game representation of genomic sequences and ANN. In: Proceedings of the 10th National Conference on Technological Trends (NCTT09), pp. 243–247, 6–7 Nov 2009
5. Sandberg, R., Winberg, G., Branden, C.I., Kaske, A., Ernberg, I., Coster, J.: Capturing Whole —Genome characteristics in short sequences using a naive Bayesian classifier. Genome Res. 11, 1404–1409 (2001)
6. Zanoguera, F., De Francesco, M.: Protein classification into domains of life using Markov chain models. In: Proceeding of the Computational Systems Bioinformatics Conference, pp. 517–519 (2004)
7. Song, C., Shi, F.: Prediction of protein subcellular localization based on Hilbert-Huang transform. Wuhan Univ. J. Nat. Sci. 17(1), 48–54 (2012)
8. Adetiba, E., Olugbara, O.O.: Lung cancer prediction using neural network ensemble with histogram of oriented gradient genomic features. Sci. World J. 2015, id. 786013, 1–17 (2015)
9. Voss, R.F.: Evolution of long-range fractal correlations and 1/f noise in DNA base sequences. Phys. Rev. Lett. 68(25), 3805 (1992)
10. Anastassiou, D.: Frequency-domain analysis of biomelecular sequences. Bioinformatics 16(12), 1073–1081 (2000)
11. Anastassiou, D.: Genomic signal processing. IEEE Signal Process. Mag. 18(4), 8–20 (2001)
12. Nair, A.S., Sreenadhan, S.P.: A coding measure scheme employing electron-ion interaction pseudopotential (EIIP). Bioinformation 1(6), 197–202 (2006)

13. Pirogova, E., Simon, G.P., Cosic, I.: Investigation of the applicability of dielectric relaxation properties of amino acid solutions within the resonant recognition model. IEEE Trans. Nanobiosci. **2**, 63–69 (2003)
14. Akay, M.: Biomedical Signal Processing, pp. 113–135. Academic Press (2012)
15. Oppenheim, A.V., Schafer, R.W.: Digital Signal Processing. Englewood Cliffs, Prentice-Hall (1975)
16. Thakur, S., Adetiba, E., Olugbara, O.O., Millham, R.: Experimentation using short-term spectral features for secure mobile internet voting authentication. Math. Prob. Eng. **2015**, id. 564904, 1– 21 (2015)
17. Adetiba, E., Ekeh, J.C., Matthews, V.O., Daramola, S.A., Eleanya, M.E.U.: Estimating an optimal backpropagation algorithm for training an ANN with the EGFR Exon 19 nucleotide sequence: an electronic diagnostic basis for non-small cell lung cancer (NSCLC). J. Emerg. Trends Eng. Appl. Sci. **2**(1), 74–78 (2011)
18. Kurban, T., Beşdok, E.: A comparison of RBF neural network training algorithms for inertial sensor based terrain classification. Sensors **9**(8), 6312–6329 (2009)
19. Oyang, Y.J., Hwang, S.C., Ou, Y.Y., Chen, C.Y., Chen, Z.W.: Data classification with radial basis function networks based on a novel kernel density estimation algorithm. IEEE Trans. Neural Netw. **16**, 225–236 (2005)
20. Lee, C.C., Chung, P.C., Tsai, J.R., Chang, C.I.: Robust radial basis function neural networks. IEEE Trans. Syst. Man Cybern. Part B: Cybern. **29**(6), 674–685 (1999)
21. Derks, E.P.P.A., Pastor, M.S., Buydens, L.M.C.: Robustness analysis of radial base function and multi-layered feed-forward neural network models. Chemometr. Intell. Lab. Syst. **28**(1), 49–60 (1995)
22. Pickett, B.E., Greer, D.S., Zhang, Y.: Virus pathogen database and analysis resource (ViPR): a comprehensive bioinformatics database and analysis resource for the coronavirus research community. Viruses **4**, 3209–3226 (2012)
23. Zou, K.H., O'Malley, A.J., Mauri, L.: Receiver-operating characteristic analysis for evaluating diagnostic tests and predictive models. Circulation **115**(5), 654–657 (2007)

An Aphid Inspired Evolutionary Algorithm

Michael Cilliers and Duncan Coulter

Abstract This paper proposes an evolutionary algorithm based on the reproduction cycle of aphids. The proposed algorithm will alternate between multiple reproduction operators based on the fitness of the population. Through the alternation of reproduction strategies the balance between exploration and exploitation can be manipulated to achieve faster convergence. Two variations on the proposed algorithm are implemented and compared to the standard evolutionary algorithm and clonal expansion. The comparison of converging times of the algorithms show that both variations of the proposed algorithm can be effective.

Keywords Aphid lifecycle · Genetic algorithm · Optimization

1 Introduction

When working with evolutionary algorithms the trade-off between exploration and exploitation has to be considered. If a specific algorithm focuses too much on exploration, it will keep on searching the whole problem space instead of converging on an optimal (sufficient) solution. On the other hand if the algorithm focuses too much on exploitation it is possible that it may get stuck on sub-optimal solutions. The correct choice for the specific balance between exploration and exploitation is problem and search space specific [1].

The first step needed to balance the exploration and exploitation of an algorithm is based on the choice of selection method used during the reproduction process. Different selection methods will have varying effects on the emergent properties of the implemented algorithm and poor quality choices can lead to the algorithm being

M. Cilliers · D. Coulter (✉)
University of Johannesburg, Johannesburg, South Africa
e-mail: dcoulter@uj.ac.za

M. Cilliers
e-mail: cilliersm@uj.ac.za
URL: http://www.uj.ac.za/

© Springer International Publishing Switzerland 2016
N. Pillay et al. (eds.), *Advances in Nature and Biologically Inspired Computing*,
Advances in Intelligent Systems and Computing 419,
DOI 10.1007/978-3-319-27400-3_26

unable to find a sufficient solution. In addition the problem space characteristics are not always fixed and therefore even with the contribution of the selection method a static trade-off between exploration and exploitation is not always optimal [3].

Varying the magnitude of changes made during the application of the mutation operation (mutation size) by the algorithm over the lifetime of a particular run, allows for a dynamic balance between exploration and exploitation [2]. A large mutation size will cause large jumps to be made by candidate solutions within in the search space which will benefit exploration. Small mutation sizes will however cause small jumps which will help exploit the current position in the search space. Varying the mutation size provides more focus towards exploration during the initial section of the algorithm and moves towards exploitation during the latter portion of the algorithm. To try and develop a better solution, inspiration can be drawn from nature. Aphids are a very successful group of insects that has developed interesting methods to be able to quickly adapt to changes in their environment [5]. By trying to emulate these methods that have been developed over many years of evolution, it could be possible to achieve better results than what is possible with the methods mentioned above.

The magnitude of the mutation can also be lowered and combined with an increased mutation rate. The combination will cause frequent small changes to individuals within in the population. In most cases the small changes can have a positive effect on the exploitation of the algorithm by moving the candidate solution up the local fitness gradient, but a negative effect on the exploration by reducing the likelihood the solution will cross an adjacent valley in the fitness landscape. It should be acknowledged however that it is possible for small changes to have profoundly positive or negative effects (for example by converting a valid solution in to an invalid solution) in most cases the magnitude of the effect will correlate with the magnitude of the change. The small mutation size should not contribute to the exploration of the algorithm, so exploration will be largely dependent on the reproduction operation. If the mutation rate is too high it can have a negative effect on the exploitation of the algorithm. Too much mutation will cause individuals to jump around their current positions within the fitness landscape, this is not an effective method of covering the local search space.

2 The Aphid Reproduction Cycle

Aphids, also known as plant lice, are insects that live on and consume the sap of a host plant. Aphids tend to be found in large groups and such infestations can cause substantial damage to the host plant. One of the factors that contribute to the success of the aphid group is due to unique aspects of their lifecycle. The species group has developed a lifecycle that makes use of multiple reproduction methods based on current environmental conditions. It should be noted however that not all aphids exhibit such flexibility in reproductive strategy selection [7].

During spring and summer these aphids reproduce asexually. A single aphid parent produces clonal offspring through parthenogenesis. Parthenogenesis is a form of asexual reproduction where an embryo is formed without fertilization. During parthenogenesis there is no genetic material contributed by a male, so the female has to fill in the absent portions. The female fills in the missing portions by copying her own genes across to the child. The offspring will receive all genetic material from the single parent and will thus be a clone of the parent. It should be noted however that due to the distinction between an individuals genotype and its subsequent expression as a phenotype it is possible for the fitness of these clones to differ in some evolutionary models. For example the contents of a protein strand are determined based on genotype alone however the function of a protein is also determined based on its tertiary structure (how the protein folds in three dimensional space) which can be influenced by environmental factors such as temperature.

In evolution the success of a parent is measured by its ability pass on its genes. In this regard clonal reproduction is ideal. If sexual reproduction is used the child will receive genetic material from both parents. The genes of both parents will be diluted in the next generation. Clonal reproduction on the other hand allows the parent to pass on all its genes to the child. Asexual reproduction allows a large number of aphids to be produced to increase population numbers while the environment is favourable. The increased birthrate can be achieved due to the efficiency of asexual reproduction. Asexual reproduction eliminates the finding of a suitable partner which can be a resource intensive activity. A disadvantage of asexual reproduction is that it limits the adaptability of the population due to the lack of genetic transfer. A lack of adaptation is however not a problem during these seasons when the population is prospering due to favourable environmental conditions such as high food availability and moderate temperatures.

During autumn the aphid population switches over to sexual reproduction. At this point the conditions deteriorate due to changing weather conditions. The changing environment and inherent increase in selection pressure requires the aphid population to adapt. The asexual reproduction that was effectively used during the spring and summer will not allow the aphids to adapt fast enough. Sexual reproduction on the other hand is ideal for allowing the population to adapt. Sexual reproduction allows gene exchange to occur which in turn allows desirable genes to be distributed throughout the population [5].

The life cycle of aphids can be compared to the production of B cells in the human immune system. The immune system produces a large number of varying B cells. The B cells are circulated throughout the body to identify antigens that are present. If a B cell has a high affinity to a corresponding antigen, it starts cloning itself to increase the number of appropriate B cells. The increased number of B cells facilitates the production of the required antibodies to deal with the detected pathogen [6].

The sexual reproduction portion of the aphid lifecycle is similar to the initial production of B cells. In both cases the focus is on exploring the environment (search space) and reaching an acceptable state. With regards to the aphids, an acceptable state is when the population is able to survive in the current conditions. In the immune system an acceptable state is when all non self proteins in the body have been elicit an immune response while self proteins do not.

The asexual reproduction of aphids and the cloning of B cells have a similar role. An acceptable state has been reached and needs to be exploited. Aphids exploit the conditions by increasing the population numbers through asexual reproduction. The immune system on the other hand takes advantage by producing the appropriate antibodies. The antibody production is achieved through the increase of B cells due to the cloning.

Aphids are not the only species that makes use of multiple reproduction methods [8]. Rotifers are very small aquatic creatures that can be found in a variety of water bodies. As with aphids, some rotifer species make use of both sexual and asexual reproduction. During spring the rotifers reproduces asexually to boost population size. When the conditions deteriorate and the water body can't sustain the population, the rotifers switch to sexual reproduction. Both aphids and rotifers make use of similar reproductive strategies, but for the purpose of this paper aphids will be used as the prototypical example.

3 Proposed Algorithm

The presented algorithm draws its inspiration from the aphid lifecycle described previously. The algorithm will take the standard genetic algorithm and adapt it to make use of multiple reproduction operators based on a measure of the harshness of the selection pressure experienced by candidate solutions. Sexual and asexual reproduction will be used to conform to the reproduction strategies used by aphids. By making use of both sexual and asexual reproduction, advantages can be drawn from both.

Sexual reproduction will allow the population to cover large portions of the search space to try and find the optimal areas. Once the optimal areas of the search space have been identified, asexual reproduction will allow the algorithm to find the optimal solution within these areas.

As with the standard genetic algorithm shown in Fig. 1, the proposed algorithm will start with a randomly generated population of potential solutions. From the population a number of individuals will be selected to reproduce. The standard genetic algorithm will then make use of a single reproduction operator to perform the reproduction. The presented algorithm will replace the reproduction portion of the standard genetic algorithm with multiple reproduction operators and a method to switch between them. Once the new solutions have been created, mutation is applied to them. These new solutions are then used to form the next generation. At this point the stopping condition of the algorithm is checked. If the stopping condition is met the algorithm terminates, otherwise it loops back to the selection of parent solutions. Figure 2a, b illustrates two variations and the changes made to the standard genetic algorithm.

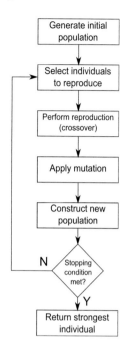

Fig. 1 Standard genetic algorithm

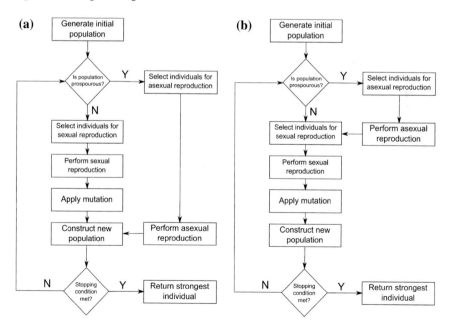

Fig. 2 Proposed algorithms (**a**—generational) (**b**—gradual)

Similar results can be obtained from techniques based on simulated annealing. Initially the simulated annealing algorithm will focus on exploration to try and cover the majority of the search space. As time passes the focus of the algorithm will shift towards exploitation to find the optimal solution in the areas identified in the first phase. Unlike the proposed algorithm, simulated annealing is time based. Simulated annealing will thus not be effective for problems with dynamic search spaces. The problem with time based algorithm is that it requires knowledge regarding the duration of the algorithm. This information is required to select an appropriate decay function/rate. The proposed algorithm will however update its behaviour based on the prosperity (fitness) of the population as a measure of the harshness of the environment. A change in the search space will affect the prosperity of the population which in turn could cause the population to adjust its strategy.

As mentioned in the previous section the algorithm is based on the standard genetic algorithm. The only change that was made is in the crossover (reproduction) step. During the crossover step a decision is added to determine whether sexual or asexual reproduction should be applied or a combination of the two. Two separate methods of applying the reproduction were implemented. The first method, henceforth referred to as the generational method, evaluates the average fitness of the population and then performs the same reproduction operator on the whole population. This approach will cause all the individuals in the population to perform sexual reproduction until the majority of the population starts to converge on possible optimal solutions. At this point the whole population will switch to asexual reproduction and start exploiting the current positions. The optimal point to switch reproduction operators will differ depending on the problem and will have to be adjusted accordingly. The check will also differ depending on whether the problem is a minimization or maximization problem.

The second method, henceforth referred to as the gradual/adaptive method, that was implemented does not use an absolute point where reproduction switches from sexual to asexual. This approach allows one portion of the population to reproduce asexually and the other to reproduce sexually. By increasing the number of individuals that reproduce asexually, as the population improves, a gradual switch from sexual to asexual reproduction can be achieved. As with the first implantation the average fitness is used to determine the fitness of the population as a whole. The average fitness is then converted to a probability that will help determine whether each individual will form part of the asexually or sexually reproducing sections of the population. As the average fitness get closer to the switch point the asexual group size increases and the sexual reproduction group size decreases. During the reproduction step of the algorithm each individual is randomly placed into one of the two reproduction groups. Once the appropriate reproduction operator has been applied to each of these groups the children are combined to form the next generation. As with the generational method, the optimal point to switch reproductive operator will be problem dependent.

4 Comparison

The prototype system will be applied to a function fitting problem. The chromosome of each solution stores the coefficients of the function being represented. The implemented algorithm will fit a cubed function and will thus store four separate values. To evaluate the fitness of each function it will be compared to the points being matched. The distance between the specific function and each point is added together to determine the quality of the function. A function with a smaller fitness will match the points closer than a function with a higher fitness. Rank based selection is used to select the individuals who will participate in sexual reproduction. Two parents are selected and one-point crossover is applied to two child solutions. Once the new individuals have been created mutation is applied to them. Individuals who reproduce asexually are only mutated. To generate the test points for the function to match, Bezier curves are used. The Bezier curve is taken and converted into a number of points. Each of the points is taken a set distance along the curve. These points are then used as the points that the algorithm attempts to fit. To compare the two implemented algorithms to the standard genetic algorithm, all three will be applied to match the same curve multiple times. For each generation the maximum fitness in the population will be averaged with the other executions of the same algorithm. These values can then be compared to determine whether one of the algorithms converges faster than the others. A second variation of the above mentioned procedure will also be applied. The variation will replace the Bezier curve being fitted once an acceptable state has been reached in order to introduce change into the environment. The algorithm will have to adapt to the new environment. The dynamic substitution of the fitness function poses a different problem to the initial adaptation of the population. The state of the population at the moment of substitution embed information from the previous problem space. Some of this embedded information will be correct and therefore beneficial while other parts will be incorrect and must therefore be unlearned. These considerations are referred to as transfer learning [4]. The time that it takes each algorithm to converge on the new solution will be compared to determine the adaptability of each algorithm.

The prototype algorithms will also be compared to a clonal selection algorithm. As with the genetic algorithm, the clonal selection algorithm will be applied multiple times to match the same curve. The maximum fitness at each generation will be averaged out across all of the runs. The values produced by the runs of the prototype algorithms can then be compared to the values produced by the clonal selection.

5 Results

The Aphid Inspired Evolutionary algorithms and the standard genetic algorithm were implemented in a prototype system using C# as the implementation language [9]. The implementation allows the problem-space defining curve to be manipulated during execution.

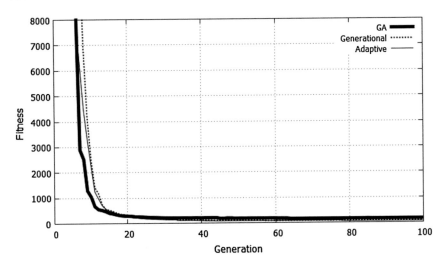

Fig. 3 Comparison overall

Figure 3 represents the convergence of the different algorithms. The graph demonstrates that the standard genetic algorithm starts off converging faster than the other variations on the presented algorithm namely generational and adaptive methods. This is expected due to the high focus on exploration that is present in sexual reproduction.

The generational algorithm should be able to match the standard genetic algorithm closer by adjusting the point at which it switches to sexual reproduction. After the initial burst in the standard genetic algorithm it slows down and both variations of the proposed algorithms start to outperform it. At this point in the algorithm exploitation starts producing better results than exploration. In Fig. 4 we can see that the gradual implementation slightly outperforms the generational implementation. This indicates that even though the sexual reproduction initially contributes less, it still does make a contribution over time.

From Fig. 5 we can see that the results are similar to the previous results. The standard genetic algorithm initially improves faster than the prototype algorithms. As the algorithms start to converge on the new solution, the two prototype algorithms become more effective and converge faster.

Testing of the clonal selection algorithm revealed that the algorithm has a problem with getting stuck on local minima. The algorithm focuses too much on exploitation without providing enough exploration. All the runs that got stuck on local minima had a negative effect on the average convergence time of the clonal selection algorithm as can be seen in Fig. 6.

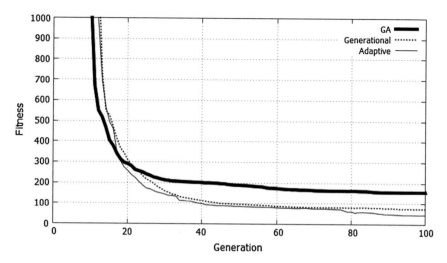

Fig. 4 Comparison (focused on convergence)

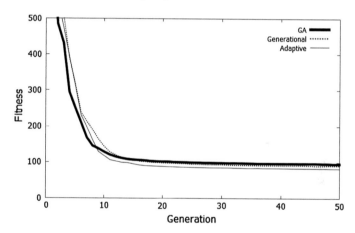

Fig. 5 Comparison of methods (dynamic environment)

Due to the sub optimal results produced by the clonal selection algorithm, a second variation was implemented. The second implementation has a larger mutation size to help avoid getting stuck on local minima. The second variation performed better than the first clonal selection algorithm as can be seen in Fig. 6. Even with the improved results the prototype algorithms still performed better initially. Once the prototype algorithms switches to asexual reproduction similar results can be observed between the prototype algorithms and the clonal selection. The adaptive method even performs better due to the smaller mutation size providing more focus on exploitation.

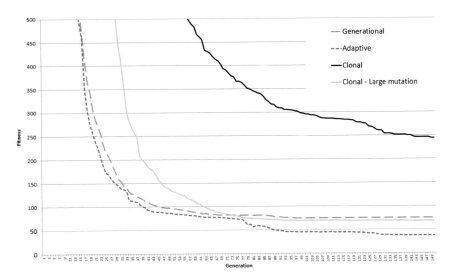

Fig. 6 Clonal selection

6 Conclusion

Both variations of the proposed algorithm were able to outperform the standard genetic algorithm in the final convergence phase. This indicated that the proposed algorithm can be effective for some problems. On the other hand the proposed algorithms where also able to outperform the clonal selection algorithm during the exploration phase. The effectiveness of the algorithm will be dependent on the specific problem and on the related search space. As the algorithm makes use of the fitness of the population to make decisions and adjust population strategy, it should work for problems with dynamic search spaces. Any major changes in the search space will affect the fitness of the population which will allow the algorithm to update the reproduction strategy accordingly.

References

1. Eiben, A.E., Schippers, C.A.: On evolutionary exploration and exploitation. Fundamenta Informaticae **35**(1), 3550 (1998)
2. Eiben, A.E., Hinterding, R., Michalewicz, Z.: Parameter control in evolutionary algorithms. IEEE Trans. Evol. Comput. **3**(2), 124141 (1999)
3. Engelbrecht, A.P.: Computational Intelligence: An Introduction, 2nd edn. Wiley (2007)
4. Pan, S.J., Yang, Q.: A survey on transfer learning. IEEE Trans. Knowl. Data Eng. **22**(10), 13451359 (2010)
5. Rispe, C., Pierre, J.-S., Simon, J.-C., Gouyon, P.-H.: Models of sexual and asexual coexistence in aphids based on constraints. J. Evol. Biol. **11**(6), 685701 (1998)

6. Sablitzky, F., Wildner, G., Rajewsky, K.: Somatic mutation and clonal expansion of B cells in an antigen-driven immune response. EMBO J. **4**(2), 345 (1985)
7. Simon, J.C., Rispe, C., Sunnucks, P.: Ecology and evolution of sex in aphids. Trends Ecol. Evol. **17**(1), 3439 (2002)
8. Snell, T.W.: Effect of temperature, salinity and food level on sexual and asexual reproduction in Brachionus plicatilis (Rotifera). Marine Biol. **92**(2), 157162 (1986)
9. C# Language Specification (C#), http://msdn.microsoft.com/en-us/library/aa645596(v=vs.71).aspx

A Neural Network Model for Road Traffic Flow Estimation

Ayalew Belay Habtie, Ajith Abraham and Dida Midekso

Abstract Real-time road traffic state information can be used for traffic flow monitoring, incident detection and other related traffic management activities. Road traffic state estimation can be done using either data driven or model based or hybrid approaches. The data driven approach is preferable for real-time flow prediction but to get traffic data for performance evaluation, hybrid approach is recommended. In this paper, a neural network model is employed to estimate real-time traffic flow on urban road network. To model the traffic flow, the microscopic model Simulation of Urban Mobility (SUMO) is used. The evaluation of the model using both simulation data and real-world data indicated that the developed estimation model could help to generate reliable traffic state information on urban roads.

Keywords Neural network · State estimation · In-vehicle mobile phone · Simulation of urban mobility (SUMO)

1 Introduction

Real-time road traffic information is important for road traffic management activities like incident detection, traffic monitoring and so on. Road traffic state estimation is a fundamental work in Intelligent Transport System (ITS). It is applicable for

A.B. Habtie (✉) · D. Midekso
Department of Computer Science, Addis Ababa University, Addis Ababa, Ethiopia
e-mail: ayalew.belay@aau.edu.et

D. Midekso
e-mail: dida.midekso@aau.edu.et

A. Abraham
Machine Intelligence Research Labs (MIR Labs), Washington 98071, USA
e-mail: ajith.abraham@ieee.org

A. Abraham
IT4Innovations—Center of Excellence, VSB—Technical University of Ostrava, Ostrava, Czech Republic

© Springer International Publishing Switzerland 2016
N. Pillay et al. (eds.), *Advances in Nature and Biologically Inspired Computing*,
Advances in Intelligent Systems and Computing 419,
DOI 10.1007/978-3-319-27400-3_27

dynamic vehicle navigation, intelligent transport management, traffic signal control, vehicle emission monitoring and so on. The two major components of ITS, Advanced Traveller Information System (ATIS) and Advanced Traffic Management System (ATMS) particularly need accurate current road traffic sate estimation and short term prediction of the future for smooth traffic flow [1].

Existing road traffic state estimation techniques can be grouped into model based approach, which are used for offline traffic state estimation and data driven approach for online state estimation [2]. Model based traffic state estimation approaches use the analytical traffic model Lighthill-Whitham-Richards (LWR) model [3] or simulation based models which are more suitable to model complex road traffic flows [4] or hybrid approaches. However, data driven approaches are preferable for real-time road traffic state estimation [2].

From all data-driven methods, Artificial Neural Networks have been applied extensively in short term traffic forecasting field and acknowledged to be a promising approach because of its superiority in modelling complex nonlinear relationships [5, 6]. Artificial Neural Networks have other advantages over other methods that make researchers choose them as road traffic modelling tool [7]. The first is the strong adaptability of Artificial Neural Networks, which enabled them to learn from past data. As they are data driven models, their transferability is strong and also need little experience when applied to different road traffic networks. Moreover, Artificial Neural Networks are very flexible in producing accurate multiple step-ahead forecast with less effort. Artificial Neural Networks have also some limitations like their "black box" nature, there are so many types of Artificial Neural Networks and as existing researches proved, appropriate network topologies and configurations can greatly affect performance of Artificial Neural Networks models [8].

Road traffic state estimation is a complex activity which cannot be done using a single forecasting method [9]. To use the data-driven approach, for example neural network, deployment of data collection infrastructure on road segments need high investment [10]. To train the neural network, model based traffic state estimation methods like microscopic or macroscopic simulators can be used [11]. Hence, application of hybrid model based and data-driven road traffic state estimation reduces computational delay and also increases forecasting accuracy [12]. For example, Habtie et al. [13] proposed hybrid method of combining Neural Network (data driven approach) and Microscopic simulator SUMO (model based approach) to estimate road traffic flow at real-time on urban road networks.

The traffic data used for estimation can be gathered from several sources like loop detectors, microwave radars and more recently from mobile users [14]. These days, road traffic information estimation from cellular networks has received much attention because of the widespread of the cellular network, its low cost, all-weather traffic information collection and with large number of mobiles to be used as location probes covering the whole road network. In order to estimate road traffic information from cellular network, the basic steps like location data collection, cell phone mobility classification, map matching, route determination and road traffic condition estimation are included [15]. In this study in-vehicle mobile phone, as it is proposed in [13], is used as road traffic data source.

In this article, a Neural Network Model is developed to estimate real-time road traffic state estimation. The model performance is evaluated through simulation. To represent road traffic flow the microscopic simulator SUMO is utilized. The reminder of this paper is organized as follows. Section 2 discusses the Artificial Neural Network Model. Section 3 presents the model application and the conclusion is shown in Sect. 4.

2 Artificial Neural Network Model for Urban Arterial Road Traffic State Estimation

Different types of artificial neural networks (ANN) have been proposed in the past few years for estimation purpose. The most popular connected multilayer perceptron (MLP) neural network architecture is chosen in this study as it is extensively applied in transportation applications [16].

In-vehicle mobile phones are used as a probe to collect traffic data and the data collected contain vehicle position, time stamp and vehicle speed on the road link. Hence position, time stamp and speeds are used as input data in the ANN model and the structure of the ANN model, which is experimentally proved with different traffic demand (free flow, 20 % demand increase, 50 % demand increase, 100 % demand increase) is adapted from Zheng and Zuylen [26] as shown on Fig. 1.

The mathematical model for the input layer, hidden layer and output layer is as follows. Input layer:

$$x(i) = \begin{bmatrix} x_1(i) \\ \vdots \\ x_n(i) \end{bmatrix} = \begin{bmatrix} p(i) \\ r(i) \\ t(i) \\ v(i) \end{bmatrix}, p(i) = \begin{bmatrix} p_1(i) \\ \vdots \\ p_n(i) \end{bmatrix}, r(i) = \begin{bmatrix} r_1(i) \\ \vdots \\ r_n(i) \end{bmatrix},$$

$$t(i) = \begin{bmatrix} t_1(i) \\ \vdots \\ t_n(i) \end{bmatrix}, v(i) = \begin{bmatrix} v_1(i) \\ \vdots \\ v_n(i) \end{bmatrix} \tag{1}$$

where $p(i)$ is position vector, $r(i)$ is link id vector, $t(i)$ is time stamp vector and $v(i)$ is speed vector. Hidden layer:

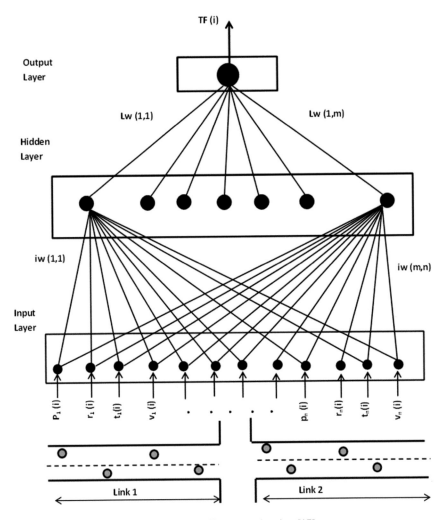

Fig. 1 Artificial neural network for road traffic state estimation [17]

$$H(i) = \begin{bmatrix} h_1(i) \\ \vdots \\ h_m(i) \end{bmatrix} = \begin{bmatrix} \varphi(\sum_{j=1}^{N} w_{j,1} x_j(i) + b_1) \\ \vdots \\ \varphi(\sum_{j=1}^{N} w_{j,m} x_j(i) + b_m) \end{bmatrix} \tag{2}$$

where $h_m(i)$ denotes the value of the mth hidden neuron, $w_{j,m}$ represent the weight connecting the jth input neuron and the mth hidden neuron, b_m is bias with fixed value for the mth hidden neuron and φ is the transfer function.

Output layer

$$y(i) = TV(i) = \varphi\left(\sum_{k=1}^{m} w_k h_k(i) + b\right) \qquad (3)$$

where $y(i)$ and $TV(i)$ denote estimated traffic speed of probe vehicle i on the link under consideration, w_k represent the weight connecting the kth hidden neuron and the output neuron, b is bias for the output and φ is the transfer function.

3   Model Application

3.1   Sample Road for Testing

The sample road, depicted on Fig. 2, used for testing the model is taken from Addis Ababa city road network. The OpenStreetMap (OSM) xml file of the selected road network is edited using Java OpenStreetMap (JOSM) [18] to remove road edges that can't be used by vehicles like road ways to pedestrian etc., and also for simplicity all road edges are set to one-way. The simplified road network consists of 13 nodes, 12 links with length range from 169 to 593 m and 4 traffic lights.

Fig. 2 Sample road network for simulation

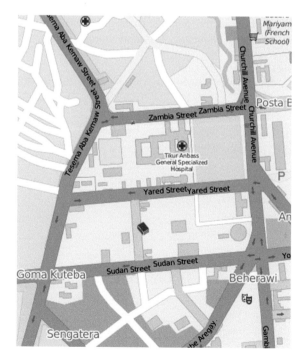

To model the arterial road traffic, a microscopic simulation package "Simulation of Urban Mobility" (SUMO) [19] is employed. To generate road network and vehicle route, the SUMO packages NETCONVERT and DFROUTER are used. From the 1 h SUMO simulation, the simulation output file, FCD output, generates large amount of simulated mobile probes at real time. At every second location updates (in terms of x, y or longitude, latitude) for the mobile probes is recorded. There are in total 437 probes and the time they spent range from 10 to 202 s. The FCD output file contains detail information of each vehicle/mobile and grows extremely large. Hence converting this location data into more compressed one is necessary [20]. Accordingly, to degrade location data, one can set up a specified percentage of simulated vehicles/mobiles to be traffic probe. Previous studies suggest that for arterial roads reliable speed estimation, a minimum penetration rate of 7 %, i.e. at least 10 probe vehicles traversing a road section (every road link) successfully is required [21–23] although factors like road type, link length and sample frequency affects the minimum sample size. In this research work, the sample is taken considering the road link length as indicated in Table 1 and sampling frequency is based on Pinpoint-temporal method [17].

3.2 Neural Network Training

A training process is needed before the ANN model can be applied to estimate traffic state. In the process, three procedures including training, testing and validation were conducted. The total training data set (110 probe data) were divided

Table 1 Number of probes taken for the sample from the FCD output based on road link length

Link # (street name)	Link length (m)	Number of sample probes	Number of location updates per probe
1 (Tesema Aba Wekaw Street)	274.6	10	15
2 (Tesema Aba Wekaw Street)	233.4	10	15
3 (Tesema Aba Wekaw Street)	258.8	10	15
4 (Tesema Aba Wekaw Street)	194.2	10	15
5 (Sudan Street)	194.2	10	15
6 (Sudan Street)	695.8	20	15
7 (Churchill Avenue)	593.9	15	15
8 (Churchill Avenue)	233.3	10	15
9 (Churchill Avenue)	531.4	15	15
10 (Zambia Street)	484.3	15	15
11 (Nigeria Street)	69.5	10	15
12 (Yared Street)	601.7	20	15
13 (Ras Damtew Street)	214.6	10	15

into three subsets [24] which are 88 probe data (80 %) for training, 11 probe data (10 %) for testing and 11 probe data (10 %) for validation. During the training process different hidden neurons like 10, 15, 20, 25 were chosen. During testing the performance in terms of Mean square error (MSE) for the case of 10, 15, 20 and 25 neurons is compared and 15 hidden neurons were used to build the network. Levenberg-Marquardt algorithm [25] was chosen for training. The trained ANN model is applied to estimate link traffic state under free flow but proved even in over saturated condition [26].

3.3 Evaluation

To evaluate how the ANN model performs, the performance indicators Root Mean Square Error (RMSE) and Mean Absolute Percentage Error (MAPE) are used and defined as follows.

$$RMSE = \sqrt{\frac{1}{n} \sum_{k=1}^{n} \left(v_{pv,k} - v_{true,k}\right)^2} \tag{4}$$

$$MAPE = 100 * \frac{1}{n} \sum_{k=1}^{n} \left| \frac{v_{pv,k} - v_{true,k}}{v_{true,k}} \right| \tag{5}$$

where $v_{pv,k}$ the estimated travel speed of the kth is probe vehicle and $v_{true,k}$ is the true link speed of the kth probe vehicle recorded by data collection points.

3.3.1 Results Based on Simulation Data

The trained ANN model is used to estimate link travel speed with a simulation data input. A correlation between the estimated link travel speed based on ANN model and the true link travel speed which is computed from the simulation data using the integration method (IM) of calculating vehicle speed is performed [17]. As depicted in Fig. 3, the estimated link travel speed has very high correlation with the true link travel speed ($R^2 > 99$ %). The linear regression between the estimated and true (simulated) link speed that predicts the best performance among these values has an equation y = 0.997*x + 0 indicating the trained ANN model performs reasonably well, where x represents true speed and y for estimated link travel speed.

The performance of the estimation method in terms of RMSE and MAPE is 0.029325 and 0.127 % respectively with average speed of 7.191 m/s.

Fig. 3 Correlation between
estimated link travel speed
and true link travel speed

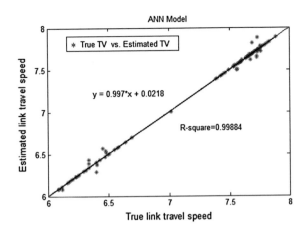

3.3.2 Results Based on Real A-GPS Data

The trained ANN model was also applied to estimate travel speed based on real
A-GPS data. To collect field based data, a car with A-GPS based mobile phone
travelled on the sample road network and location updates in terms of longitude,
latitude, timestamp, speed and accuracy is recorded at every 3 s for about 45 min
(see Fig. 4).

Fig. 4 Real A-GPS based
moving vehicle location data

Fig. 5 Correlation between estimated link travel speed and true link travel speed using real A-GPS data

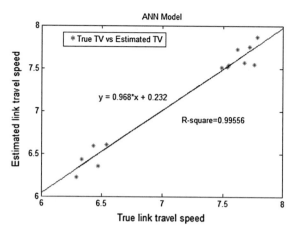

A sample using Pinpoint-Temporal method [17] with 10 s time interval is taken and at every road link for 15 location points i.e. total of 195 A-GPS based vehicle location points are taken. The estimation result is shown on Fig. 5. Each point represents travel speed on each trip i.e., at every road link. From the regression formula of Fig. 5, it can be seen that the trained ANN model performs very good. The RMSE and MAPE are 0.101034 and 1.1877 % respectively, which show the possible application of the ANN model to real link travel speed measurements.

4 Conclusions

This paper presents a neural network model used for estimating urban road traffic flows at real-time. To evaluate the performance of the model a simulation data using the microscopic simulator SUMO was employed. Besides a real world data gathered using in-vehicle A-GPs enabled mobile phone was also used to validate the trained neural network performance. The performance of the developed model based on the performance matrices RMSE and MAPE are 0.029325 and 0.127 % based on simulation data and 0.101034 and 1.1877 % using the real A-GPS based data respectively.

References

1. Kumar, S.V., Vanajakshi, L., Subramanian, S.C.: Traffic State Estimation and Prediction under Heterogeneous Traffic Conditions (2011)
2. Aydos, J., Hengst, B., Uther, W., Blair, A., Zhang, J.: Stochastic Real-Time Urban Traffic State Estimation: Searching for the Most Likely Hypothesis with Limited and Heterogeneous Sensor Data (2012)

3. Lighthill, M.J., Whitham, G.B.: On kinematic waves, II. a theory of traffic flow on long crowded roads. Proc. R. Soc. Lond. A **229**, 317–345 (1955)

4. Gartner, N.H., Messer, C.J., Rathi, A.K.: Traffic flow theory: A state-of-the-art report. In: Committe on Traffic Flow Theory and Characteristics (AHB45) (2001)

5. Zhang, H.: Recursive prediction of traffic conditions with neural network models. J. Transp. Eng. **126**, 472–481 (2000)

6. Wei, C., Lin, S., Li, Y.: Empirical validation of freeway bus travel time forecasting. Transp. Planning J. **32**, 651–679 (2003)

7. Vlahogianni, E.I., Golias, J.C., Karlaftis, M.G.: Short term traffic forecasting: overview of objectives and methods. Transp. Rev. **24**, 533–557 (2004)

8. Ishak, S., Alecsandru, C.: Optimizing traffic prediction performance of neural networks under various topological, input, and traffic condition settings. J. Transp. Eng. **130**, 452–465 (2004)

9. You, J., Kim, T.J.: Development and evaluation of a hybrid travel time forecasting model. Transp. Res. Part C: Emerg. Technol. **8**, 231–256 (2000)

10. van Lint, H.: Reliable travel time prediction for freeways: Netherlands TRAIL Research School. Delft University of Technology (2004)

11. Anderson, J., Bell, M.: Travel time estimation in urban road networks. In: IEEE Conference on Intelligent Transportation System, 1997. ITSC'97, pp. 924–929 (1997)

12. You, J., Kim, T.J.: Development and evaluation of a hybrid travel time forecasting model. Transp. Res. Part C: Emerg. Technol. **8**, 231–256 (2000)

13. Habtie, B., Ajith A., Dida, M.: In-Vehicle mobile phone-based road traffic flow estimation: a review. In: Journal of Network and Innovative Computing (JNIC), vol. 2, pp. 331–358 (2013)

14. Lv, W., Ma, S., Liang, C., Zhu, T.: Effective data identification in travel time estimation based on cellular network signaling. In: Wireless and Mobile Networking Conference (WMNC), 2011 4th Joint IFIP, 2011, pp. 1–5

15. Gundlegard, D., Karlsson, J.M.: Route classification in travel time estimation based on cellular network signaling. In: 12th International IEEE Conference on Intelligent Transportation Systems, 2009. ITSC'09, pp. 1–6 (2009)

16. TOPUZ, V.: Hourly Traffic Flow Predictions by Different ANN Models, Sept (2010)

17. Habtie, A.B., Ajith A., Dida M.: Road Traffic state estimation framework based on hybrid assisted global positioning system and uplink time difference of arrival data collection methods. In: AFRICON, 2015. IEEE, (2015) in press

18. JOSM: Java OSM Editor. URL: https://josm.openstreetmap.de/

19. Behrisch, M., Bieker, L., Erdmann, M.J., Krajzewicz, D.: SUMO-simulation of urban mobility-an overview. In: SIMUL 2011, The Third International Conference on Advances in System Simulation, pp. 55–60 (2011)

20. Tao, S., Manolopoulos, V., Rodriguez Duenas, S., Rusu, A.: Real-time urban traffic state estimation with A-GPS mobile phones as probes. J. Transp. Technol. **2**, 22–31 (2012)

21. Ferman, M.A., Blumenfeld, D.E., Dai, X.: An analytical evaluation of a real-time traffic information system using probe vehicles. In: Intelligent Transportation Systems, pp. 23–34 (2005)

22. Manolopoulos, V., Tao, S., Rodriguez, S., Ismail, M., Rusu, A.: MobiTraS: A mobile application for a smart traffic system. In: NEWCAS Conference (NEWCAS), 2010 8th IEEE International, pp. 365–368 (2010)

23. Zhao, Q., Kong, Q.-J., Xia, Y., Liu, Y.: Sample size analysis of GPS probe vehicles for urban traffic state estimation. In: 2011 14th International IEEE Conference on Intelligent Transportation Systems (ITSC), pp. 272–276 (2011)

24. Adeoti, O.A., Osanaiye, P.A.: Performance analysis of ANN on dataset allocations for pattern recognition of bivariate process. Math. Theory Model. **2**, 53–63 (2012)

25. Ranganathan, A.: The levenberg-marquardt algorithm. Tutorial on LM Algorithm, pp. 1–5, (2004)

26. Zheng, F., Van Zuylen, H.: Urban link travel time estimation based on sparse probe vehicle data. Transp. Res. Part C: Emerg. Technol. **31**, 145–157 (2013)

Optimization of a Static VAR Compensation Parameters Using PBIL

Dereck Dombo and Komla Agbenyo Folly

Abstract Static VAR Compensators (SVCs) are part of the family of the Flexible Alternating Current Transmission Systems (FACTS) devices that are predominantly used for quick and reliable line voltage control. Under contingency conditions, they can be used to provide dynamic, fast response reactive power. In addition, SVCs can be used to increase power transfer capability, and damp power oscillations. This paper is concerned with damping of power oscillations using SVC. To perform the function of damping controller, supplementary control is required for the SVC. In addition, the control parameters for the SVC need to be tuned adequately. In this paper, Population Based Incremental Learning (PBIL) is used to tune the parameters of SVC in a multi machine power system. Population-Based Incremental Learning (PBIL) is a technique that combines Genetic Algorithm and simple competitive learning derived from Artificial Neural Network. It has recently received increasing attention due to its effectiveness, easy implementation and robustness. To show the effectiveness of the approach, simulations results are compared with the results obtained using simple Genetic Algorithm (SGA) and the conventional SVC under different operating conditions.

Keywords Static VAR compensator · Low frequency oscillations · Population-based incremental learning

D. Dombo (✉) · K.A. Folly
Department of Electrical Engineering, University of Cape Town, Private Bag,
Rondebosch, Cape Town 7701, South Africa
e-mail: Dereck.Dombo@alumni.uct.ac.za

K.A. Folly
e-mail: Komla.Folly@uct.ac.za

© Springer International Publishing Switzerland 2016
N. Pillay et al. (eds.), *Advances in Nature and Biologically Inspired Computing*,
Advances in Intelligent Systems and Computing 419,
DOI 10.1007/978-3-319-27400-3_28

1   Introduction

Low frequency oscillations have become the main problems for small signal stability due to the increased interconnection of power systems and the transfer of bulk
power across the various regions. These oscillations are in the range of 0.2–3.5 Hz
[1–3]. Some of the causes of low frequency oscillations are heavy power transmission on weak transmission lines and exciter's high gain. To damp these oscillations, Power System Stabilizers (PSSs) have been traditionally used. Although
PSSs are the most cost effective devices to damp low frequency oscillations, they
may not necessarily provide adequate damping to the system, in particular when the
operating conditions of the system are varying over a wide range [4, 5]. Recently,
Flexible Alternating Current Transmission Systems (FACTS) devices such as
Thyristor Controlled Series Compensator (TCSC), Static Synchronous compensator
(STATCOM) and Static VAR Compensators (SVCs) have been proposed to damp
low frequency oscillations [6, 7]. In terms of cost, SVC is the cheapest among the
FACTS devices and therefore the most attractive to be used in developing countries
such as South Africa.

This paper is concerned with damping of power oscillations using SVC. To
perform the function of damping controller, supplementary control is required for
the SVC. In addition, the control parameters for the SVC need to be tuned adequately. Several alternative design techniques such as pole placement, phase gain
margin technique and H∞ technique have been used to tune the parameters of SVC
[1]. In this paper, Population Based Incremental Learning (PBIL) is used to tune the
parameters of SVC in a multi machine power system. Population-Based Incremental Learning (PBIL) is a technique that combines aspects of Genetic Algorithms
and simple competitive learning derived from Artificial Neural Networks [8, 9]. It
has recently received increasing attention due to its effectiveness, easy implementation and robustness. To show the effectiveness of the approach, simulations
results are compared with those obtained using standard Genetic Algorithm
(GA) and the conventional SVC.

2   Overview of Genetic Algorithms

GAs are search methods based on the mechanics of natural selection, genetics and
emulating species evolution through generations [8, 12]. They produce powerful
tools for system optimization and other applications by combining the principle of
survival of the fittest with information exchange among individuals [9, 11]. John
Holland came up with the ideas of GAs in the 1960s [12]. When Holland came up
with the algorithm, ideally the goal was not really to solve certain identified
problems. The main goals were to study the process of adaptation formally as it
happens in natural systems and to improve ways to import mechanisms of natural
systems into computer systems [10–12]. The way in which the GAs operate is such

that there is a change from one population of chromosomes to a new population with the use of some kind of natural selection. The fitter individuals have the ability to produce more offspring in the next generation than the less fit ones [10, 11]. Over the years, GAs have been used for different kinds of applications because of their computational simplicity [11, 13]. Compared to conventional optimization methods, GAs are more flexible in that they search from multiple points and not a single point, rely on payoff information not derivatives or continuity and use probabilistic transition rules not deterministic rules.

In this paper, the Simple Genetic Algorithm (SGA) was implemented to tune the parameters of a SVC and the results were compared with the ones using Population-Based Incremental learning (PBIL). For more details on SGA the reader is referred to [8, 10–12]. The Pseudo code of GA is given below [14].

3 Overview of Population-Based Incremental Learning

Population-based incremental learning (PBIL) is a technique that combines aspects of Genetic Algorithms and simple competitive learning derived from Artificial Neural Networks [15, 16]. Unlike GAs which performance depends on crossover operator, PBIL performance depends on the learning process of the probability vector. The probability vector guides the search, which produces the next sample point from which learning takes place [16, 17]. The learning rate determines the speed at which the probability vector is shifted to resemble the best (fittest) solution vector [17]. Initially, the values of the probability vector are set to 0.5 to ensure that the probability of generating 0 or 1 is equal. As the search progresses, these values are moved away from 0.5, towards either 0.0 or 1.0. Like in GA, mutation is also used in PBIL to maintain diversity. In this paper, the mutation is performed on the probability vector; that is, a forgetting factor is used to relax the probability vector toward a neutral value of 0.5 [18]. The pseudo code for the standard PBIL is shown in Figs. 1 and 2, [15–17].

4 Overview and Placement of SVC

4.1 SVC Placement

SVC is a shunt connected static var absorber by which the output is adjusted to exchange inductive or capacitive current so as to control the bus voltage. The SVC is known to improve the dynamic voltage control, thus increasing the system loadability. Furthermore, an additional stabilizing signal and supplementary control placed on the voltage control loop of the SVC controller can provide oscillation damping of the system.

Fig. 1 Pseudocode for SGA
[14]

//**initialize generation 0**;
k:=0;
P_k:= a population n randomly-generated
        individuals; initialize population

// **Evaluate** P_k:
Compute fitness (i) for each $i \in P_k$:
do
{   // **Create generation** k **+1;**
      //**Crossover**
      //**Mutation**
      // **Evaluation**
      // **Increment**
      k:: +1;
}
while fitness of fittest individual in P_k:is not
        satisfied the requirement ;
return the fittest individual from P_k:

Fig. 2 Pseudocode for
standard PBIL

Begin
g:= 0;
//initialize probability vector
for i:=1 to l, do $PV_i^0 = 0.5$;
endfor;
while not termination condition do
generate sample $S(g)$ from (PV(g) , pop.)
Evaluate samples $S(g)$
Select best solution $B(g)$
// update probability vector $PV(g)$ toward best solution according to (1)
//mutate $PV(g)$
Generate a set of new samples using the new
probability vector
    g=g+1
end while // e.g., $g>G_{max}$

There are different forms of SVC. In its simplest form, SVC is a controllable inductor in parallel with a switchable capacitor [1].

In terms of improving small signal stability, the effectiveness of the SVCs mainly depends on their location, signals used as input and the method used to perform controller design. Since SVCs are mainly used for voltage control, they should be located at the bus that needs voltage control.

The power system model used in this paper is the IEEE two-area system, 4-machine power system as shown in Fig. 3. It can be seen from Fig. 3 that area 1

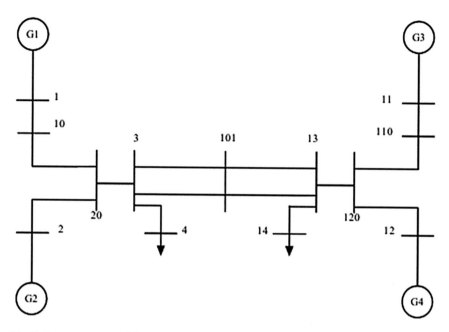

Fig. 3 Power system model

contains generator 1 (G1) and generator 2 (G2), while area 2 contains generator 3 (G3) and generator 4 (G4). In this paper, each generator is modelled by the detailed six order differential equations. The machines are equipped with simple exciter systems. For more information on this system, the reader is referred to [1].

For this system, the SVC should be placed at the middle of the interconnection between the two areas (at bus 101) where voltage swings are the greatest [1]. Current magnitude was used as input to the SVC.

4.2 Problem Formulation and Objective Functions

Three operating conditions were considered during the design of SVC controllers. These operating conditions are varying with the amount of power transferred from area 1 to area 2 and are listed in Table 1. It is well known that the system exhibits two local modes one in area 1 and the other in area 2 and one inter-area mode. The

Table 1 Selected open-loop operating conditions including eigenvalues and damping ratios

Case	P_e [MW]	Eigenvalue (ζ)
1	200	$0.0216 \pm j4.34$ (−0.005)
2	300	$0.0333 \pm j4.12$ (−0.008)
3	400	$0.0219 \pm j4.02$ (−0.055)

inter-area modes are the difficult modes to control, hence in this paper only inter area modes will be considered.

The purpose of the design is to optimize the parameters of the supplementary control for the SVC such that the small signal stability is improved.

The transfer function of the supplementary control is given by:

$$K(s) = K_r \left(\frac{T_w}{1+sT_w}\right)\left(\frac{1+sT_c}{1+sT_b}\right) \tag{1}$$

where, K_r is the gain, T_b and T_c represent suitable time constants. T_w is the washout time constant needed to prevent steady-state offset of the voltage. The value of T_w is not critical for the supplementary control and was set to 10 s.

Since most of oscillation modes considered in this paper are unstable and dominate the time response of the system, it is expected that by maximizing the minimum damping ratio, a set of system models could be simultaneously stabilized over a wide range of operating conditions [17, 18]. The following objective function was used to design the supplementary control of the SVC.

$$J = \max\left(\min\left(\zeta_{i,j}\right)\right). \tag{2}$$

where $i = 1, 2, \ldots n$, and $j = 1, 2, \ldots m$ and $\zeta_{i,j} = \frac{-\sigma_{i,j}}{\sqrt{\sigma_{i,j}^2 + \omega_{i,j}^2}}$ is the damping ratio of the ith eigenvalue in the jth operating condition. σ_{ij} is the real part of the eigenvalue and the ω_{ij} is the frequency. n denotes the total number eigenvalues and m denotes the number of operating conditions.

4.3 Application of GA to Controller Design

The configuration of the standard GA is as follows:

Encoding: Binary
Population: 100
Generations: 100
Crossover: 0.7
Mutation: 0.1.

4.4 Application of PBIL to Controller Design

The configuration of the PBILs is as follows:

Encoding: Binary
Population: 100
Generations: 100

Learning rate (LR): 0.1
Mutation (Forgetting factor-FF): 0.005.

4.5   Design of Conventional SVC

The conventional SVC was designed based on the steps detailed in [19] using a
MATLAB toolbox called Power System Analysis Toolbox (PST) [20].

5   Simulation Results

5.1   Eigenvalue Analysis

A MATLAB toolbox called PST was used to carry out the simulations [21]. Table 2
shows the eigenvalues of the system with the conventional SVC, GA-SVC and
PBIL-SVC. It can be seen that with the introduction of the SVCs, the damping of
the system was improved compared to the open-loop system. For all the operating
conditions considered, the conventional SVC gives the least damping followed by
the GA-SVC. The PBIL-SVC is observed to outperform both the conventional SVC
and the GA-SVC by providing the best damping under all the operating conditions.

5.2   Time Domain Simulations

In all the simulations described in the Figs. 4, 5, 6 and 7 below, a 10 % step change
in V_{ref} was considered as disturbance. The performances of the SVCs were
investigated when the system was operating under different operating conditions.
However, for simplicity, only the results for operating condition 1 are shown here.
   Figures 4 and 5 show the changes in the active power of G1 and G2, respectively
for the above disturbance. From the responses in Figs. 4 and 5, it can be seen that in
the absence of SVC (No SVC) the system is poorly damped. The inclusion of SVC
with supplementary control damped the oscillations. However, both the PBIL based
SVC and the GA based SVC give better responses with slightly smaller overshoots
compared to the conventional SVC. It can be seen that the performances of PBIL

Table 2 Closed-system eigenvalues and damping ratio

Case	SVC	GA-SVC	PBIL-SVC
1	−0.336 ± j4.34 (0.077)	−1.877 ± j4.19 (0.41)	−2.93 ± j4.12 (0.58)
2	−0.366 ± j4.29 (0.08)	−0.59 ± j3.44 (0.17)	−1.26 ± j2.08 (0.23)
3	−0.334 ± j4.12 (0.42)	−2.36 ± j3.87 (0.52)	−4.73 ± j4.94 (0.69)

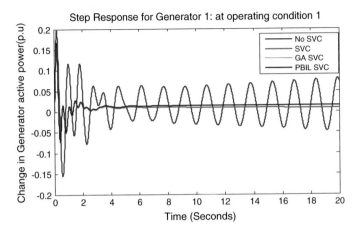

Fig. 4 Responses in active power of G1 for a 10 % step response in V_{ref}

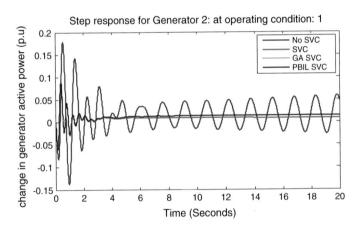

Fig. 5 Responses in active power for G2 for a 105 step response in V_{ref}

based SVC and the GA based SVC are very similar. However, this is difficult to see from the graphs since the PBIL step response is on top of the GA step response.

Figures 6 and 7 show the active power responses for G3 and G4, respectively. It can be seen that without SVC there are growing oscillations and the system is unstable. The introduction of SVC controllers damped the oscillations. However the PBIL based SVC and GA based SVC perform better than the conventional SVC as their responses settle much quicker compared to those of the conventional SVC.

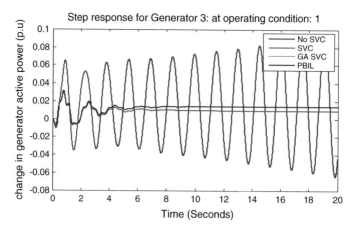

Fig. 6 Responses in active power for G3 for a 10 % step response in V_{ref}

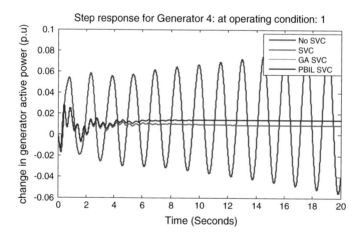

Fig. 7 Responses in active power for G4 for a 10 % step response in V_{ref}

6   Conclusions

SVC parameter optimization based PBIL has been presented in this paper. To show the effectiveness of the approach, simulations results based on PBIL are compared with those obtained using standard Genetic Algorithm (GA) and the conventional SVC. Frequency and time domain simulations show that the introduction of SVC with of supplementary control can improve the system damping. Although the conventional SVC was able to improve the damping of the system, its performance was not as good as those of the GA-SVC and the PBIL-SVC. Eigenvalue analysis shows that PBIL based SVC provides the best damping to the system followed by

GA based SVC. The conventional SVC provided the least damping. More works need to be done to improve the time domain performance of the PBIL-SVC.

Acknowledgment This work is based on the research supported in part by the National Research Foundation of South Africa, UID 83977 and UID 85503

References

1. Kundur, P.: Power system stability and control. In: Bali, N.J., Lauby, M.G. (eds.) Mc Graw-Hill, Inc, (1994)
2. Kundur, P., Paserba, J., Ajjarapu, V., Anderson, G.: Definition and classification of power system stability IEEE/CIGRE joint task force on stability terms and definitions. IEEE Trans. Power Syst. **19**(3), 1387–1401 (2004)
3. Kundur, P., Paserba, J., Vitet, S.: Overview on definition and classification of power system stability. Qual. Secur. Electr. Power Deliv. Syst. **2003**, 1–4 (2003)
4. Folly, K.A.: Design of power system stabilizer: a comparison between genetic algorithms (GAs) and population-based incremental learning (PBIL). In: Proceedings of the IEEE PES 2006 General Meeting, Montreal, Canada (2006)
5. Jiang, P., Yan, W., Gu, W.: PSS parameter optimization with genetic algorithm. In: Third International Conference on Electric Utility Deregulation and Restructuring and Power Technologies—DRPT (2008)
6. Mhaskar, U.P., Kulkarni, A.M.: Power oscillations damping using FACTS devices: modal controllability, observability in local signals, and location of transfer function zeros. IEEE Trans. power Syst. **21**(1) (2006)
7. Abd-Elazim, S.M., Ali, E.S.: Bacteria Foraging optimization algorithm based SVC damping controller design for power system stability enhancement. Electr. Power Energy Syst. **43**, 933–940 (2012)
8. Whitley, D.: A Genetic Algorithm Tutorial. Statistics and Computing, pp. 65–85. http://www.cs.colostate.edu/~genitor/MiscPubs/tutorial.pdf (1994)
9. Sheetekela, S., Folly, K.A.: Power system controller design: a comparison between breeder genetic algorithm (BGA) and population-based incremental learning (PBIL). In: Proceedings of the International Joint Conference on Neural Networks (IJCNN) (2010)
10. Goldberg, D.E.: Genetic Algorithms in Search. Optimization & Machine Learning, Addison-Wesley (1989)
11. Davis, L.: Handbook of Genetic Algorithms. International Thomson Computer Press (1996)
12. Holland, J.H.: Adaptation in Natural and Artificial Systems: An Introductory Analysis with Applications to Biology, Control and Artificial Intelligence. University of Michigan Press, Ann Arbor (1975)
13. Yao, J., Kharma, N., Grogono, P.: Bi-objective multipopulation genetic algorithm for multimodal function optimization. IEEE Trans. Evol. Comput. **14**(1), 80–102 (2010)
14. Genetic Algorithms. http://www.cs.ucc.ie/~dgb/courses/tai/notes/handout12.pdf
15. Baluja, S.: Population-Based Incremental Learning: A Method for Integrating Genetic Search Based Function Optimization and Competitive Learning. Technical Report, CMU-CS-94-163, Carnegie Mellon University (1994)
16. Baluja, S., Caruana, R.: Removing the Genetics from the Standard Genetic Algorithm. Technical Report, CMU-CS-95-141, Carnegie Mellon University (1995)
17. Folly, K.A.: Performance evaluation of power system stabilizers based on population-based incremental learning (PBIL) algorithm. Int. J. Power Energy Syst. **33**(7), 1279–1287 (2011)
18. Folly, K.A.: An improved population-based incremental learning algorithm. Int. J. Swarm Intell. Res. (IJSIR) **4**(1), 35–61 (2013)

19. Ayres, H.M., Kopcak, I., Castro, M.S., MIlano, F., Da Costa, V.F.: A didactic procedure for designing power oscillation dampers of FACTS devices. Simul. Model. Pract. Theory **18**(6), 896–909 (2010)
20. Milano, F.: Power System Analysis Toolbox. Documentation for PSAT version, 2.1.2 (2008)
21. Power System Toolbox, version 3. Copyright Joe Chow/Graham Rogers (1991–2008)

ImmunoOptiDrone—Towards Re-Factoring an Evolutionary Drone Control Model for Use in Immunological Optimization Problems

Kevin Downs and Duncan Coulter

Abstract The work explores the potential for re-purposing an existing evolutionary model, originally intended for the generation of flight control software for low cost search and rescue drones. Freed from real world physical constraints inherent. The re-purposed model is suitable for use in clonal expansion based optimization contexts. An argument is then made for greater adoption of re-usable component oriented evolutionary systems.

Keywords Clonal expansion · Gene expression programming · Drone control

1 Introduction

In this article a discussion of the re-purposing of an existing evolutionary model for application in a different problem domain is presented. Nature and biologically inspired systems are a relatively recent approach to software engineering and as such are prone to repeating some of the mistakes made, and addressed, in other fields of endeavour. To adopt a biological metaphor one could compare the evolution of these approaches to the, now discredited, biological phenomenon of *ontogeny recapitulating phylogeny* in which a developing embryo appears to pass through several phases which recapture the features of its evolutionary past beneficial to its final form or not.

The work presented here details an existing model for the evolutionary generation of flight control software for drone platforms. The structure and generation of the flight control software is discussed as well as the simulation environment in which the fitness of candidate control systems is evaluated. The fitness evaluation of the

K. Downs · D. Coulter (✉)
University of Johannesburg, Johannesburg, South Africa
e-mail: dcoulter@uj.ac.za
URL: http://www.uj.ac.za/

K. Downs
e-mail: 201125187@student.uj.ac.za

© Springer International Publishing Switzerland 2016
N. Pillay et al. (eds.), *Advances in Nature and Biologically Inspired Computing*,
Advances in Intelligent Systems and Computing 419,
DOI 10.1007/978-3-319-27400-3_29

327

control software via simulation is a necessity since such evaluation by any means other than *in silico* would be prohibitively expensive.

The following section outlines the evolutionary approach adopted in general which is then followed by a discussion of the drone system itself. The modification of the model to utilize immunological computation is then discussed by first briefly describing the immune algorithm used followed by a presentation of the new model.

2   Gene Expression Programming

In traditional evolutionary computation approaches to problem solving populations of individuals are created and evaluated according to a function that determines how fit the individual is for the task at hand. The individuals are then chosen based on their fitness and then undergo reproduction. This reproductive phase introduces genetic variation in the children of the selected parent individuals by way of genetic operators [6]. As such this fitness function drives the evolutionary processes which continues until the fitness of an individual reaches an acceptable level and the program terminates.

The principle differences between the approaches to evolutionary computation that exist are found in either the representation of an individual within the population or the reproduction method which is used to introduce genetic variety [1]. In traditional Genetic Algorithms, GAs, these individuals within the population are expressed as linear genotypes, that is, some fixed string representation of the behaviour of the individual. Genetic Programming differs in that the genotype is expressed into a tree structure which is executed. The tree structure represents a candidate solution to the problem and the tree structure is subject to evolution directly. These tree structures are referred to as the chromosomes' phenotype, and the application of genetic operators to these representations is therefore referred to as phenotypic evolution [1]. These genetic operators evolve the phenotype representations of the individual directly.

Gene Expression Programming, GEP, as defined by Ferreira [5], is an approach to evolutionary computation. The principle difference between GEP and other evolutionary computation approaches lies in the representation of the behaviour of an individual and in the approach to evolution. Chromosomes are represented by fixed length strings which, like in GA's, are referred to as the genotype representing that individual. This linear representation allows for easier application of genetic operators such as recombination, transposition, replication and mutation. These genotypes may then be expressed as phenotypes in order to be evaluated. These expression trees are exclusively used for the expression of the behaviour associated with the genotype, and as such no phenotypic evolution takes place [4]. The phenotype is simply evaluated and assigned a fitness.

GEP chromosomes are divided into one or more genes, which themselves are divided into both head and tail sections. The head of the gene is composed of both functions, of a certain arity, and arguments to the functions. The tail of gene is composed of only arguments to the functions present in the head of the gene. This representation mimics open reading frames when considered from a biological perspective. Open reading frames in biology are coding sequences for a gene, where the start of the gene is identified by a starting codon. The gene is then comprised of a number of amino acid codons before ending with a termination codon. In GEP, this start codon is always the start of the gene. The entire symbolic string representing the genotype is therefore considered the open reading frame and in GEP is referred to as a K-expression by Ferreira [4]. The length of these genes is calculated as the sum of the length of the tail, t, and that of the head. The length of the tail of gene is a function of the length of head, h, and the maximum arity of the functions present in the head, n. The length of the tail can therefore be defined as follows in Eq. 1:

$$t = h \times (n - 1) + 1 \tag{1}$$

One such benefit of this representation is that during phenotypic expression, all invalid regions of the genotype are discarded. This eliminates the need to search the potentially infinite problem space consisting of invalid programs. Genotypes are expressed through the creation of expression trees and as such each argument that applies to a function within the head of the gene is placed in a position that corresponds to the arity of that function. Chromosomes may therefore contain some values which do not contribute to the candidate solution that they represent. These expression trees are also guaranteed to be correct with regards to the functions that appear within.

3 Proposed Model

The system presented in this paper uses an approach to evolutionary computation inspired by GEP called multiple-output GEP, moGEP, as defined by Mwaura and Keedwell [7]. MoGEP was developed as a method of creating control systems for robotic behaviour, specifically in problems such as avoidance and guidance. Chromosomes within MoGEP are comprised of a number of genes which are expressed individually as a number of sub expression trees. Other approaches exist, to which MoGEP is similar, such as Cartesian GP, Parisian GP and Multi Expression Programming (MEP). MoGEP differs from these other approaches in the way in which the chromosomes are evaluated. Cartesian GP used both directed graphs, instead of tree-like structures, and also utilizes a separate, potentially identical fitness function that is used to evaluate each gene within the chromosome [7]. MoGEP also differs from Parisian GP in that in MoGEP, chromosomes code for the entire candidate solution to a problem, where in Parisian GP, the chromosome forms a sub-solution to the entire problem [7].

MoGEP has been chosen as the initial model for adaption owing to the contribution provided by each gene within the chromosome to the solution to the problem. Initially, the MoGEP approach focussed on reading the values associated with

Table 1 Table containing arguments for functions appearing in the head of a gene

Function arguments

Symbol	Returned value	True	False
l	Ascent vector	Ascend	Descend
f	Facing angle	Rotate left	Rotate right
k	Velocity	Accelerate	Decelerate
a	X coordinate	–	–
b	Y coordinate	–	–
c	Altitude (Z coordinate)	–	–
d	Power remaining for drone	–	–
e	Structural integrity of the drone	–	–
g	Angle to highest detected temperature	Rotate	–
h	Angle to movement detected	Rotate	–
i	Distance to highest detected temperature	Rotate	–
j	Distance to detected movement	Rotate	–
x	Double constant	Constant	Constant

sensors placed in positions around a simulated robot and applying forces to simulated motors through the use of various kinematic models. Unlike Mwaura and Keedwells system, in this model the state variables representing the drone in 3d space are simply updated directly based on the evaluation of their phenotype representations. As such, drones maintain state about their position within an environment. These maintained state variables include the acceleration of a drone, the facing angle, as well as their ascent vector. Removing the kinematic models required for the simulation of motors ensures that the actions taken by a drone based on the evaluation of its' phenotype are deterministic. These simplifications allow for the problem domain being modelled to be extended to n-dimensions and provide the basis for the adaptions proposed to this system.

Each of the chromosomes within this system are, as in MoGEP, divided into a number of genes, where each gene is itself comprised of both a head an a tail. Genes are, as in GEP, fixed length strings where the head is comprised of both functions and arguments, as defined in Table 1. Functions used within the described system, referred to as I and G and presented in Eqs. 2 and 3 respectively, are both functions with arity 4. I, also referred to as the "If less than or equal" function executes the *true* action of the node that appeared as argument 3 if the sensor value returned from the argument in 1 is less than or equal to the value returned from the argument in position 2. Otherwise, I, executes the *false* action of node 4. G, also referred to as the "If greater than or equal" which executes the *true* action of the node appearing in argument 3 if the sensor value returned from argument 1 is greater than the sensor value returned from argument 2. Otherwise, G performs the *false* action of the node provided as argument 4. The addition of the function to MoGEP has been performed in order to provide for greater genetic diversity with regards to the head portions of genes. This cause is further supported through the use of multiple execution options for a given node within the expression tree. Within the equations presented below, the value of c(*true*) represents the execution of the true branch of the argument represented by c. The same is true for the second case, where d(*false*) represents the execution of the false branch of the argument represented by d.

$$
I(a,b,c,d) = \begin{cases} c(\textit{true}), & \text{if } \textit{value}(a) \leq \textit{value}(b) \\ d(\textit{false}), & \text{otherwise} \end{cases} \tag{2}
$$

$$
G(a,b,c,d) = \begin{cases} c(\textit{true}), & \text{if } \textit{value}(a) \geq \textit{value}(b) \\ d(\textit{false}), & \text{otherwise} \end{cases} \tag{3}
$$

Arguments correspond to both sensor output and an action that is to take place depending on the position in the expression tree at which they are encountered. When an argument is encountered and evaluated within the expression tree for a gene, the value represented by the argument is always returned. This is to ensure that a value tree is executed. In the case of an argument not having an associated action, this ensures that the value is able to be utilized further towards the root of the tree, thereby ensuring that each value contributes to the candidate solution.

Within the arguments presented in Table 1, the entries which appear in the *true* column represent the action which takes place when the functions at the root of 4 nodes within a subtree of the decision tree evaluates to true based on the 1st and 2nd nodes. The entries within the *false* column take place when the function evaluates to false.

Based on the discussed arguments and functions, the example phenotype representation presented below (Fig. 1) codes for behaviour where the drone will rotate to face the angle of movement detected if the distance to the moving object, based

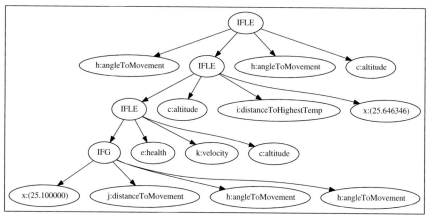

IhIIG(25.1)jhhekccixhcifkgcfab

Fig. 1 Example phenotype representation for the individual, IhIIG(25.1)jhhekccixhcifkgcfab

Fig. 2 Rendered
environment represented as a
3-d heightmap

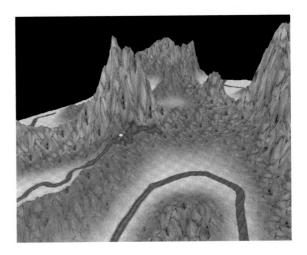

on results from the relevant sensors, is greater than the constant value, 25.1. Otherwise, the drone will also rotate to the angle to movement. The value returned from the function being executed will then be compared to the structural integrity of the drone. If the structural integrity of the drone is greater than the returned value, the drone will increase accelerate, otherwise it will return the ascent vector of the drone. This returned value is then compared with the current altitude, and if it is less, the drone will return the distance to the highest detected temperature. Otherwise the function will return the constant value 25.646346. This value is finally compared to the angle to movement that is detected. If the angle to movement is less than this returned value, the drone will rotate to face this angle, otherwise it will simply return it's altitude to the root of the decision tree.

In this way, the drone's state is updated based on the decisions taken throughout the execution of the various decision trees which represent these genes within the chromosome. The environment within which these chromosomes are evaluated is represented as a 3-dimensional height map in jMonkeyEngine3, an example of which is provided in Fig. 2. Height maps for the proposed system may either be rendered from images or created with use of arrays of values. This allows the height map to be adapted to represent values from a function being modelled.

The individual genes are then evaluated in order, with each evaluation leading to a change in the state maintained by the drone based on the result of the execution of the phenotype structure. The proposed model evaluates the fitness of a given drone within the population based on a number of criterion. This problem is therefore considered to be a multi-objective problem where the criteria used to evaluate each drone is the distance to the object of interest, the fuel used by the drone, the total number of control cycles used, as well as the damage that has occurred to the drone during the simulation. This system has also been designed to be extensible, where the fitness function may simply be adapted given the problem domain being explored. In this way, the fitness could be considered to be a maximum or minimum value thereby allowing for the adaptation to multiple domains.

4 Experimental Setup

In order to evaluate the viability of the initial system to the problem at hand, populations were initialized based on a number of parameters. These parameters included the number of individuals, the number of genes that each individual is comprised of, as well as the length of the head for each gene. These individuals were randomly generated and allowed to evolve according to the canonical GEP algorithm as presented by Ferreira [4]. The replacement strategy of the population for subsequent generations is performed between parents and children, where the weakest parent is always replaced by the strongest child. This takes place by way of a Parent-offspring competition [1]. The object to be located was placed in a fixed position for testing purposes and the algorithm was allowed to progress for a total of 50 generations, after which the average and best fitnesses were recorded. Roulette Wheel Selection was implemented for the selection process among the parents of the next generation [1].

In order to contrast between the various genetic operators so that the best configuration given the problem domain may be found, a number of operators have been implemented. These operators are then configured at the start of each execution of the simulation. In terms of Recombination, both One Point and Two Point Recombination were implemented at a selectable rate. Both Gene and Root Transposition were likewise implemented according to Ferreria's definitions of these operators [4]. Owing to the variety of genetic material within each chromosome representing a solution, mutation is implemented in such a way that each allele within each gene is subject to mutation at a fixed rate. The mutation operator in use examined each allele and randomly selected either a function node, or argument node to replace the value contained within. In cases where floating point valued nodes were subject to mutation, either the value was regenerated or the preexisting value was simply modified by a fixed value, namely ± 0.1. In terms of the fitness function, presented in Eq. 4, the weights were fixed such that $dWeight = 10.0$, $hWeight = 2.0$, $pWeight = 1.0$, and $cWeight = 0.5$. Therefore, more emphasis was placed on the distance to the object being searched for. The problem can therefore be considered to be a minimization problem.

$$fitness(x_i) = (dWeight * distance(x_i, p)) + (hWeight * health(x_i))$$
$$+(pWeight * power(x_i)) + (cWeight * controlCycles(x_i)) \tag{4}$$

5 Results

Results, as presented in Table 2, are classified according to the parameters that were selected for the execution of the algorithm. Let P be the tuple comprised of the parameters for the simulation, such that:

$P = \{nGenes, hLength, mRate, rOp, rRate, tOp, tRate\}$

Table 2 Table containing the results of various configurations utilizing Roulette Wheel Selection over the course of 50 generations of a population comprised of 30 Drones

Results									
Test	nGenes	hLength	mRate	rOp	rRate	tOp	tRate	Best	Avg
1	20	5	0.04	One point	1.00	Gene	0.01	1836.56	3204.14
2	20	5	0.02	One point	1.00	Gene	0.01	1905.25	2204.65
3	20	6	0.02	One point	0.7	Gene	0.01	2456.15	3360.75
4	30	5	0.02	One point	1.00	Gene	0.01	2304.54	2980.5
5	30	5	0.04	One point	1.00	Gene	0.01	1950.15	2845.32
6	30	5	0.02	Two point	1.00	Gene	0.01	1450.31	2745.34
7	30	5	0.04	Two point	1.00	Gene	0.01	2285.31	3047.51
8	30	5	0.02	One point	1.00	Root	0.01	1845.65	2403.54
9	30	6	0.04	One point	1.00	Root	0.01	2206.45	3014.68
10	20	6	0.04	Two point	0.8	Root	0.02	2050.49	3405.95

Where *nGenes* is the number of genes that each drone is comprised of, *hLength* is the length of the head of each gene, *mRate* is the mutation rate, *rOp* is the recombination operator, *rRate* is the rate at which recombination occurs, and finally *tOp* and *tRate* are the transposition operator and transposition rate respectively. Presented within these results are both the average and best fitnesses at the end of 50 generations.

As can be seen within Table 2, the algorithm is tunable in order to favour either exploration or exploitation within the simulation. This is evident in both the average and best fitnesses found in Test 3 compared to other configurations. This configuration, where each chromosome was comprised of 20 genes, with a head length of 6, resulted in poor convergence with the selected operations and rates, instead favouring exploration. This is in comparison with Test 6, presented in Fig. 3, where the best

Fig. 3 Average versus Best Fitness over the course of 50 generations of 30 Drones comprised of 30 genes and a head length of 5, using Roulette Wheel Selection, Uniform Mutation at a rate of 0.02, Two Point Recombination, and Gene Transposition at a rate of 0.01

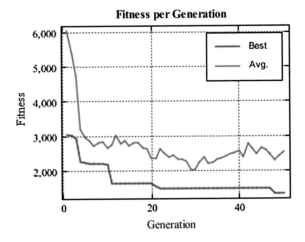

fitness of any configuration is observed. The configuration presented in Test 6 was able to generate the best solution observed throughout the testing process, owing to the increased tendency towards exploitation during the simulation.

6 ImmunoOptiDrone via Clonal Expansion

The adaptive immune response is in truth not a single response but rather a multitude of interrelated responses each dealing with a different sub-problem each governed by differing cells and effector molecules. The problem of self / non-self recognition is mediated by T-cell lymphocytes through negative selection while the problem of providing an optimal response is handled by B-cell lymphocytes via clonal expansion's positive selection [2].

In general clonal expansion is most similar to a mutation-heavy evolutionary algorithm. Figure 4 details the steps involved in clonal expansion which has been adapted to ImmunoOptiDrone. Initially immature B-cells are produced en mass in

Fig. 4 UML Activity Diagram— ImmunoOptiDrone

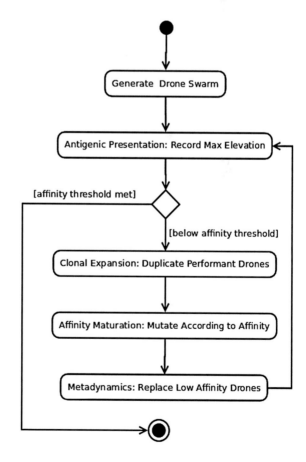

the bone marrow. Each B-Cell then undergoes somatic hyper-mutation in a specialized variable region of its DNA. This region controls which peptide sequence the lymphocyte will bond to i.e. its affinity to a specific antigen. During maturation clonal expansion is used in order to filter out those lymphocytes which have affinity to peptide sequences which are not present and amplify the response of those lymphocytes whose affinity is for those antigens which are present. In order to do so antigen presenting cells provide samples on their surface (via the major histo-compatibility complex molecule) of those peptides which are present within the organism together with danger signals such as those produced when tissues are damaged. Those lymphocytes which happen to bond with the presented antigens are preferentially duplicated via the process of clonal expansion and a limited subset of those clones have their life-spans increased to serve as memory cells and thereby increase the rapidity of future responses to those antigens [3].

In order to adapt the evolutionary drone control mechanism discussed in this model to the task of function optimisation the following observations need to made. Firstly as the new problem domain is purely virtual physical constraints on drones need not apply. The original evolutionary system would have to heavily penalize drone control software which lead to crash however in the new problem domain this is not such a dire consequence. Additionally drones are not limited to their initial number and be dynamically created and destroyed by the system as needed. The second observation is that the fitness function of the evolutionary drone control system can be adapted so that instead of favouring drones which come closest to victims in need of rescue it instead favours those which maintain a fixed distance above the terrain's surface yet achieve a high elevation. In this way the function being optimised can be substituted for the terrain of the environment. The system can then be mapped onto the clonal expansion algorithm by having the fitness function of the underlying GEP algorithm substitute for affinity in an evolutionary algorithm layered on top of it. Individual lymphocytes are then realized by the drones themselves with new instances brought in and out according to their affinity at fixed intervals. Somatic hyper-mutation does not need to be expressly implemented instead the pre-existing mutation function of the underlying GEP can be used.

References

1. Engelbrecht, A.: Computational Intelligence. Wiley, Chichester (2007)
2. Dasgupta, D., Nino, F.: Immunological Computation: Theory and Applications. CRC Press, Boca Raton (2008). ISBN 978-1420065459
3. de Castro, L.N., Timmis, J.: Artificial Immune Systems: A New Computational Intelligence Approach. Springer, London (2002). ISBN 978-1-85233-594-6
4. Ferreira, C.: Gene expression programming: a new adaptive algorithm for solving problems. Complex Syst. 13(2), 87–129 (2001)
5. Ferreira, C.: Gene expression programming in problem solving. Soft Computing and Industry, pp. 635–653. Springer, London (2002)
6. Mitchell, M.: An introduction to genetic algorithms. MIT Press, Cambridge (1998)
7. Mwaura, J., Keedwell, E.: On using gene expression programming to evolve multiple output robot controllers. In: 2014 IEEE International Conference on Evolvable Systems (2014)

A Generative Hyper-Heuristic for Deriving Heuristics for Classical Artificial Intelligence Problems

Nelishia Pillay

Abstract A recent direction of hyper-heuristics is the automated design of intelligent systems with the aim of reducing the man hours needed to implement such systems. One of the design decisions that often has to be made when developing intelligent systems is the low-level construction heuristic to use. These are usually rules of thumb derived based on human intuition. Generally a heuristic is derived for a particular domain. However, according to the no free lunch theorem different low-level heuristics will be effective for different problem instances. Deriving low-level heuristics for problem instances will be time consuming and hence we examine the automatic induction of low-level heuristics using hyper-heuristics. We investigate this for classical artificial intelligence. At the inception of the field of artificial intelligence search methods to solve problems were generally uninformed, such as the depth first and breadth first searches, and did not take any domain specific knowledge into consideration. As the field matured domain specific knowledge in the form of heuristics were used to guide the search, thereby reducing the search space. Search methods using heuristics to guide the search became known as informed searches, such as the best-first search, hill-climbing and the A* algorithm. Heuristics used by these searches are problem specific rules of thumb created by humans. This study investigates the use of a generative hyper-heuristic to derive these heuristics. The hyper-heuristic employs genetic programming to evolve the heuristics. The approach was tested on two classical artificial intelligence problems, namely, the 8-puzzle problem and Towers of Hanoi. The genetic programming system was able to evolve heuristics that produced solutions for 20 8-puzzle problems and 5 instances of Towers of Hanoi. Furthermore, the heuristics induced were able to produce solutions to the instances of the 8-puzzle problem which could not be solved using the A* algorithm with the number of tiles out of place heuristic and at least one admissible heuristic was evolved for all 25 problems.

Keywords Heuristics · Informed search · Generative hyper-heuristic

---

N. Pillay (✉)
School of Mathematics, Statistics and Computer Science,
University of KwaZulu-Natal, Pietermartizburg, South Africa
e-mail: pillayn32@ukzn.ac.za

© Springer International Publishing Switzerland 2016
N. Pillay et al. (eds.), *Advances in Nature and Biologically Inspired Computing*,
Advances in Intelligent Systems and Computing 419,
DOI 10.1007/978-3-319-27400-3_30

1 Introduction

A recent advancement in the field of optimisation is hyper-heuristics. These are methods that select or generate low-level construction and perturbative heuristics [1]. Generative hyper-heuristics create new low-level heuristics comprised of existing low-level heuristics or variables representing problem characteristics. The heuristics evolved can be disposable, i.e. created for a particular instance of a problem, or reusable, i.e. created for one problem instance and applied to other instances. Genetic programming and variations thereof are generally employed by generative hyper-heuristics to create new heuristics [1–4]. Genetic programming [5] is an evolutionary algorithm that explores a program space rather than a solution space by iteratively refining an initial population of programs by means of fitness evaluation, selection and regeneration. Genetic operators commonly used for regeneration are reproduction, mutation and crossover [6]. Each program is represented as a parse tree constructed by selecting elements from the function and terminal sets specified for the application domain. In generative hyper-heuristics each program represents a heuristic.

An emerging research direction of hyper-heuristics is the automated design of intelligent systems to reduce the man hours involved in designing these system. The study presented focuses on the design decision of inducing low-level problem specific construction heuristics which are usually based on human intuition. This study implements and evaluates a generative construction hyper-heuristic (GCHH) for the induction of disposable heuristics for use with informed searches in the domain of artificial intelligence. The heuristics used by informed searches are usually derived by humans. Heuristics enabling a search to produce a solution with a minimum cost are referred to as admissible heuristics [7]. The generative hyper-heuristic was tested on two classical artificial intelligence problems, namely, the 8-puzzle problem and Towers of Hanoi. The GCHH generated heuristics that produced solutions for 20 8-puzzle problem instances of differing difficulty and 5 instances of the Towers of Hanoi problem with 3, 4, 5, 6 and 7 discs. Furthermore, the generated heuristics were able to solve five of the problems which could not be solved using the A* algorithm with the number of tiles out of place heuristic.

The following section describes the generative hyper-heuristic. Section 3 gives details of the experimental setup used to evaluate the hyper-heuristics including an overview of the problem domains. The performance of the heuristics induced by the hyper-heuristic is discussed in Sect. 4. Section 5 provides a summary of the findings of the study and proposes future extensions of this work.

2 Generative Construction Hyper-Heuristic

The generative hyper-heuristic uses genetic programming to explore the space of programs representing heuristics. The generational genetic programming algorithm used is depicted in Algorithm 1.

Algorithm 1 Genetic programming algorithm [5]

Create an initial population
while Termination criterion is not met **do**
    Evaluate the population
    Select parents
    Apply genetic operators to the selected parents
end while

Fig. 1 An example of an individual

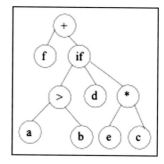

An example of an element of the population, called an individual, is illustrated in Fig. 1. Each individual is created by choosing elements from the function and terminal set. The grow method [5] is used to create each tree. The function set for this study is comprised of arithmetic operators, arithmetic logical operators and the *if* function. The arithmetic operators perform the standard addition (+), subtraction (−) and multiplication (∗) operators. A protected division operator is used which returns a value of 1 if the denominator evaluates to zero. The arithmetic logical operators include <, >, <=, >=, == and ! =. The *if* function takes three arguments and performs the function of an if-then-else statement, i.e. if its first argument evaluates to true the output of its second argument is returned else the value of the third argument is returned. The elements of the terminal set are variables representing characteristics of the problem domain. These are a to f in Fig. 1 and are problem specific.

Each element of the population is evaluated by using it with an informed search to solve the particular problem. The informed search is the best first search using the heuristic function $f(n) = h(n) + g(n)$. $h(n)$ is the heuristic evolved, i.e. the value the individual in the population evaluates to and $g(n)$ is an estimate of the cost of getting from an initial state to the particular state and is the level in the tree representing the state space representation for the problem. Informed searches solve a problem by identifying the steps of getting from the initial state specified for the problem to the goal state. The initial state is at level zero. The heuristic is used to decide which state to go to next, i.e. which move or operator to apply. The state resulting from applying an operator/move to the initial state is at level 1, so $g(n)$ will be 1 at this level. Similarly, the state resulting from application of an operator/move to any of the states at level 1 are at level 2. $h(n)$ is an estimate of the cost from the state to the goal state. If the heuristic $h(n)$ always underestimates the actual cost of getting from a state to the goal it is said to be admissible and the best first search implemented with an admissible heuristic is the A* algorithm [7]. The A* algorithm is guaranteed to find the minimum cost which for the problem domains used in this study is the shortest solution path. The best-first search algorithm used in this study is that presented in Luger et al. [7].

A Pareto fitness function is used by the selection method to choose parents. This involves a vector comparison over n dimensions, with each dimension representing an objective of the problem. In this study two objectives are considered, firstly finding a solution and secondly minimizing the length of the solution path. The tournament selection method is used [5] to choose parents of successive generations. The selection method chooses t individuals randomly from the population and returns the fittest of the t individuals. Each comparison is performed using the Pareto fitness function. This function firstly compares accuracy, if one individual is closer to a solution than another the former is considered to be fitter. However, if both individuals have the same value for accuracy, the length of the "solution" paths are considered. The individual with the shorter "solution" path is treated as fitter.

The mutation and crossover operators are used to create each generation. The mutation operator evokes tournament selection to choose a parent. The mutation operator performs standard parse tree mutation [5]. A mutation point is randomly selected in a copy of the parent and the subtree rooted at this point is replaced with a randomly created subtree. The crossover operator is parse tree crossover defined in [5] and is applied to two parents selected using tournament selection. Crossover points are randomly selected in copies of each of the parents and the subtrees rooted at these points are swapped to create offspring.

The genetic programming algorithm terminates when either an admissible heuristic, i.e. a heuristic producing a solution with the known minimum number of moves is found, or the maximum number of generations specified has been performed

3 Experimental Setup

The generative hyper-heuristic was tested on two classical artificial intelligence problems, namely, the 8-puzzle problem and the Towers of Hanoi [7]. The 8-puzzle problem involves moving 8 tiles on a 3×3 board from timesan initial configuration to a goal configuration. The board contains 8 tiles, numbered 1–8 and a space allowing for the movement of one tile at a time. An example of an initial and goal state is illustrated in Fig. 2. One tile is moved into the space at a time to get from the initial state to the goal state. This can be visualised in terms of moving the space up, down, left or right.

Heuristics that have been defined for this domain are the number of tiles out of place, Manhattan distance and tile reversal. The number of tiles out of place for the initial state in Fig. 2 is 7 as 5 is the only tile in place. There are no tile reversals, i.e. two tiles next to each other vertically or horizontally in the state which are swapped in the goal state. The Mahanttan distance is the sum of the minimum distance of each

Fig. 2 An example of the
8-puzzle problem

Initial State			Goal State		
1	2	3	2	8	1
8		4	4	6	3
7	6	5		7	5

Table 1 8-puzzle problem instances

Problem instance	Initial state	Goal state	Known optimum
Puzzle 1	123804765	134862705	5
Puzzle 2	123804765	281043765	9
Puzzle 3	123804765	281463075	12
Puzzle 4	134805726	123804765	6
Puzzle 5	231708654	123804765	14
Puzzle 6	231804765	123804765	16
Puzzle 7	123804765	231804765	16
Puzzle 8	283104765	123804765	4
Puzzle 9	876105234	123804765	28
Puzzle 10	867254301	123456780	31
Puzzle 11	647850321	123456780	31
Puzzle 12	123804765	567408321	30
Puzzle 13	806547231	012345678	31
Puzzle 14	641302758	012345678	14
Puzzle 15	158327064	012345678	12
Puzzle 16	328451670	012345678	12
Puzzle 17	035428617	012345678	10
Puzzle 18	725310648	012345678	15
Puzzle 19	412087635	123456780	17
Puzzle 20	162573048	123456780	10

tile from its position in the goal state. The Manhattan distance of the initial state in
Fig. 2 is 10. The tile reversal heuristic is usually used with the Manhattan distance
heuristic.

The generative hyper-heuristic has been tested on the twenty problem instances
listed in Table 1. These instances were obtained from previous work and online
assignments.

An example of the Towers of Hanoi problem with 3 discs is illustrated in Fig. 3.

The discs from the first pole must be moved to the third pole without placing
a larger disc on a smaller disc at any point. Only one disc can be moved at a time.
The five problem instances of the Towers of Hanoi problem together with the known
minimum moves for solving these instances are listed in Table 2.

Fig. 3   3 disc Towers of
Hanoi

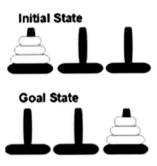

Problem instance	Known optimum
3 discs	7
4 discs	15
5 discs	31
6 discs	63
7 discs	127

Table 2   Towers of Hanoi problem instances

The terminal set for the 8-puzzle problem was comprised of 20 elements. The first 9 elements are variables representing the content of each position on the board in the state that the heuristic is calculated for, e.g. a represents the content at position $(0,0)$. Given the initial state in Fig. 6, a will evaluate to 1. The second 9 elements correspond to the content of each position on the board of the goal state. The remaining two terminals correspond to the row and column index of the space on the board of the state being evaluated.

Six terminals are used for the Towers of Hanoi problem. The first three elements are variables that evaluate to the topmost ring on each of the three poles in the state that the heuristic is being calculated for. The remaining three elements evaluate to the pole that each of the three discs is on in the state that the heuristic is being calculated for.

For both the 8-puzzle problem and Towers of Hanoi the fitness is calculated to be $f(n) = h(n) + g(n)$, with $g(n)$ defined as in Sect. 2. For the 8-puzzle problem $h(n)$ is the Manhattan distance. $h(n)$ for Towers of Hanoi is calculated to be the number of discs on the first and second poles multiplied by two.

The parameter values used for the genetic programming generative hyper-heuristic for the 8-puzzle and Towers of Hanoi problems are listed in Table 3. These values were determined empirically by conducting trial runs.

The genetic programming hyper-heuristic was implemented in Java and all simulations were run on an Intel Core i7 with Windows 7. Due to the stochastic nature of genetic programming, ten runs were performed for each problem instance, each with a different random number generator seed to prevent any distortion in results caused by random noise. The performance of the heuristics induced by the generative hyper-heuristic are discussed in the following section.

Table 3 Parameter values

Parameter	8-puzzle	Towers of Hanoi
Population size	500	200
Tournament size	4	4
Maximum tree depth	10	4
Mutation depth	3	3
Maximum number of generations	50	50
Crossover application rate	90 %	90 %
Mutation application rate	10 %	10 %

4 Results and Discussion

This section examines the performance of the generative hyper-heuristic in inducing heuristics for the 8-puzzle and the Towers of Hanoi problems. Section 4.1 discusses the results for the 8-puzzle problem and Sect. 4.2 for the Towers of Hanoi.

4.1 8-Puzzle Problem

In order to measure the performance of the heuristics induced by the generative hyper-heuristic, the A* algorithm with the Manhattan distance and number of tiles out of place were applied to the 20 problem instances. The results are listed in Table 4. The values listed are the number of moves needed to obtain a solution. A hyphen indicates that a solution could not be found in the maximum allocated time of 5 min. The A* algorithm with the Manhattan distance heuristic was able to find solutions for all 20 problem instances with the known minimum number of moves. The A* algorithm with the number of tiles out of place heuristic did not perform as well and was not able to induce a solution within the 5 min time limit for five of the problems. For the problem instances for which the A* algorithm together with the number of tiles out of place heuristic produced the optimal solution, this was achieved in less than a second. The same runtimes were obtained for the A* algorithm with the Manhattan distance heuristic except for puzzles, 9, 10, 11 and 13 which took 8, 29, 26 and 27 s respectively.

The results obtained by the generative hyper-heuristic are listed in Table 5. The second column lists the number of the 10 runs that evolved heuristics that produced solutions as a percentage. Column 3 presents the percentage of the 10 runs producing admissible heuristics. The last column specifies the average runtime of the runs producing solutions.

Table 4 Results for the A* Algorithm

Problem instance	Manhattan distance	No. of tiles out of place
Puzzle 1	5	5
Puzzle 2	9	9
Puzzle 3	12	12
Puzzle 4	6	6
Puzzle 5	14	14
Puzzle 6	16	16
Puzzle 7	16	16
Puzzle 8	4	-
Puzzle 9	28	-
Puzzle 10	31	-
Puzzle 11	31	-
Puzzle 12	30	-
Puzzle 13	31	-
Puzzle 14	14	14
Puzzle 15	12	12
Puzzle 16	12	12
Puzzle 17	10	10
Puzzle 18	15	15
Puzzle 19	17	17
Puzzle 20	10	10

From Table 5 it is evident that the heuristics induced were able to produce solutions for all 20 problem instances, with at least one of the heuristics produced being admissible. For the more difficult problems, namely, Puzzle 9, 10 and 13 fewer admissible heuristics were produced. Future work will examine this further and look at other mechanisms besides Pareto fitness that can be used to evolve admissible heuristics. Each of the heuristics evolved was disposable and different heuristics were produced for each of the ten runs. It is hypothesised that this is in response to selection noise resulting from the stochastic nature of genetic programming. Each seed produces a different set of random numbers resulting in the search following a different route, thereby generating a different heuristic to solve the problem.

4.2 Towers of Hanoi

The A* algorithm with h(n) defined in Sect. 3 was able to produce optimal solutions for all five problem instances in less than a second. Table 6 lists the percentage runs for which GCHH produced a solution, the percentage runs producing an admissible

Table 5 Results for GCHH

Problem instance	% of successful runs	% of admissible runs	Average runtime (s)
Puzzle 1	100	100	<1 s
Puzzle 2	100	100	<1 s
Puzzle 3	100	100	5.3
Puzzle 4	100	100	<1 s
Puzzle 5	100	100	98.4
Puzzle 6	100	100	2.5
Puzzle 7	1100	100	2
Puzzle 8	100	100	<1 s
Puzzle 9	100	100	20
Puzzle 10	100	20	14
Puzzle 11	60	10	140
Puzzle 12	100	100	2
Puzzle 13	100	1008	42
Puzzle 14	100	100	47
Puzzle 15	100	100	<1 s
Puzzle 16	100	100	1
Puzzle 17	100	100	<1 s
Puzzle 18	100	100	4
Puzzle 19	100	1100	134
Puzzle 20	100	100	<1 s

Table 6 Results for GCHH

Problem instance	% of successful runs	% of admissible runs	Average runtime (s)
3 discs	100	100	< 1 s
4 discs	100	100	< 1 s
5 discs	100	100	1.1
6 discs	100	100	6
7 discs	100	100	< 1 s

heuristic and the average runtime over the 10 runs. The GCHH was able to produce admissible heuristics for all 10 runs for each problem instance. As in the case of the 8-puzzle problems different heuristics were generated on each of the runs for each problem instance.

5 Conclusion

This study forms part of an initiative using hyper-heuristics for automated design and has focussed on the automatic induction of domain specific heuristics to be used by artificial intelligence search methods. The hyper-heuristic implemented used genetic programming to evolve heuristics for the 8-puzzle and the Towers of Hanoi problems. The function set consisted of standard arithmetic, arithmetic logic and conditional statements and the terminal set variables representing different characteristics/aspects of the problem domain. The heuristics generated were able to produce solutions to all 25 problems and at least one of the evolved heuristics was admissible for each problem. The heuristics induced were also able to produce solutions to five problems that the A* algorithm using the number of tiles out place heuristic was not successful in solving. For three of the problem instances for the 8-puzzle problem the number of admissible heuristics produced was lower than that obtained for the other problem instances and reasons for this together with mechanisms for improving this will be studied as future extensions of this work. Future research will also focus on applying the use of generative hyper-heuristics in other domains that rely on the use of heuristics to find solutions, specifically combinatorial optimization problems such as educational timetabling, travelling salesman and nurse rostering, which use a human derived heuristic to create an initial solution which is further optimized using optimization techniques such as metaheuristics.

References

1. Burke, E.K., Gendreau, M., Hyde, M., Kendall, G., Ochoa, G., Ozcan, E., Qu, R.: Hyper-heuristics: a survey of the state of the art. J. Oper. Res. Soc. **64**, 1695–1724 (2013)
2. Burke, E.K., Hyde, M., Kendall, G., Ochoa, G., Ozcan, E., Woodard, J.: Exploring hyper-heuristic methodologies with genetic programming. Comput. Intell. **6**, 177–201 (2009)
3. Sabar, N., Ayob, M., Kendall, G., Qu, R.: A dynamic multiarmed bandit-gene expression programming hyper-heuristic for combinatorial optimization problems. IEEE Trans. Cybern. **45**(2), 217–228 (2015)
4. Sosa A., Ochoa G., Terashima-Marín, Conant-Pablos, S.E.: Grammar-based generation of variable-selection heuristics for constraint satisfaction problems. Genetic Programming and Evolvable Hardware, September 2015, doi:10.1007/s10710-015-9249-1 (2015)
5. Koza, J.R.: Genetic Programming: On the Programming of Computers by Means of Natural Selection. MIT Press, Cambridge (1992)
6. Banzhaf, W., Nordin, P., Keller, R.E., Francone, F.D.: Genetic Programming—An Introduction—On the Automatic Evolution of Computer Programs and Its Applications. Morgan Kaufmann Publishers, Inc. (1998)
7. Luger, G.F., Stubblefield, W.: Artificial Intelligence: Structures and Strategies for Complex Problem Solving. Addison-Wesley Longman (1998)

A Priority Rate-Based Routing Protocol for Wireless Multimedia Sensor Networks

Loini Tshiningayamwe, Guy-Alain Lusilao-Zodi and Mqhele E. Dlodlo

Abstract The development of sensor hardware have made it possible to transmit real time multimedia data over a wireless medium using tiny resource constrained sensors. However, in current wireless sensor networks, multimedia traffic which has stringent bandwidth and delay requirements is not distinctively differentiated from other data types during transmission which makes it difficult to meet its service requirements. Next generation wireless sensor networks are predicted to deploy a different model where service is allocated depending on the nature of data to be transmitted. Applying traditional wireless sensor routing algorithms to wireless multimedia sensor networks may lead to high delay and poor visual quality for multimedia applications. In this paper, we propose a priority based rate routing protocol that assigns priorities to traffic depending on their service requirements. We study, the performance of our proposed routing algorithm for real time traffic when mixed with three non real time traffic but with different priorities: high, medium and low priority. Initial results from the simulation show that the proposed algorithm performs better compared to two existing algorithms PCCP and CCF in terms of delay, loss and throughput.

Keywords Wireless sensor network · Multimedia · Priority · Routing

L. Tshiningayamwe (✉) · M.E. Dlodlo
Department of Electrical Engineering, University of Cape Town,
Rondebosch 7701, South-Africa
e-mail: tshloi001@uct.ac.za

M.E. Dlodlo
e-mail: Mqhele.Dlodlo@uct.ac.za

G.-A. Lusilao-Zodi
Department of Computer Science, Polytechnic of Namibia, Windhoek, Namibia
e-mail: gzodi@polytechnic.edu.na

© Springer International Publishing Switzerland 2016
N. Pillay et al. (eds.), *Advances in Nature and Biologically Inspired Computing*,
Advances in Intelligent Systems and Computing 419,
DOI 10.1007/978-3-319-27400-3_31

347

1 Introduction

A wireless multimedia sensor network (WMSN) consists of small sensor nodes used to monitor certain physical phenomena [1]. The sensor nodes are equipped with cameras and microphones that cancapture audio, videos, images and scalar data. Wireless sensor nodes suffer from device specific challenges such as low memory, low battery life and low power consumption. The nodes are connected to a sink that serves as a base station.

Due to the fact that sensed data travels almost always upstream, congestion appears more in the upstream direction. Two types of congestion exist in WMSN. Link level congestion occurs when more than one wireless node using carrier sense multiple access (CSMA-like) protocols tries to access the transmission medium at the same time resulting in collisions. The second type of congestion occurs as a result of buffer overflow and is called node level congestion. Buffer overflow happen when the incoming rate is higher than the rate responsible for forwarding a packet out of the buffer. Both types of congestion impacts the scarce resources and quality of service (QoS) of the nodes [2].

A number of routing protocols [3–5] were proposed in the past decade to address the issue by adjusting the transmission rate of nodes so as to alleviate congestion. Most of these routing algorithms are built on the assumption that the data collected by sensor nodes are all of the same type and require a similar service from the network. Wireless multimedia sensor networks, however relies heavily on multimedia content; and multimedia content has different characteristics. Multimedia is often divided into three categories, namely streaming of stored multimedia content, streaming of live multimedia and real time interactive applications [6]. Real time applications have got unique characteristics such as bounded delay, delay variation, packet loss, and guaranteed throughput with limited fluctuations. Thus, data collected by the sensors in WMSN differ in terms of their importance. Real time data is considered to be the most important. This is because real time data has strict service requirements and is therefore considered to have the highest priority. Deploying single service routing in wireless multimedia sensor networks may lead to high delay and poor visual quality for delay sensitive multimedia applications.

In this paper, we propose a Priority based Rate Routing Protocol (PRRP) that distributes network resources in a way that real time data are transmitted with very little delay as opposed to other traffic types. The proposed PRRP protocol divides the traffic collected by the sensor nodes into four categories based on their service requirements: real time high priority (RT-HP), non real time high priority (NRT-HP), medium priority non real time (NRT-MP) and low priority non real time (NRT-LP). High priority traffic refers to reliable and delay sensitive data [5]. Medium and low priority traffic are not delay sensitive and can tolerate loss. The priorities of the traffic may vary and are set depending on the network applications. While most of the existing protocols use packet service time [3] and Random Early Detection (RED) based techniques for [5] for congestion detection, an Adaptive Random Early Detection (A-RED) queue management scheme is used in PRRP to detect the congestion at nodes. A-RED has the advantage over RED of automatically tuning its parameters depending on the application behaviour; this helps reducing the possibility of buffer overflows, packet loss and retransmission. Upon detecting the congestion, new rates are calculated for each traffic in order to ensure that the service requirements of each traffic is met. We study, the performance of the proposed routing algorithm for real time traffic when competing for network resources with low, medium and high priority non real time traffic. Initial results from the simulation indicate that the proposed algorithm performs better compared to two existing algorithms PCCP and CCF in terms of delay, loss and throughput.

The remainder of the paper is organized as follow. Section 1 provides a review of related work followed by the proposed PRRP system design in Sect. 3. The next section provides detailed information of the PRRP rate adaptation. The results of PRRP are presented and discussed in Sect. 4. A conclusion of the paper is covered in Sect. 5.

2 Related Work

Congestion impacts the efficiency and performance of application in WMSN. It is for this reason that congestion control techniques are developed to alleviate the effects of congestion. In recent years, several congestion control protocols have been proposed for WMSN [7–9]. These protocols can be divided into four categories based on how they detect the presence of congestion and react to it. The categories include: (1) Queue assisted protocols, (2) Priority aware protocols, (3) Resource control protocols and (4) Content aware protocols. A general congestion control protocol consists of three units, namely: congestion detection, congestion notification and rate adaptation. The congestion detection unit is responsible with detecting the presence of congestion in the network. Several methods of detecting congestion proposed so far includes packet service time and inter arrival time [3, 4], queue length [10, 11], channel load [12] or a combination of these parameters [13]. Congestion notification unit is concerned with disseminating congestion related information once congestion has been detected. Two congestion notification techniques are often used, which are implicit and explicit notification. Implicit notification piggybacks congestion information in a packet header while explicit notification make use of a special extra packet to carry congestion information to other connected nodes.

Queue based protocols are concerned with keeping queue levels at the sensor nodes low. Examples of queue based protocols are [9, 10, 12]. Congestion detection and avoidance (CODA) [12] is an energy efficient congestion control protocol for wireless sensor networks (WSN). Buffer occupancy and channel load are used to measure congestion in CODA. Once congestion is detected, broadcast backpressure messages are sent towards the source nodes in order for them to adjust their rates. Upon receiving the backpressure message, a node can decide whether or not to send this message further to its neighbour nodes. Similar to the traditional transmission control protocol (TCP), additive increase multiplicative decrease (AIMD) is used to adjust rates which often lead to packet loss which is undesirable in WSN.

Fusion [11] detects congestion by using queue length. Three techniques, namely hop by hop flow control, prioritized MAC and rate limiting that operates at different layers of the protocol stack are integrated as a way to control congestion. Hop by hop flow control deals with congestion detection by monitoring the queue size of the nodes. In addition to this, it also deals with controlling congestion after it has been detected. This is done by setting a congestion bit in the header of every packet being sent from the congested node. Due to the broadcast nature of wireless networks, other nodes in the network will overhear the congestion bit and will not transmit to it. Rate limiting is concerned with lowering the transmission rates of sensor nodes while the prioritized MAC uses back off techniques in order to allow congested node to transmit their data. Despite the fact that fusion achieves good throughput and fairness, its non-smooth rate adjustment harm link utilization and fairness.

Compared to queue based protocols, priority based protocols assigns priorities to sensor nodes to ensure that certain nodes receive priority dependent throughput. PCCP [7] is a priority based protocol which addresses both link level and node level congestion in single path and multipath network. PCCP uses packet interarrival time and packet service time to infer congestion. One of the known weaknesses of PCCP is that during low congestion, PCCP increases the transmission rate for all transmitting nodes regardless of their priorities.

Another example of a priority based protocol is PBRC-SD [14] which detects congestion by observing the queue length of the node. PBRC-SD weakness is derived from the RED strategy which uses static threshold values to monitor the behaviour of the queue. When static threshold values are used, the behaviour of the queue becomes too sensitive to the congestion level making it difficult to predict parameters such as the delay and throughput. CCF [4] uses packet service time to detect the presence of congestion in the network. It uses implicit notification to notify other nodes about the presence of congestion. CCF mostly focuses on achieving fairness among the sensor nodes. A known weakness of CCF is low utilization of the available capacity when certain nodes do not have any data to send or when they become idle for a long period.

A different type of congestion control protocol was proposed in [16]. This protocol considers the content of the multimedia during congestion whereby packets consisting of I-frames do not get dropped during congestion. This is because the I-frame is considered to be the most important frames in multimedia content and a loss of such packets have a big impact on the quality of the video.

Motivated by the limitations of PBRC-SD [14], an adaptive priority based routing protocol for WMSN has been proposed. Unlike in PBRC-SD, the threshold values used in congestion detection are not predetermined but are rather being automatically calculated depending on the network status.

Fig. 1 Network model

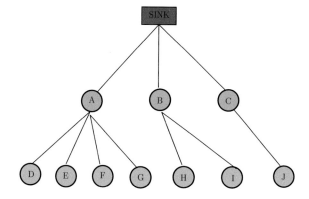

3 Proposed PRRP System Design

3.1 Network Model

The network model considered in this study is formulated as a set of sensor nodes $\mathcal{N}$ and wireless links $\mathcal{L}$. Every sensor node is capable of sensing multimedia content from the environment and route it to the sink node S along a fixed route $l \in \mathcal{L}$ as depicted in Fig. 1. In the diagram, node D till node J are called children nodes while node A, B and C are called parent nodes. A node can serve as both a child and a parent node. Every node can transmit two types of traffic, the source traffic that has been locally generated by a node and transit traffic which has been generated by children node and is routed through a parent node. For example, node B receives transit traffic from sensor node H and I and at the same time may generate its own traffic which is referred to as source traffic. The network model considered in this paper is shown below.

3.2 Node Queueing Model

Figure 2 below presents the queuing model considered at each node. When traffic arrives at every child and parent node, it goes through a classifier were the traffic is separated depending on its service requirements. This means that high priority real time traffic, high priority non real time traffic, medium priority non real time and low priority non real time are all send to their own separate queues. From the queue, every traffic is send to a scheduler. In our protocol, weighted round robin is used for scheduling. This means that every queue is assigned a weight and therefore queues are serviced based on their weights. The queue with the higher weight, in this case the high priority real time queue gets the highest bandwidth allocation while the low priority non real time traffic gets the lowest bandwidth.

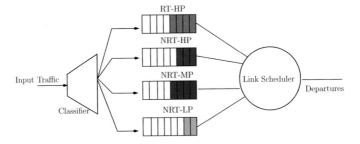

Fig. 2 Node queueing model

3.3 Congestion Detection and Notification

Similar to existing models, PRRP consist of three (3) components, namely congestion detection unit, congestion notification unit and rate adaptation unit.The congestion detection unit is responsible for monitoring the amount of traffic in the queues
of every node in the network. A technique similar to the RED active queue management is used to indicate congestion in PRRP. Thus, for each queue k of a sensor node
i, two values the minimum threshold $(min_{th}^k(i))$ and maximum threshold $(max_{th}^k(i))$
are used to estimate the congestion level of the queue. Assuming $q^k(i)$ is the queue
size for traffic class k in node i, the congestion indicator $I_X^k(i)$ for the queue is defined
as follows.

$$I_X^k(i) = \begin{cases} 0, & q^k(i) \leq min_{th}^k(i) \\ \frac{q^k(i)-min_{th}^k(i)}{max_{th}^k(i)-min_{th}^k(i)}, & min_{th}^k(i) \leq q^k(i) \leq max_{th}^k(i) \\ 1, & q_k(i) \geq max_{th}^k(i) \end{cases} \tag{1}$$

When the queue length is less than $min_{th}^k(i)$, then there is no congestion and the
congestion index $I^k(i)$ is set to zero, so there is no need for a child node to decrease
its transmission. On the other hand, when the queue length is greater then $max_{th}^k(i)$,
then there is severe congestion and the congestion index is set to 1. In this case, a
child should decrease its transmission rate to reduce the probability of packet loss
at the parent node. When the queue length is between the $min_{th}^k(i)$ and $max_{th}^k(i)$ then
$I^k(i)$ will vary between 0 and 1, depending on the queue length. Unlike in RED active
queue management, the values for $min_{th}^k(i)$ and $max_{th}^k(i)$ are not fixed. They are instead
calculated automatically depending on the transmission rate of each traffic and the
current buffer occupancy as shown in Eqs. 2 and 3.

$$min_{th}^{k}(i) = \frac{min\left(\lambda^{k}(i)\right)}{\left(min\,\lambda^{k}(i) + max\,\lambda^{k}(i)\right)} \times q^{k}(i) \tag{2}$$

$$max_{th}^{k}(i) = \frac{max\left(\lambda^{k}(i)\right)}{\left(min\,\lambda^{k}(i) + max\,\lambda^{k}(i)\right)} \times q^{k}(i) \tag{3}$$

where $\lambda^{k}(i)$ refers to the input rate of class k to node i and $q^{k}(i)$ the queue length at queue k of node i. To reduce the effect of abrupt changes of the transmission rate have on the overall calculated threshold values, we average both threshold values using the exponential weighted moving average with parameter 0.9.

The congestion notification component is responsible for informing other nodes about the state of congestion in the network. Once congestion has occurred,other nodes need to be notified about the occurence of congestion so that they can perform rate adjustment. In our study, we make use of implicit notification to reduce overhead observed when using explicitt of congestion notification. The rate controller is activated depending on the congestion level.

3.4 Proposed PRRP Rate Adaptation

The PRRP protocol is motivated by the limitations found in existing widely accepted protocols. In [8], it is clearly indicated that during low congestion, the PCCP algorithm will increase the scheduling rate and source rate of all traffic sources without paying attention to their priority index. PCCP also decrease the sending rate of all traffic sources based on their priority index in the case of high congestion. Instead in our algorithm, exact rate adjustment is performed at each node. Like every congestion control technique, PRRP consist of congestion detection, congestion notification and rate controller. The system design of PRRP is shown in Fig. 4 below.

The congestion index of a parent sensor node i is used to asign and adjust the rate of each of its child nodes as well as the rate of local traffic generated by itself. Assume a parent sensor node i has N_i child nodes. Then, it has $N_i + 1$ queues, N_i queues for its child nodes and one queue for its local traffic source. For each queue k in sensor node i, the congestion index $I^{k}(i)_X, k \in 1, 2, \ldots, N_i + 1$ is computed using Eq. 1. Then the average $\bar{I}_{w}^{k}(i)$ is obtained as follows.

$$\bar{I}_{w}^{k}(i) = \frac{\sum_{j=1}^{N_i+1} I_{X}^{j}(i) - I_{X}^{k}(i)}{N_i \sum_{j=1}^{N_i+1} I_{X}^{j}(i)} \tag{4}$$

Since each node i has its own priority. We denote $SP(i)$ as the source priority at sensor node i. We define the total priority, $TP(i)$, as the sum of priorities of all nodes in the subtree rooted at node i. Let $C(i)$ be the set of node is child nodes. Then the total priority, $TP(i)$, is calculated as:

$$TP(i) = \sum_{j \in C(i)} TP(j) + SP(i) \tag{5}$$

In each queue k in node i, the weight w_i^k and the input rate r_i^k are calculated as:

$$w_i^k = \frac{p_i^k \bar{I}_X^k(i)}{\sum_{j=1}^{N_i+1} p_i^j \bar{I}_X^j(i)} \tag{6}$$

$$r_i^k = w_i^k \times r_i \tag{7}$$

where p_i^k is obtained as follows. $p_i^1 = \frac{SP(i)}{TP(i)}, p_i^k = \frac{TP(k)}{TP(i)}, k \in \{2, 3, \ldots, N_i + 1\}$. Note that the rate r_i is calculated for all active sources. The congestion index is set to infinity for an inactive sensor node, yielding to a transmission rate equal to zero. Thus, only sensor node i, the output rate r_i is shared only between active traffic.

In an event of congestion,active low priority traffic may suffer since PRRP will attempt to provide more resources to high priority traffic, that is RT-HP and NRT-HP. To guarantee a certain level of fairness, we define a minimum sending rate value for all active traffic below which the sensor nodes will not reduce the transmitting rate. This way the bandwidth is not only occupied by a few traffic while others are starving. If an active traffic reaches the minimum sending rate, it cannot decrease its sending rate any more. In our experiment, the minimum sending rate is set to 0.1, however a better value can be obtained from experiments.

4 Simulation and Results

The evaluation of the PRRP protocol is done using a packet event-oriented simulation written in C++. The simulator is applied to the network model presented in Fig. 1. The exponential and constant bit rate traffic is chosen in order to test the throughput, delay and loss of the proposed model PRRP protocol in comparison to two existing models, namely PCCP and CCF.The simulation lasted 60 s and the results are presented below.

4.1 Throughput and Delay for Constant Bit Rate Traffic

In Fig. 3, sensor node 3 was switched off between 30 and 70 s. As it can be seen in the Fig. 3, at 30 s, the throughput for CCF dropped and it rises again at 70 s.The drop is due to the fact that CCF cannot effectively allocate the remaining capacity and use work conservation scheduling algorithm.However, during the same period, PCCP and PRRP has got high throughput.

Figure 4 shows the delay for PRRP, CCF and PCCP. From the results it is observed that the proposed PRRP protocol achieves lower delay than CCF and PCCP. This is

Fig. 3 Throughput for fixed traffic

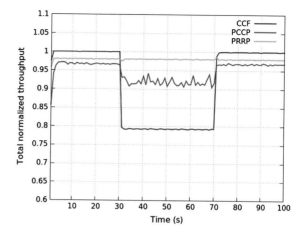

Fig. 4 Delay

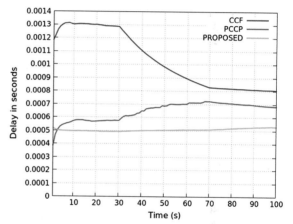

because in addition to service priority provided to real-time traffic, the RED queue can handle delay better than the FIFO queue which is implemented at each queue of the nodes.

4.2 Throughput and Delay for Exponential Traffic

The same simulation was run using exponential traffic and the results are shown in Figs. 5 and 6. As it can be seen in Fig. 5, the proposed protocol and PCCP achieves a higher throughput compared to CCF. CCF uses packet service time to detect congestion and therefore the rate adjustment is based on packet service time. This leads to incorrect allocation of rates among child nodes and thus the drop in throughput between 30 and 70 s in CCF. PCCP solves this by setting thresholds which are auto

Fig. 5 Network throughput

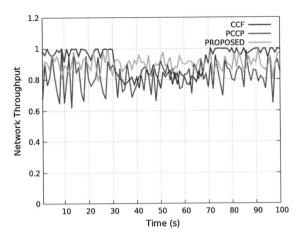

Fig. 6 Delay

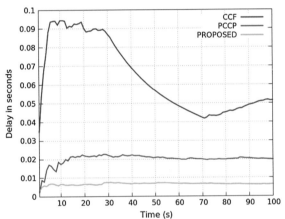

tuned depending on the current network load which results in higher throughput. It can also be seen that PCCP and proposed were able to correctly share the available capacity.

Figure 6 reveals that proposed and PCCP has lower delay while CCF has higher delay. This is attributed to the fact that CCF is a non-work conserving protocol therefore nodes need to wait to be attended to. Figure 5 shows that proposed has a lower loss probability when compared with CCF and PCCP. The queue size in PCCP is often kept short and therefore there is no packet loss experienced due to buffer overflow.

5 Conclusion and Future Work

In this paper, we proposed an upstream congestion control protocol for WMSN called PRRP. PRRP detects congestion by using adaptive RED mechanism and a rate controller is implemented that will give higher rates to real time traffic in the event of congestion.We have demonstrated through simulation results that PRRP achieves higher throughput depending on traffic priority and lower packet losses.Future work will investigate the performance of the rate adaptation in a multipath network and we will also implement an energy efficient protocol that will check the energy level of a node before it can transmit traffic.

References

1. Safai, F., Mahzoon, H., Talebi, M.S.: A simple-priority based scheme. Int. J. Wireless Mobile **4**(1), 165–175 (2012)
2. Lin, Q., Wang, R., Guo, J., Sun, L.: Novel congestion control approach in wireless multimedia sensor networks. J. China Univ. Posts Telecommun. **18**(2), 1–8 (2011)
3. Wang, C., Li, B., Sohraby, K., Daneshmand, M., Hu, Y.: Upstream congestion control in wireless sensor networks through cross-layer optimization. IEEE J. Sel. Areas Commun. **25**(4), 786–795 (2007)
4. Ee, C.T., Bajcsy, R.: Congestion control and fairness for many-to-one routing in sensor networks. In: ACM SenSys, pp. 148–161 (2004)
5. Yaghmaee, M.H., Adjeroh, D.A.: Priority-based rate control for service differentiation and congestion control in wireless multimedia sensor networks. Comput. Netw. **53**(11), 1798–1811 (2009)
6. Rakocevic, V.: Congestion control for multimedia applications in the wireless Internet. Int. J. Commun. Syst. **17**, 723–734 (2004)
7. Wang, C., Li, B., Sohraby, K., Daneshmand, M., Hu, Y.: Upstream congestion control in wireless sensor networks through cross-layer optimization. Sel. Areas Commun. IEEE J. **25**(4), 786–795 (2007)
8. Ee, C.T., Bajcsy, R.: Congestion control and fairness for many-to-one routing in sensor networks. In: Proceedings of the 2nd International Conference on Embedded Networked Sensor Systems, pp. 148–161 (2004)
9. IFRC 2006.pdf
10. Wan, C.-Y., Eisenman, S.B., Campbell, A.T., Crowcroft, J.: Siphon: overload traffic management using multi-radio virtual sinks in sensor networks. In: Proceedings of the 3rd International Conference on Embedded Networked Sensor Systems, pp. 116–129 (2005)
11. Hull, B., Jamieson, K., Balakrishnan, H.: Mitigating Congestion in Wireless Sensor Networks, pp. 134–147 (2004)
12. Wan, C.-Y., Eisenman, S.B., Campbell, A.T.: CODA: congestion detection and avoidance in sensor networks. In: Proceedings of the 1st International Conference on Embedded Networked Sensor Systems, pp. 266–279 (2003)
13. Nath, B.: TARA: Topology-aware resource adaptation to alleviate congestion in sensor Networks. IEEE Trans. Parallel Distrib. Syst. **18**(7), 919–931 (2007)
14. Yaghmaee, M., Adjeroh, D.: A new priority based congestion control protocol for wireless multimedia sensor networks. In: International Symposium on a World of Wireless, Mobile and Multimedia Networks, 2008. WoWMoM 2008, p. 18 (2008)

15. Sergiou, C., Vassiliou, V., Paphitis, A.: Hierarchical Tree Alternative Path (HTAP) algorithm for congestion control in wireless sensor networks. Ad Hoc Netw. **11**(1), 257–272 (2013)

16. Mahdizadeh Aghdam, S., Khansari, M., Rabiee, H.R., Salehi, M.: WCCP: A congestion control protocol for wireless multimedia communication in sensor networks. Ad Hoc Netw. **13**(Part B), 516–534 (2014)

17. Wan, C.-Y., Eisenman, S.B., Campbell, A.T.: Congestion detection and avoidance in sensor networks. In: Proceedings of the 1st International Conference on Embedded Networked Sensor Systems, pp. 266–279 (2003)

18. Wan, C.-Y., Eisenman, S.B., Campbell, A.T., Crowcroft, J.: Siphon: overload traffic management using multi-radio virtual sinks in sensor networks. In: ACM Proceedings of the 3rd International Conference on Embedded Networked Sensor Systems, pp. 116–129 (2005)

19. Yer, Y.G., Gandham, S., Venkatesan, S.: STCP: a generic transport layer protocol for wireless sensor networks. In: Proceedings of 14th International Conference on Computer Communications and Networks, pp. 449–454, Oct 2005

20. Akyildiz, I.F., Vuran, M.C., Akan, O.B.: A cross-layer protocol for wireless sensor networks. In: 40th Annual Conference on Information Sciences and Systems, pp. 1102–1107, March 2006

21. Lee, D., Chung, K.: Adaptive duty-cycle based congestion control for home automation networks. IEEE Trans. Consum. Electron. **56**(1), 42–47 (2010)

22. Aghdama, S.M., Khansarib, M., Rabieec, H.R., Saleh, M.: WCCP: A congestion control protocol for wireless multimedia communication in sensor networks. Ad Hoc Netw. **13**, 516–534 (2014)

Newcastle Disease Virus Clustering Based on Swarm Rapid Centroid Estimation

Fatma Helmy Ismail, Ahmed Fouad Ali, Saleh Esmat
and Aboul Ella Hassanien

Abstract Newcastle disease is considered to be one of the most important serious infectious poultry disease. The work introduced in this paper addresses the problem of clustering Newcastle disease dataset obtained from the National Center for Biotechnology Information GenBank (NCBI). A lightweight swarm clustering algorithm called Rapid Centroid Estimation (RCE) is applied in the clustering task. However, the best number of clusters is selected using silhouette measure. Hence, RCE is compared with K-means for the same number of clusters. The experiment shows that the external quality measures (purity, entropy, rand index, precision, recall and F-measure) of the RCE clustering technique outperform the ones of the K-means.

Keywords Particle swarm clustering · Clustering · K-means · RCE · Swarm optimization

1 Introduction

The main aim of clustering is to define parts of the data that has high degree of similarity with other parts of the data. The similar parts are grouped together as clusters. Similarity can measured in terms of Euclidean distance and other measures

F.H. Ismail (✉)
Faculty of Computer Science, Member of Scientific Research Group in Egypt,
Misr International University, Cairo, Egypt

A.F. Ali
Faculty of Computers and Information, Member of Scientific Research Group in Egypt,
Suez Canal University, Ismailia, Egypt

S. Esmat
Faculty of Veterinary Medicine, Member of Scientific Research Group in Egypt,
Cairo University, Giza, Egypt

A.E. Hassanien
Faculty of Computers and Information, Chair of Scientific Research Group in Egypt,
Cairo University, Giza, Egypt

© Springer International Publishing Switzerland 2016
N. Pillay et al. (eds.), *Advances in Nature and Biologically Inspired Computing*,
Advances in Intelligent Systems and Computing 419,
DOI 10.1007/978-3-319-27400-3_32

[1]. Particle Swarm Optimization (PSO) is a recent stochastic optimization approach proposed by Kennedy and Eberhant in 1995 [2]. PSO suffers from stagnation when particles prematurely converge on particular regions of the potential solution space. An approach called Regrouping Particle Swarm Optimization (RegPSO) was proposed by Evers and Ghalia to overcome stagnation problem [3]. Clustering data using Particle Swarm Optimization (PSC) was proposed by Cohen and de Castro in 2006 [4]. In their experiments on benchmark datasets, they reported that PSC is superior to K-means. A Modified PSC was proposed by Szabo in 2010 [5]. MPSC eliminates the need for initial weight and velocity during the update procedure. It reduced the computation time but it still suffers from a long optimization time. In their work, Mitchell and Steven introduced a lightweight modification for MPSC called Rapid Centroid Estimation (RCE) [6]. The modification has significantly reduced the time complexity and the efficiency of each position update. The work introduced in this paper is a comparison between RCE and K-means clustering algorithms.

The rest of this paper is organized as follow. Section 2 demonstrates the main clustering concepts and the details of RCE algorithm. Section 3 shows an explanation of the system model. Section 4 introduces the clustering evaluation measures. Section 5 demonstrates the experiment results. The conclusion and the future work are discussed in Sect. 6.

2 Rapid Centroid Estimation (RCE)

Clustering is a division of data into groups (clusters) of similar objects. Each cluster consists of objects that are similar between themselves and dissimilar to objects of other cluster. Rapid centroid estimation algorithm (RCE) [6] is one of the clustering algorithms based on swarm optimization. The main RCE steps are presented in Algorithm 1 and Fig. 1.

The main steps of the RCE algorithm sre summarized as follow.

- **Step 1**. RCE algorithm starts by setting the initial parameters of it such as number of particles n_c, maximum number of iterations, maximum number of stagnation number s_{max} and stagnation threshold ε.
- **Step 2**. The iteration counter is initialized and initial population of article is generated randomly.
- **Step 3**. The personal best position p and the best global position g are assigned and the Euclidean distance for each particle and datum is calculated. The distance matrix, each particle position and the data position are updated.
- **Step 4**. The best position is assigned by calculating the global minimum an the process is stopped when a long stagnation is detected.
- **Step 5**. The Euclidian distance between each particle and an input pattern is calculated
- **Step 6**. The inertia weight is updated
- **Step 7**. The process is repeated until algorithm reaches to maximum number of iterations and maximum number of stagnation.
- **Step 8**. Produce the best found solution so far.

Algorithm 1 Rapid Centroid Estimation

1: Set the initial value of nc, max_{itr}, s_{max} and ϵ.
2: Set $t := 0$.
3: **for** $i = 1 : nc$ **do**
4:     Generate the initial population $x_i^{(t)}$ randomly.
5:     Calculate the distance of p, g for each particle and each datum.
6: **end for**
7: **while** $(t < max_{itr} \&\& s_c < s_{max})$ **do**
8:     Set $t := t + 1$.
9:     Update distance matrix, $D(t) = d(y_j, x_i) \ \forall i, j$
10:     For each particle, Find the closest data point, $\lfloor Dx^{min}(t) Ix^{min} \rfloor = min(D, i)$
11:     For each data point, Find the closest particle, $\lfloor Dy^{min}(t) Iy^{min} \rfloor = min(D, j)$
12:     **if** $Dx_i^{min}(t) < Dx_i^{min}(t - 1)$ **then**
13:         $p_i(t + 1) = y_{Ix^{min}}(t)$
14:     **else**
15:         $p_i(t + 1) = p_i(t)$.
16:     **end if**
17:     **if** $Dy_j^{min}(t) < Dy_j^{min}(t - 1)$ **then**
18:         $g_j(t + 1) = y_{Iy^{min}}(t)$
19:     **else**
20:         $g_j(t + 1) = g_j(t)$.
21:     **end if**
22:     **if** $f(x_i(t) < f(M(t))$ **then**
23:         $M(t + 1) = x(t)$
24:     **else**
25:         $M(t + 1) = M(t)$.
26:     **end if**
27:     **if** $f(M(t + 1) - f(M(t) > -\epsilon)$ **then**
28:         $s_c = s_c + 1$
29:     **else**
30:         $s_c = 0$.
31:     **end if**
32:     **for** $i = 1 : nc$ **do**
33:         Calculate the Euclidean distance between each particle and an input pattern $d(p_i(t) - x_i(t))$
34:         Find the winning particle $x_{most_w in}(t) = x(t)$
35:         $y_i^{cluster} = \forall y \in x_i(t)$
36:         $N_i = size(y_i^{cluster})$
37:     **end for**
38:     **if** $(N_i > 0)$ **then**
39:         $x_i(t + 1) = x_i(t) + \Delta x_i(t + 1)$
40:     **else**
41:         $x_i(t + 1) = x_i(t) + \phi \otimes (x_{most_w in}(t) - x_i(t))$
42:     **end if**
43:     $w(t + 1) = 0.95 w(t)$
44: **end while**
45: Produce the best particle.

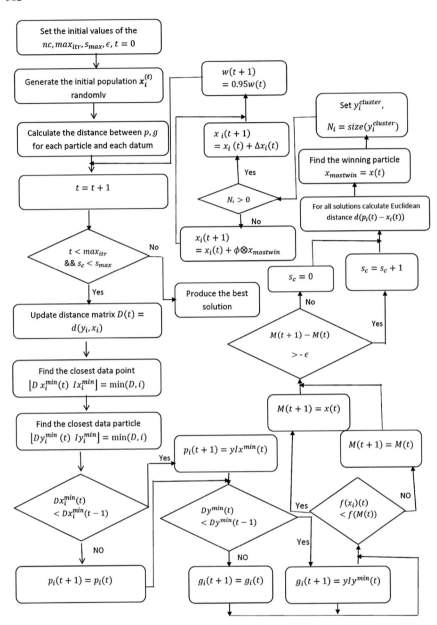

Fig. 1 Flowchart of RCE algorithm

3 Proposed System Components

RCE and K-means algorithm are applied on the Newcastle dataset.

3.1 Dataset Features

Newcastle disease is considered to be one of the infectious poultry disease. The dataset is obtained from the National Center for Biotechnology Information Gen-Bank (NCBI). It is a multi-class data, consists of 213 records with six features in each.

3.2 Algorithms Parameters Setting

RCE is applied in a supervised manner with the following parameters setting:

1. Number of clusters (K): The important parameters in K-means [7] is defining the distance measure between two data points and defining the number of clusters. silhouette measure [8] is used to define the number of clusters K. The silhouette value for each point from the input data points is a measure of how similar that point is to points in its own cluster, when compared to points in other clusters. Its value for the ith point, s_i, is defined as

$$s_i = \frac{b_i - a_i}{max(b_i, a_i)} \tag{1}$$

   where a_i is the average distance from the ith point to the other points in the same cluster as i, and b_i is the minimum average distance from the ith point to points in a different cluster, minimized over clusters.
   The silhouette value range extends from -1 to $+1$. the positive silhouette value indicates that i is well-matched to its own cluster, and poorly-matched to neighboring clusters. the negative value indicates that the clustering solution may have either too many or too few clusters. RCE algorithm is tested with K equals two, three and four. The silhouette value was positive for $K = 2$ and started to be negative starting from $K = 3$. That is why the value of $K = 2$ is chosen to run the K-means algorithm. Figures 2 and 3 show the silhouette value for $K = 2$ and $K = 3$.
2. Number of particles (n_c): n_c is set to 3.
3. Maximum number of iterations (max_iter):max_iter is set to 100.
4. Stagnation threshold (ε): ε is set to zero.

Fig. 2 Silhouette value for $K = 2$

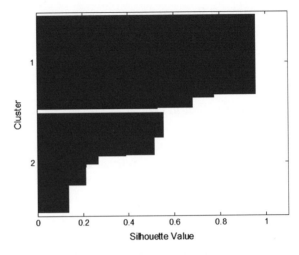

Fig. 3 Silhouette value for $K = 3$

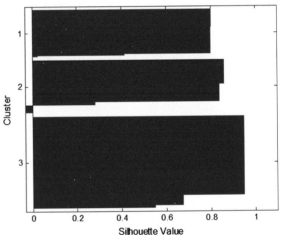

4 Evaluation Methodology

Many measures are used to judge the clustering quality. The external quality measures allow evaluate the working of clustering by comparing the groups produced by clustering method with known classes [9]. This section introduces four external quality measures which are entropy, purity, rand index and F-measure.

4.1 *Entropy*

Entropy of clustering indicates how various semantic classes are distributed within each cluster. For cluster j, the class distribution p_{ij} is calculated using Eq. (2).

$$p_{ij} = \frac{m_{ij}}{m_j} \tag{2}$$

where m_j is the number of values of cluster j and m_{ij} is the number of values of class i in cluster j. Then using this class distribution, the entropy of each cluster j is calculated using Eq. (3)

$$e_j = \sum_{i=1}^{L} P_{ij} \log_2 p_{ij} \tag{3}$$

where L is the number of classes. The total entropy for a set of clusters is calculated as the sum of the entropies of each cluster weighted by the size of each cluster using Eq. (4)

$$e = \sum_{j=1}^{k} \frac{m_j}{m} e_j \tag{4}$$

where m_j is the size of cluster j, K is the number of clusters and m is the total number of data points. Smaller entropy indicates better clustering solution.

4.2 Purity

One of the ways of measuring the quality of a clustering solution is cluster purity. The purity of cluster j is given by Eq. (5)

$$purity_j = \max(p_{ij}) \tag{5}$$

The overall purity of clustering is calculated as in Eq. (6)

$$purity = \sum_{j=1}^{k} \frac{m_j}{m} * purity_j \tag{6}$$

where K is the number of clusters. In general, the larger the purity value, the better the clustering solution.

4.3 Rand Index

During clustering, two input vectors are assigned to the same cluster if and only if they are similar. A true positive (TP) decision assigns two similar inputs to the same cluster; a true negative (TN) decision assigns two dissimilar inputs to different clusters. There are two types of errors can be committed. A (FP) decision assigns

Table 1 The contingency table

	Relevant	Non relevant
Retrieved	True positive (TP)	False positive (FP)
Not retrieved	False negative (FN)	True negative (TN)

two dissimilar inputs to the same cluster. A (FN) decision assigns two similar inputs to different clusters. Rand Index measures (Eq. 7) the accuracy of clustering result in terms of percentage of decision that is correct. This notion can be made clearer with the help of the contingency Table 1.

$$RI = \frac{TP + TN}{TP + FP + FN + TN} \tag{7}$$

4.4 F-Measure

Precision and Recall Eqs. (8) and (9) are the two basic and most frequent measures. A single measure that trades off precision versus recall is F-measure which is the weighted harmonic mean of precision and recall.

$$P = \frac{TP}{TP + FP} \tag{8}$$

$$R = \frac{TP}{TP + FN} \tag{9}$$

A balanced F-measure (Eq. 10) since it equally weights precision and recall is given by

$$F = \frac{2PR}{P + R} \tag{10}$$

5 Experiments and Discussion

Table 2 shows a comparison between RCE and K-means algorithm from the point of view of entropy, purity, precision, recall, rand index and F-measure. Entropy of RCE is lower than K-means which indicates better clustering using RCE. Meanwhile, the purity, precision, recall, rand index and F-measure of RCE are higher than the ones of K-means which confirms a better clustering quality of RCE.

Table 2 The external quality measures values: Comparative analysis where CTs is Clustering technique and (K = 2)

CTs	Entropy	Purity	Precision	Recall	RI	F-measure
K-means	0.66	0.755	0.61	0.61	0.63	0.61
RCE	0.035	0.98	0.99	0.99	0.99	0.99

6 Conclusion and Future Works

This paper evaluates RCE clustering algorithm on numeric dataset which describes the features of Newcastle poultry disease. Rapid Centroid Estimation shows better external clustering quality measures than K-means. Many Swarm optimization techniques are adopted to perform clustering. In the future work, we intend to expand the comparison to include many other swarm clustering techniques. Also, it is planned to increase the size of the dataset to check the performance optimization obtained by the RCE. One of the major challenges to be faced is to determine the number of clusters automatically.

References

1. Peterson, M.R., Doom, T.E., Raymer, M.L.: Ga-facilitated knn classifier optimization with varying similarity measures. In: Proceedings IEEE Congress on Evolutionary Computation, pp. 2514-2521 (2005)
2. Eberhart, R., Kennedy, J.: A new optimizer using particle swarm theory. In: IEEE Sixth International Symposium on Micro Machine and Human Science, pp. 39-43 (1995)
3. Evers, G.I., Ghalia, M.B.: Regrouping particle swarm optimization: a new global optimization algorithm with improved performance consistency across benchmarks. In: Proceedings International Conference on Systems, Man, and Cybernetics 2009, pp.3901-3908, San Antonio, TX, USA (2009)
4. Cohen, S.C.M., de Castro, L.N.: Data clustering with particle swarms. In: Proceedings 2006 IEEE Congress on Evolutionary Computations, pp.1792-1798 (2006)
5. Szabo, A., Prior, A.K.F., de Castro, L.N.: The proposal of a velocity memoryless clustering swarm. In: Proceedings 2010 IEEE Congress on Evolutionary Computation (CEC), pp.1-5 (2010)
6. Yuwono, M., Su, S.W., Moulton, B., Nguyen, H.: Method for increasing the computational speed of an unsupervised learning approach for data clustering. In: Proceedings 2012 IEEE Congress on Evolutionary Computation, pp.2957-2964 (2012)
7. Witten, I.H., Frank, E.: Data Mining: Practical Machine Learning Tools and Techniques. Morgan Kaufmann, San Francisco (2005)
8. Kaufman, L., Rousseeuw, P.J.: Finding Groups in Data: An Introduction to Cluster Analysis. Wiley, Hoboken (1990)
9. Bisht, S., Paul, A.: Document clustering: a review. Int. J. Comput. Appl. **73**(11), 7 (2013). http://citeseerx.ist.psu.edu/viewdoc/summary?doi=10.1.1.403.993

Design of Nature Inspired Broadband Microstrip Patch Antenna for Satellite Communication

Pushpendra Singh, Kanad Ray and Sanyog Rawat

Abstract This paper presents the design and synthesis of a nature-inspired micro-strip patch antenna based on the sunflower structure. The antenna structure was based on the Fibonacci pattern found in a sunflower, with the antenna elements in the position of the seeds. Simulation is done using computer simulation technology Microwave studio and the geometry offers impedance bandwidth of 5.28 GHz with enhanced radiation parameters. The geometry has simple structure, therefore can be used for satellite communication applications.

Keywords Microstrip antenna · Sunflower spiral antenna · Golden angle · Nature inspired

1 Introduction

Wireless telecommunication is developing very fast, and rapid changes to the antenna system are often desired to have room for the system needs. The classic primer antenna structures might not meet up the feat needed for the space given. The continuous technological advancement of wireless communication systems has led to increased constraints on the antennas utilized in terms of cost, size, bandwidth, gain, polarization, and many other antenna performance parameters [1]. Communication systems are becoming compact antennas with improved performance. Micro-strip antennas may be proved very useful structure for this communication system. [2] Micro-strip patch antennas are preferred in many

P. Singh · K. Ray (✉) · S. Rawat
Department of Electronics and Communication, Amity University, Rajasthan, India
e-mail: kanadray00@gmail.com

P. Singh
e-mail: singhpushpendra548@gmail.com

S. Rawat
e-mail: sanyog.rawat@gmail.com

© Springer International Publishing Switzerland 2016
N. Pillay et al. (eds.), *Advances in Nature and Biologically Inspired Computing*,
Advances in Intelligent Systems and Computing 419,
DOI 10.1007/978-3-319-27400-3_33

applications due to their advantage over conventional antennas such as low profile, light weight, small size and compatibility. But the major drawback of micro-strip antenna is its narrow bandwidth and low gain [3, 4]. Several methods like stacked patches, defected ground, use of active devices have been reported in literature [5–8]. Here we are exploring a new method for building such antennas that could have several times improved radiation performance. It might result in a much longer range, lower power operation, better gain, better use of available bandwidth and even enhanced speeds. The key to this new design is a mathematical concept that is capable of defining many complex shapes found in nature. Efficiency can be hugely increased by improvement of transmission and quality of reception by finding such formula-based antennas. To the best of our knowledge the structure we have developed has not been explored before.

In this work a new geometry of nature inspired antenna is proposed. The main objective of the present investigation is to provide a frame work for solving the optimization problems related to antenna application in satellite communication [9]. Nature inspired microstrip patch antennas based on the sunflower structure are more suitable in the field of the satellite applications at higher frequency. The proposed antenna may find use in aircraft because it can be easily- mounted on the skin of aircraft or spacecraft. A simple construction, adjustable and moderate gain, broad bandwidth are some of the key features of this design. Sunflower based microstrip patch antenna structure shows a non-uniformly spaced, direct radiating array which is responsible for reducing complexity and cost compared with conventional antenna.

This paper is structured in five sections. Inspiration description has been proposed in Sect. 2. Nature—inspired antenna design is described in Sect. 3. Discussion of simulation result has been shown in Sect. 4. Conclusion is drawn in Sect. 5.

2 Inspiration

The sunflower captures light in an efficient manner, due to the orientation and distribution of its seeds. The idea for this work started with the following hypothesis. If we build an antenna in which each element follows the distribution of the seeds in a sunflower (Fig. 1), the arrangement has a good performance in capturing the electromagnetic wave. One of the mathematical patterns that can be found frequently in nature is the Fibonacci sequence. [10] The angle, distribution of the seeds and spiral form are some example of these patterns. The Fibonacci sequence is defined as-

Fig. 1 Fibonacci sequence
found in the number of seeds
in a sunflower [13]

$$F = \{F_0, F_1, F_2\}$$
Such as $\quad F_0 = 0$
$$F_1 = 1$$
$$F_k = F_{k-1} + F_{k-2}$$

3 Antenna Design and Analysis

Spiral elements are chosen in this antenna because they are easy to design and their shape is similar to a sunflower. The basic structure of the spiral is shown Fig. 2.

We calculate the position of the elements in the proposed design by the following cartesian Eqs. 1 and 2

$$Xn = Sn\sqrt{n/\pi}\, \cos\theta \cdot \beta \tag{1}$$

$$Yn = Sn\sqrt{n/\pi}\, \sin\theta \cdot \beta \tag{2}$$

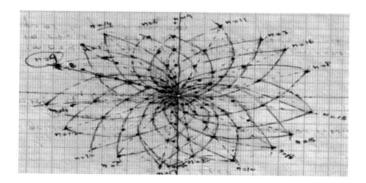

Fig. 2 Basic structure of proposed design with spiral elements

Fig. 3 Seeds distribution in
the sun flower [13]

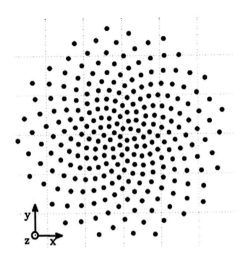

Where

n = 2, 3, 4........
β = Phase contol parameter
S (golden mean) = 1.695
$Sn = Sn - 1\sqrt{2}$

The distribution of the antenna elements is achieved using a structure, which typically follows the pattern of the seeds in sunflower as shown in Fig. 3.

In many cases, the head of the flower is made up of small seeds which are produced at the centre, and then migrate towards the outside filling eventually all the space. Each new seed appears at a certain angle in relation to the proceeding one. In order to optimize the filling of the seeds, it is necessary to choose the most irrational number which is the least well approximated by a fraction [10]. This number is called golden mean. The corresponding golden angle is 137.5°. With this angle, one can obtain the optimal filling spacing between the seeds as shown in Fig. 3 in text.

Golden ratio:
    Golden ratio may be defined as

$$\delta = 1 + 1/\delta$$

After solving it

$$\delta = 1.618$$

Fibonacci number series is
{2, 3, 5, 8, 13, 21, 34, 55, 89...................}

If we take the two successive Fibonacci numbers [14]

P	Q	P/Q
2	3	1.5
3	5	1.666666666...
5	8	1.6
8	13	1.625

Golden Angle: To choose the irrational part of the golden mean (0.695 aprox), the rotation is about 222.5° angle. In the other direction it is about **137.5°**, called the Golden Angle [11].

Analysis of proposed design:
The length and the position of the spiral elements in the proposed antenna is obtained from Eqs. 1 and 2. For example the position of the 5 elements can be calculated as follows:

Case 1: $(\beta = 1)$

	(X_2, Y_2)	$(\theta = 137.5 \times 1)$
$n = 2$	$(0.068, -0.780)$	137.5^0
$n = 3$	$(0.117, -1.344)$	....
$n = 4$	$(0.192, -2.20)$	....
$n = 5$	$(0.302, -3.47)$	....
$n = 6$	$(0.469, -4.70)$	....

Similarly we can obtain another coordinate of the points-

$$(\beta = 2, 3, 4, 5, 6 \ldots\ldots\ldots\ldots\ldots\ldots\ldots)$$

The maximum value of β is taken as 20 in order to maintain the design of the antenna simple.

4 Simulation Results

The initial design consists of FR-4 substrate and the size of the ground plane is considered as 20×20 mm, substrate height is 1.59 mm with microstrip line feed as shown in Fig. 4.

The variation of simulated reflection coefficient with frequency is shown in Fig. 6a. It clearly indicates that the antenna is resonating at 13 GHz and provides a

Fig. 4 Front view of initial
design

Fig. 5 Rear view of proposed
antenna

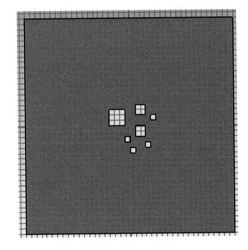

bandwidth of nearly 3.1 GHz (12.4–15.5 GHz). The variation of gain with respect
to frequency is shown in Fig. 6b.

In order to improve bandwidth and other radiation properties the antenna is
modified by etching slots in the ground plane as shown in Fig. 5. The variation of
simulated reflection coefficient with frequency is shown in Fig. 7a which shows that
the antenna is now resonating at resonant frequencies of 12.4, 14.8 and 16.5 GHz
and has operating bandwidth of 5.28 GHz. The variation of gain of the antenna with
respect to frequency is shown in Fig. 7b. It may be seen that from frequency 12 to
15 GHz, i.e. in the frequency range having first two resonance frequencies, gain of

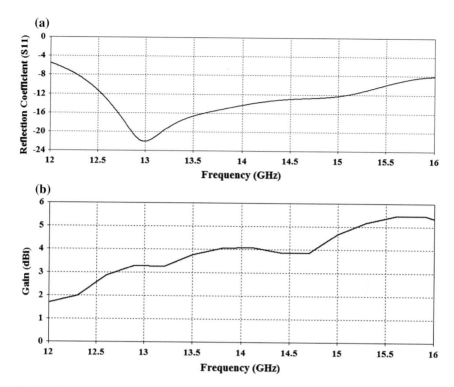

Fig. 6 a Variation of reflection coefficient (S11) as a function of frequency, **b** Variation of gain with frequency

antenna is nearly 3.5 dBi. There after for a range of frequency, i.e. from 15 to 17.3 GHz, improvement in gain is realized and gain of antenna at the third resonance frequency of 16.5 GHz is close to 5 dBi.

The narrow bandwidth of micro-strip antenna is one of the important features that restrict its wide usage. In the present work the bandwidth of sunflower based micro-strip antenna is increased by cutting rectangular slots in the ground material. The proposed antenna has resonance at 3 different frequencies 12.4, 14.8 and 16.5 GHz which is suitable for many applications. The fractional bandwidth of proposed antenna is found to be 36.23 %.

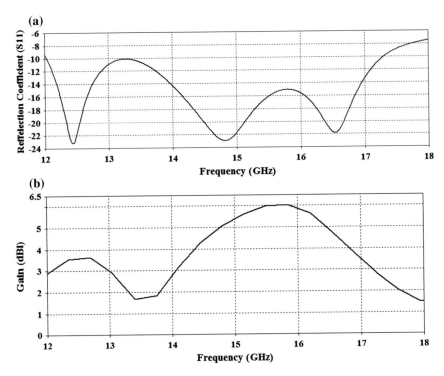

Fig. 7 **a** Variation of reflection coefficient (S11) as a function of frequency, **b** Variation of gain with frequency for proposed antenna

The elevation radiation patterns of proposed antenna at three frequencies within the impedance bandwidth region are shown in Fig. 8. It may be observed that the radiation patterns at all these frequencies are stable and maximum radiations are directed normal to the patch geometry. The antenna could serve as a potential candidate for Ku band of satellite communication and other suitable applications having better radiation performance as compared to the recently reported works [14–16].

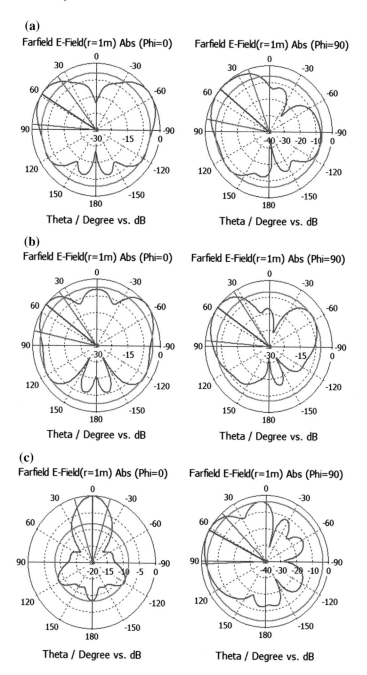

Fig. 8 Variation of E and H plane elevation patterns at **a** 12.4, **b** 14.8, and **c** 16.5 GHz

5 Conclusion

The Sunflower structure shows non-uniformly spaced, direct radiating elements which are responsible for reducing complexity and may reduce the cost compared with conventional antenna. The antenna based on sunflower structure satisfy the requirements of satellite communication in form of the reduced directivity at several spot beam areas and reduced lobe level at several values. Reducing cost and improving the antenna performance is important in the field of satellite communication. The proposed antenna may be useful in above application area. The antenna gain can be further improved by increasing the spiral elements (seeds) which lead to increase in the complexity.

References

1. Ramat-samii, Y., M. Kovita, J., Rajagopalan, H.: Nature-inspired optimization technique in communication antenna design. Proc. IEEE **100**(7), 2132–2144 (2012)
2. Gopal Jangid, K., Kulkar, V.S., Sharma, V.: Compact circular micro-strip patch antenna with modified ground plane for broadband performance. In: International Conference on Signal Propagation and Computer Tech, pp. 25–28 (2014)
3. Kumar, P., Singh, G.: Advantage computational technique in electromagnetic. In: 4th International Conference on Communication System and Network Technologies IJECS-IJENS, vol. 2012, (2012)
4. Surjati, I., KN, Y.: Increasing bandwidth dual frequency triangular micro-strip antenna for WiMax application. Int. J. Electr. Comput. Sci. **10**(06) (2006)
5. Takaki, R., Ogiso, Y., Hayashi, M., Katsu, A.: Simulation of Sunflower Spiral and Fibonacci Numbers. Tokyo institute of technology, Japan (2003)
6. Rawat, S., Sharma, K.K.: Stacked elliptical patches for circularly polarized broadband performance. In: International Conference on Signal Propagation and Computer Technology (ICSPCT 2014), pp. 232–235 (2014)
7. Grob, V., Pfeifer, E., Rutishauser, R.: Sympodial construction of Fibonacci-type leaf rosettes in pinguicula moranensis (lentibulariaceae). Ann. Bot. **100**(4), 857–863 (2007)
8. Rawat, S., Sharma, K.K.: A compact broadband microstrip patch antenna with defected ground structure for C-band applications. Central Eur. J. Eng. Springer, pp. 287–292 (2014)
9. Puri, M., Dhanik, S.S., Mishra, P.K., Khubchandani, H.: Design and simulation of double ridged horn antenna operating for UWB application. In: IEEE India Conference (2013)
10. Carolina Vigano, M.: Sunflower array antenna for multi-beam setallite application. Laurea in Ingegneria delle Telecomunicazioni Universita di Ingegneria Firenze, Italia (Thesis Dilft university of technology) (2011)
11. Vogel, H.: A better way to construction the sunflower head. Math. Biosci. **44**(44), 179–189 (1979)
12. Delgado, J.A.V.; Mera, C.A.V.: A bio-insired patch antenna array using fibonacci sequence in Oak-Tree. Published Research Report, USA (2013)
13. Rawat, S., Sharma, K.K.: Stacked configuration of rectangular and hexagonal patches with shorting pin for circularly polarized wideband performance. Central Eur. J. Eng. Springer **4**, 20–26 (2014)
14. Delgado, J.A.V.; Mera, C.A.V.: A bio-insired patch antenna array using fibonacci sequence in Oak-Tree. Published Research Report, USA (2013)

15. Rawat, S., Sharma, K.K.: Annular ring microstrip patch antenna with finite ground plane for ultra-wideband applications. Int. J. Microwave Wirel. Technol. pp. 179–184 (2015)
16. Ahsan, M.R., Islam, M.T., Ullah, M.H.: "A simple design of planar microstrip antenna on composite substrate for Ku/K band satellite applications. Int. J. Commun. Syst. doi: 10.1002/dac.2970. (2015)

Using Headless Chicken Crossover for Local Guide Selection When Solving Dynamic Multi-objective Optimization

Mardé Helbig and Andries P. Engelbrecht

Abstract One of the major issues that should be addressed when solving dynamic problems, is a loss of diversity. In addition, when solving multi-objective optimisation problems, one of the goals is to find a diverse set of solutions. Therefore, a key component of a dynamic multi-objective optimisation algorithm (DMOA) is an approach to increase diversity of either the dynamic multi-objective optimisation algorithm DMOA's individuals or the guides that guide the search of the DMOA. This study investigates whether using the headless chicken macromutation operator for local guide selection improves the performance of the dynamic vector evaluated particle swarm optimisation (DVEPSO) algorithm. Results indicate that the operator does improve the accuracy of the set of solutions that is found by DVEPSO. However, fewer solutions are found.

Keywords Dynamic multi-objective optimisation · Dynamic vector evaluated PSO · Headless chicken macromutation

1 Introduction

Optimisation problems with more than one objective, with at least two objectives in conflict with one another, and at least one objective changing over time, are referred to as dynamic multi-objective optimisation problems (DMOOPs). When solving a DMOOP, a dynamic multi-objective optimisation algorithm (DMOA) has to track the changing set of optimal trade-off solutions over time. Therefore, after the DMOA converged to the set of optimal solutions in one environment, diversity loss has to be addressed to ensure that the DMOA can track the changing set of optimal solutions after a change in the environment has occurred. In addition, similar to a

M. Helbig (✉) · A.P. Engelbrecht
Department of Computer Science, University of Pretoria, Hatfield, South Africa
e-mail: mhelbig@cs.up.ac.za

A.P. Engelbrechtat
e-mail: engel@cs.up.ac.za

© Springer International Publishing Switzerland 2016
N. Pillay et al. (eds.), *Advances in Nature and Biologically Inspired Computing*,
Advances in Intelligent Systems and Computing 419,
DOI 10.1007/978-3-319-27400-3_34

multi-objective algorithm (MOA) solving a static multi-objective optimisation problem (MOOP), the DMOA has to find a set of trade-off solutions that are as close as possible to the optimal set of trade-off solutions, and find a diverse set of solutions.

The dynamic vector evaluated particle swarm optimisation (DVEPSO) algorithm is a co-operative particle swarm optimisation (PSO)-based algorithm where each sub-swarm solves only one objective and knowledge is then shared between the various sub-swarms [1]. The sub-swarm's search is guided by a global guide and local guides. The global guide is a gbest (best position found so far by the entire sub-swarm) of either the sub-swarm itsef or another sub-swarm. A local guide is normally the pbest (best position found so far by the particle) of a particle. After a change in the environment has occurred, diversity is introduced into the swarm by re-initialising a percentage of the particles.

In evolutionary algorithms (EAs), mutation is used to add additional diversity into the population by introducing new genetic material to an individual. In this study, the headless chicken macromutation operator [2] is incorporated into the DVEPSO algorithm to increase the diversity of the sub-swarms during the entire optimisation process while solving DMOOPs. The goal of the study is to investigate whether the introduction of the headless chicken macromution operator to produce the local guide, improves the performance of DVEPSO. In this study the headless chicken operator uses crossover on either a particle's position, a particle's pbest and a random position in the search space, or a particle's position, the gbest of a selected sub-swarm and a random position in the search space.

The rest of the paper's layout is as follows: Sect. 2 discusses background information that is required for the rest of the paper. Background information with regards to headless chicken macromutation, parent-centric crossover (PCX) and multi-objective optimisation (MOO) are discussed. DVEPSO used in this study is discussed in Sect. 3. Section 4 discusses the experimental setup for this study, including the benchmark functions, performance measures and the statistical analysis approach that was used to analyse the data. The results are presented in Sect. 5. Finally, the conclusions are presented in Sect. 6.

2 Background

This section discusses the headless chicken macromutation operator, parent-centric crossover (PCX) and multi-objective optimisation (MOO).

2.1 Headless-Chicken Macromutation Operator

A macromutation operator that combines a parent individual and a randomly generated individual using a crossover operator was proposed by Jones [2]. This operator is referred to as the headless chicken operator. Since inheritance does not take place, it cannot be referred to as a crossover operator. However, new randomly generated genetic material is introduced, and therefore it is referred to as a macromutation operator.

2.2 Parent-Centric Crossover

PCX, proposed by Deb et al. [3], is performed as follows:

1. The mean vector $\mathbf{g}$ of the μ selected parents are calculated.
2. For each offspring p one parent $\mathbf{x}_p$ is selected with equal probability and the direction vector $\mathbf{d}_p = \mathbf{x}_p - \mathbf{g}$ is calculated.
3. From each of the $(\mu - 1)$ parents perpendicular distances $\mathbf{D}_i$ to $\mathbf{d}_p$ are calculated and the average $\mathbf{D}$ is found.
4. The offspring is then created as follows:

$$\mathbf{y} = \mathbf{x}_p + w_\zeta \left| \mathbf{d}_p \right| + \sum_{i=1, i \neq p}^{\mu} w_\eta \mathbf{D} \mathbf{e}_i \tag{1}$$

where $\mathbf{e}_i$ are the $(\mu - 1)$ orthonormal bases that span the subspace perpendicular to $\mathbf{d}_p$, and w_η and w_ζ are zero-mean normally distributed with variance σ_η and σ_ζ respectively.

2.3 Multi-objective Optimisation

A MOOP does not have a single optimum due to the conflict between its objectives, i.e. improving one objective leads to the worsening of at least one other objective. Therefore, when an algorithm solves MOOPs, it has to find a diverse set of trade-off solutions and a set of trade-off solutions that is as close as possible to the set of optimal trade-off solutions. The set of optimal trade-off solutions in the decision variable space is referred to as the Pareto-optimal set (POS) and in the objective space is referred to as the Pareto-optimal front (POF).

3 Dynamic Vector-Evaluated PSO

The vector evaluated particle swarm optimisation (VEPSO) algorithm was introduced by Parsopoulos et al. [4] and is a co-operative PSO-based algorithm. Each sub-swarm solves only one objective function and the knowledge of the sub-swarms are shared through the velocity update of the particles. Similar to the basic PSO, each particle has a local guide and a global guide that guide its search through the search space. The local guide is the particle's pbest (best position found so far by the particle) and the global guide is selected from one of the sub-swarms through a knowledge transfer strategy [1].

Greeff and Engelbrecht [1] extended VEPSO for dynamic multi-objective optimisation (DMOO). The extended algorithm, DVEPSO, used in this study works as follows:

- Each objective function is optimised by a sub-swarm that is a gbest PSO.
- If a particle's new position leads to a better objective function value than its current pbest position, the particle's new position is selected as the particle's pbest position. Only the objective function being optimised by the sub-swarm is taken into consideration for the pbest update.
- If a particle's new position is non-dominated with regards to the sub-swarm's current gbest position, one of these two positions is randomly selected as the sub-swarm's gbest position. If a particle's new position dominates the current gbest position, the particle's position is selected as the new gbest position.
- The local guide is created using the headless chicken operator (refer to Sect. 2.1), which performs PCX (refer to Sect. 2.2).
- The global guide is selected according to a random knowledge sharing topology [1], where the selected sub-swarm can be another sub-swarm or the sub-swarm itself.
- Sentry particles [5] are used to detect a change in the environment.
- If a change in the environment has occurred, 30 % of the particles of the sub-swarm whose objective function changed is randomly re-initialised. After re-initialisation, each particle's objective function value and its pbest objective function value are re-evaluated.
- If the archive is full, a solution is removed from a crowded region.

4 Experimental Setup

This section discusses the experimental setup of the experiments conducted for this study. Section 4.1 discusses the DVEPSO configurations that were evaluated on the selected benchmark functions. The benchmark functions and performance measures used to evaluate the performance of the algorithms are discussed in Sects. 4.2 and 4.3, respectively. Section 4.4 discusses the approach that was followed to analyse the performance of the various DVEPSO configurations.

4.1 Dynamic Vector Evaluated PSO Configurations

The following DVEPSO configurations were used in the study:

- Default DVEPSO using gbest PSO as discussed in Sect. 3.
- DVEPSO using headless chicken macromutation, performing PCX on:

- a particle's position, a randomly generated position in the search space and the sub-swarm's pbest position, or
- a particle's position, a randomly generated position in the search space and the sub-swarm's gbest position

to generate the local guide. The following values were used for σ_η and σ_ζ: 0.05, 0.1, 0.2 and 0.5. If the particle's position was equal to either the pbest or gbest positions, the particle's previous position was used.

Each algorithm was executed on each benchmark function for each environment for 30 independent runs of 20 environments or changes each. Swarm sizes of 20 particles were used, and c_1, c_2 and the inertia weight were set to values that lead to convergent behaviour [6], namely 1.49, 1.49 and 0.72 respectively. The source code of DVEPSO is available in the opensource library CIlib [7].

4.2 Benchmark Functions

Based on the analysis of DMOOPs in [8], nine benchmark functions were selected of various DMOOP types [9] to compare the performance of the DVEPSO configurations. These functions include: a modified version of DIMP2 [10] with a concave POF (referred to as DIMP2 in the rest of the paper), dMOP2$_{dec}$ [11], dMOP3 [12], FDA4 [9], FDA5 [9], FDA5$_{iso}$ [11], FDA5$_{dec}$ [11], HE2 [8], and HE7 [8, 11]. For each benchmark function the following severity of change (n_t) and frequency of change (τ_t) combinations were used: $n_t = 10$ and $\tau_t = 5$, $n_t = 10$ and $\tau_t = 10$, $n_t = 10$ and $\tau_t = 25$, $n_t = 10$ and $\tau_t = 50$, and $n_t = 20$ and $\tau_t = 10$.

Each of these benchmark functions were selected based on the characteristics of their POF and POS and the specific difficulties they present to a DMOA. DIMP2 is a type I DMOOP, i.e. the POS changes over time and the POF remains static. where each decision variable has its own rate of change. dMOP2$_{dec}$ is a type II DMOOP, i.e. both the POF and POS change over time. The POF changes from concave to convex and vice versa. Furthermore, the POF is deceptive, i.e. dMOP2$_{dec}$ has at least two optima, but the search space favours the deceptive local POF and not the global POF. dMOP3 is a type I DMOOP where the variable controlling the spread of the solutions changes over time. HE2 is a type III problem where the POF changes over time, but the POS remains static. It addition, it has a the POF with disconnected pieces, where each piece is continuous. HE7 is a type III DMOOP with a non-linear POS and each decision variable has a different POS. All of the previously mentioned DMOOPs have 2 objectives. FDA4 is a type I DMOOP with 3 objective functions. FDA5, FDA5$_{dec}$ and FDA5$_{iso}$ are type II problems with 3 objectives, where the spread of solutions in the POF varies over time. In addition, FDA5$_{dec}$ has a deceptive POF and FDA5$_{iso}$ has an isolated POF. If the majority of the fitness landscape is fairly flat and no useful information is provided with regards to the location of the POF, the POF is referred to as being isolated.

4.3 Performance Measures

Based on an analysis of performance measures in [13, 14], the following two performance measures were selected for this study:

- the alternative accuracy measure (acc_{alt}) [15], referred to in this article as *acc*. It measures the difference between the hypervolume [16] of the obtained POF and the hypervolume of the true POF. A low *acc* value indicates good performance.
- stability (*stab*) [15] that quantifies the effect of changes in the environment on *acc* of the DMOA. It measures the difference in *acc* of two executive environments, i.e. whether the *acc* remains relatively stable for the various environments. A low *stab* value indicates good performance.

The hypervolume value required for these measures was calculated according to [17]. The reference vector used for the hypervolume calculations was the worst objective function value for each dimension.

4.4 Statistical Analysis

For each DMOOP, for each environment, and for each performance measure the following process was followed for the statistical analysis of the results [18]: the average performance measure value over 30 runs was calculated for each time step just before a change in the environment occurred. These values obtained by the DVEPSO configurations are tested using a Kruskal-Wallis test to determine whether there is a statistical significant difference between the algorithms' performance. If the Kruskal-Wallis test indicated that there was a statistical significant difference between the performance measure values, a pair-wise Mann-Whitney U test was performed for each pair of DMOAs. If the Mann-Whitney U test indicated a statistical significant difference, the average performance measure value of each time step just before a change in the environment occurred was used to award wins and losses as follows: for each environment the average performance measure values of the two DMOAs were compared. The DMOA with the best performance measure value was awarded a win and the other DMOA was awarded a loss. In order to ensure that wins awarded to a DMOA that tracked the changing POF very well for a specific DMOOP did not skew the results, the number of wins and losses were normalised for each DMOOP. For all statistical tests a confidence level of 95 % was used.

5 Results

This section presents the results obtained by the various DVEPSO configurations. It should be noted that DVEPSO has been extensively evaluated against other DMOA on a wide range of benchmark functions and environment types [18]. Therefore, in

Table 1 Overall wins for the various DVEPSO configurations

Results	DMOO Algorithm								
	D_d	$D_{p,\sigma=0.05}$	$D_{p,\sigma=0.1}$	$D_{p,\sigma=0.2}$	$D_{p,\sigma=0.5}$	$D_{g,\sigma=0.05}$	$D_{g,\sigma=0.1}$	$D_{g,\sigma=0.2}$	$D_{g,\sigma=0.5}$
Wins	255.35	107.05	101.35	97.15	107.1	81.5	85.7	74.95	107.5
Losses	345.25	74.1	66.8	68.0	65.05	92.45	95.75	98.5	112.65
Diff	−89.9	32.95	34.55	29.15	42.05	−10.95	−10.05	−23.55	−5.15
Rank	9.0	3.0	2.0	4.0	**1.0**	7.0	6.0	8.0	5.0

this paper the effect of incorporating headless chicken macromutation into DVEPSO is investigated by comparing various DVEPSO configurations' (refer to Sect. 4.1) performances against one another.

5.1 Overall Performance

This section discusses the results that were obtained over all the environments, all the performance measures and all the benchmark functions. The overall wins and losses are presented in Table 1. In all tables in this section, D_d refers to the default DVEPSO configuration and $D_{hc,\sigma=s}$ refers to a DVEPSO configuration that uses headless chicken macromutation with the particle's position, a randomly generated position and a guide hc, and a σ value of s, where s can be either 0.05, 0.1, 0.2 or 0.5. hc is either the pbest (represented by p) or the gbest (represented by g). The results indicate that D_d performed poorly, obtaining the worst rank and being awarded 89.9 more losses than wins. The D_p configurations outperformed D_d and the D_g configurations. All D_p configurations obtained more wins than losses. However, all other configurations obtained more losses than wins. The best performing configuration was $D_{p,\sigma=0.5}$, obtaining 42.05 more wins than losses.

Table 2 presents the average *acc* values obtained by the various DVEPSO algorithms for each benchmark function. The results indicate that D_d performed the best for DIMP2. Each D_p configuration performed the best for at least one benchmark function. Two D_p configurations performed the best for two benchmark functions. $D_{g,\sigma=0.5}$ performed the best for two benchmark functions, namely dMOP2$_{dec}$ and FDA4. All other D_g algorithms did not perform the best for any of the benchmark functions.

5.2 Performance per Measure

The wins and losses that were obtained per measure, over all benchmarks and all environments, are discussed in this section. Table 3 presents the wins and losses obtained for each performance measure. For *acc* a similar trend was observed as for the overall performance. Once again, the D_p configurations outperformed D_d and the D_g configurations. All D_p configurations obtained more wins than losses. However,

Table 2 Average *acc* value for various DVEPSO configurations

DMOOP	DMOO Algorithm								
	D_d	$D_{p,\sigma=0.05}$	$D_{p,\sigma=0.1}$	$D_{p,\sigma=0.2}$	$D_{p,\sigma=0.5}$	$D_{g,\sigma=0.05}$	$D_{g,\sigma=0.1}$	$D_{g,\sigma=0.2}$	$D_{g,\sigma=0.5}$
DIMP2	**0.490**	1.150	1.150	1.401	1.332	1.048	1.224	1.134	1.341
dMOP2$_{dec}$	0.066	0.063	0.047	0.055	0.053	0.050	0.051	0.058	**0.044**
dMOP3	1.425	0.171	**0.162**	3.807	0.185	0.177	0.165	0.176	0.168
HE2	4.313	**1.785**	2.779	2.913	2.661	3.111	2.998	2.563	2.635
HE7	1.825	**0.164**	0.203	0.193	0.213	0.183	0.174	0.202	0.180
FDA4	29.277	7.117	7.415	8.308	5.628	7.343	7.350	6.034	**5.009**
FDA5	25.838	124.814	8.594	7.080	**6.613**	8.620	8.724	8.513	7.732
FDA5$_{iso}$	25.038	6.983	105.919	6.527	**5.634**	7.116	6.925	6.153	6.264
FDA5$_{dec}$	26.871	11.015	9.468	**7.845**	10.288	10.350	10.683	10.436	9.957

Table 3 Overall wins and losses per measure for various DVEPSO configurations

PM	Results	DMOO Algorithm								
		D_d	$D_{p,\sigma=0.05}$	$D_{p,\sigma=0.1}$	$D_{p,\sigma=0.2}$	$D_{p,\sigma=0.5}$	$D_{g,\sigma=0.05}$	$D_{g,\sigma=0.1}$	$D_{g,\sigma=0.2}$	$D_{g,\sigma=0.5}$
acc	Wins	74.0	95.55	89.75	85.65	95.6	74.3	79.95	69.05	96.2
acc	Losses	269.0	50.45	43.25	44.35	41.4	76.7	72.05	74.95	88.8
acc	Diff	−195.0	45.1	46.5	41.3	54.2	−2.4	7.9	−5.9	7.4
acc	Rank	9.0	3.0	2.0	4.0	**1.0**	7.0	5.0	8.0	6.0
stab	Wins	181.35	11.5	11.6	11.5	11.5	7.2	5.75	5.9	11.3
stab	Losses	76.25	23.65	23.55	23.65	23.65	15.75	23.7	23.55	23.85
stab	Diff	105.1	−12.15	−11.95	−12.15	−12.15	−8.55	−17.95	−17.65	−12.55
stab	Rank	**1.0**	5.0	3.0	5.0	4.0	2.0	9.0	8.0	7.0

all other configurations obtained more losses than wins. The best performing configuration was $D_{p,\sigma=0.5}$, obtaining 54.2 more wins than losses. D_d performed very poor, obtaining the worst rank and awarded 195 more losses than wins. For *stab*, D_d performed the best and obtained 105 more wins than losses. This indicates that even though D_d performed poorly for *acc* in comparison to the other algorithms, it performed relatively stable for *acc* in the various environments. Except for $D_{g,\sigma=0.05}$, D_p outperformed D_g with regards to stability. If only *acc* is taken into account, $D_{p,\sigma=0.5}$ is the best configuration. However, if both *acc* and *stab* are important, then $D_{p,\sigma=0.1}$ is the best configuration, obtaining the second best rank for *acc* and the third best rank for *stab*.

5.3 Performance per Environment

Table 4 presents the wins and losses over all benchmark functions and all performance measures for each of the environments. D_d performed poorly in all environments, obtaining the worst rank for all environments, except $n_t = 20$, $\tau_t = 10$ where it

Table 4 Overall wins and losses per environment for various DVEPSO configurations

$n_t - \tau_t$	Results	DMOO Algorithm								
		D_d	$D_{p,\sigma=0.05}$	$D_{p,\sigma=0.1}$	$D_{p,\sigma=0.2}$	$D_{p,\sigma=0.5}$	$D_{g,\sigma=0.05}$	$D_{g,\sigma=0.1}$	$D_{g,\sigma=0.2}$	$D_{g,\sigma=0.5}$
10–10	Wins	49.1	20.55	18.25	24.0	21.65	20.85	20.5	11.3	20.95
10–10	Losses	68.0	13.1	16.4	11.65	14.0	20.6	19.2	26.4	20.7
10–10	Diff	−18.9	7.45	1.85	12.35	7.65	0.25	1.3	−15.1	0.25
10–10	Rank	9.0	3.0	4.0	**1.0**	2.0	7.0	5.0	8.0	6.0
10–25	Wins	49.3	22.8	24.7	15.95	17.55	21.85	14.45	13.85	18.65
10–25	Losses	70.8	7.85	7.95	16.7	16.1	8.6	22.25	20.85	27.0
10–25	Diff	−21.5	14.95	16.75	−0.75	1.45	13.25	−7.8	−7.0	−8.35
10–25	Rank	9.0	2.0	**1.0**	5.0	4.0	3.0	7.0	6.0	8.0
10–50	Wins	50.1	17.7	23.55	20.55	25.65	12.1	17.8	21.0	19.65
10–50	Losses	71.0	21.95	10.1	13.1	8.0	23.35	20.9	16.7	22.0
10–50	Diff	−20.9	−4.25	13.45	7.45	17.65	−11.25	−3.1	4.3	−2.35
10–50	Rank	9.0	7.0	2.0	3.0	**1.0**	8.0	6.0	4.0	5.0
10–5	Wins	50.7	23.3	13.0	18.35	20.5	10.1	17.95	13.35	27.85
10–5	Losses	62.4	14.35	21.65	10.3	15.15	26.35	11.75	18.35	15.8
10–5	Diff	−11.7	8.95	−8.65	8.05	5.35	−16.25	6.2	−5.0	12.05
10–5	Rank	8.0	2.0	7.0	3.0	5.0	9.0	4.0	6.0	**1.0**
20–10	Wins	56.15	22.7	21.85	18.3	21.75	16.6	15.0	15.45	20.4
20–10	Losses	73.05	16.85	10.7	16.25	11.8	13.55	21.65	16.2	27.15
20–10	Diff	−16.9	5.85	11.15	2.05	9.95	3.05	−6.65	−0.75	−6.75
20–10	Rank	9.0	3.0	**1.0**	5.0	2.0	4.0	7.0	6.0	8.0

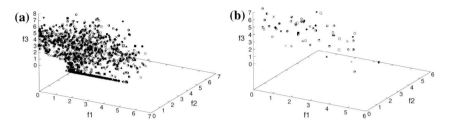

Fig. 1 Approximated POFs found by D_d, $D_{p,\sigma=0.05}$ and $D_{g,\sigma=0.1}$. **a** POF D_d. **b** POF $D_{p,\sigma=0.05}$

obtained the second worst rank. $D_{g,\sigma=0.1}$ obtained the best rank for two environments, namely $n_t = 10, \tau_t = 25$ (a slowly changing environment) and $n_t = 20, \tau_t = 10$ (a gradually changing environment). The best configuration for fast changing environments ($n_t = 10, \tau_t = 10$) was $D_{g,\sigma=0.2}$. $D_{g,\sigma=0.5}$ performed the best in very slow changing environments ($n_t = 10, \tau_t = 50$) and $D_{g,\sigma=0.5}$ performed the best in very fast changing environments ($n_t = 10, \tau_t = 5$). D_d was the only algorithm that obtained more losses than wins in all environments. In contrast, $D_{p,\sigma=0.05}$ was the only algorithm that obtained more wins than losses in all environments.

5.4 Discussion of Results

It should be noted that for problems on which the default DVEPSO struggled to converge to the true POF, the headless chicken macromutation configurations obtained solutions that were closer to the true POF. However, it found only a few solutions. This is illustrated in Fig. 1.

The default DVEPSO algorithm found more solutions that were further away from the true POF. However, the headless chicken macromuation configuration obtained only a few solutions, but these solutions were closer to the true POF. For most of the measured iterations (just before a change in the environment occurs) the headless chicken configurations obtained solutions that produced a better *acc* value.

6 Conclusions

This study investigated the effect of using the headless chicken macromuation operator, that applies crossover to either a particle's position, a randomly generated position and the pbest position, or a particle's position, a randomly generated position and the gbest position, on the performance of the dynamic vector evaluated particle swarm optimisation (DVEPSO) algorithm. The results indicated that the headless chicken macromutation operator does improve the performance of DVEPSO on problems where DVEPSO struggles to converge to the true Pareto-optimal front

(POF). However, incorporating this operator into DVEPSO leads to fewer solutions. A larger set of solutions may be produced by lowering the crossover probability of the macromutation operator (for the crossover operation).

Future work will include investigating the effect of the crossover probability on both the accuracy of the found POF, as well as the spread of solutions on the found POF. Furthermore, the effect of the following two options for parent-centric crossover (PCX) on the performance of DVEPSO will be investigated: crossover applied to the particle's position, and the particle's pbest and gbest positions; and crossover applied to the particle's pbest and gbest positions, as well as a solution from the archive. Once a robust DVEPSO algorithm has been developed taking the results of this study into account, DVEPSO will be evaluated against state-of-the-art dynamic multi-objective optimisation algorithms (DMOA) that were recently proposed [19].

References

1. Greeff, M., Engelbrecht, A.: Dynamic multi-objective optimisation using PSO. In: Nedjah, N., dos Santos Coelho, L., de Macedo Mourelle, L. (eds.) Multi-Objective Swarm Intelligent Systems. Studies in Computational Intelligence, vol. 261, pp. 105–123. Springer, Berlin (2010)
2. Jones, T.: Crossover, macromutation, and population-based search. In: Proceedings of the International Conference on Genetic Algorithms, pp. 73–80. San Francisco, U.S.A. (1995)
3. Deb, K., Joshi, D., Anand, A.: Real-coded evolutionary algorithms with parent-centric recombination. In: Proceedings of the Congress on Evolutionary Computation, vol. 1, pp. 61–66 (May 2002)
4. Parsopoulos, K., Vrahatis, M.: Recent approaches to global optimization problems through particle swarm optimization. Nat. Comput. 1(2–3), 235–306 (2002)
5. Carlisle, A., Dozier, G.: Tracking changing extrema with adaptive particle swarm optimizer. In: Proceedings of World Automation Congress, vol. 13, pp. 265–270. Orlando, Florida, U.S.A. (June 2002)
6. F.v.d.Bergh: An analysis of particle swarm optimizers. Ph.D. thesis, Department of Computer Science University of Pretoria (2002)
7. Pamparà, G., Engelbrecht, A., Cloete, T.: CIlib: A collaborative framework for computational intelligence algorithms—Part I. In: Proceedings of World Congress on Computational Intelligence, pp. 1750–1757. Hong Kong, 1-8 June 2008. http://www.cilib.net. Last accessed on: 10 December 2013
8. Helbig, M., Engelbrecht, A.: Benchmarks of dynamic multi-objective optimisation algorithms. ACM Comput. Surv. 46(3), 37 (2014)
9. Farina, M., Deb, K., Amato, P.: Dynamic multiobjective optimization problems: test cases, approximations, and applications. IEEE Trans. Evol. Comput. 8(5), 425–442 (2004). October
10. Koo, W., Goh, C., Tan, K.: A predictive gradient strategy for multiobjective evolutionary algorithms in a fast changing environment. Memet. Comput. 2(2), 87–110 (2010)
11. Helbig, M., Engelbrecht, A.: Benchmarks for dynamic multi-objective optimisation. In: Proceedings of IEEE Symposium Series on Computational Intelligence, pp. 84–91. Singapore (April 2013)
12. Goh, C.K., Tan, K.: A competitive-cooperative coevolutionary paradigm for dynamic multi-objective optimization. IEEE Trans. Evol. Comput. 13(1), 103–127 (2009). February
13. Helbig, M., Engelbrecht, A.: Issues with performance measures for dynamic multi-objective optimisation. In: Proceedings of IEEE Symposium Series on Computational Intelligence, pp. 17–24. Singapore (April 2013)

14. Helbig, M., Engelbrecht, A.: Performance measures for dynamic multi-objective optimisation algorithms. Inf. Sci. **250**, 61–81 (2013). November
15. Cámara, M., Ortega, J., de Toro, F.: A single front genetic algorithm for parallel multi-objective optimization in dynamic environments. Neurocomputing **72**(16–18), 3570–3579 (2007)
16. Zitzler, E., Thiele, L.: Multiobjective optimization using evolutionary algorithms a comparative case study, vol. 1498, pp. 292–301 (1998)
17. Fonseca, C., Paquete, L., Lopez-Ibanez, M.: An improved dimension-sweep algorithm for the hypervolume indicator. In: Proceedings of Congress on Evolutionary Computation, pp. 1157–1163 (July 2006)
18. Helbig, M., Engelbrecht, A.: Analysing the performance of dynamic multi-objective optimisation algorithms. In: Proceedings of the IEEE Congress on Evolutionary Computation, pp. 1531–1539. Cancún, Mexico (June 2013)
19. Helbig, M., Engelbrecht, A.: Results of the IEEE CEC 2015 DMOO Competition. https://sites.google.com/site/cec2015dmoocomp/

Updating the Global Best and Archive Solutions of the Dynamic Vector-Evaluated PSO Algorithm Using ϵ-dominance

Mardé Helbig

Abstract Dynamic multi-objective optimisation problems have more than one objective, at least two objectives that are in conflict with one another and at least one objective that changes over time. These kinds of problems do not have a single optimum due to the conflict between the objectives. Therefore, a new approach is required to determine the quality of a solution. Traditionally in multi-objective optimisation (MOO) Pareto-dominance have been used to compare the quality of two solutions. However, in order to increase the speed of convergence and the diversity of the found solutions, ϵ-dominance has been proposed. This study investigates the effect of using ϵ-dominance for two aspects of the dynamic vector evaluated particle swarm optimisation (DVEPSO) algorithm, namely: updating the global best and managing the archive solutions. The results indicate that applying ϵ-dominance instead of Pareto-dominance to either both of these aspects of the algorithm, or only to the global best update, does improve the performance of DVEPSO.

Keywords ϵ-dominance · Dynamic vector-evaluated pso · Dynamic moo

1 Introduction

The conflict between at least two of a multi-objective optimisation problem (MOOP)'s objectives results in a MOOP not having a single optimum, but a set of optimal trade-off solutions. The goal of a multi-objective algorithm (MOA) is to find a set of trade-off solutions that contains a diverse set of solutions and that is as close as possible to the optimal set of trade-off solutions. In the decision variable space the optimal set of trade-off solutions is called the particle swarm optimisation (POS) and in the objective space the Pareto-optimal front (POF).

Since a single optimum does not exist for MOOPs, a new approach is required to determine the quality of a solution. The most commonly used relation in multi-

M. Helbig (✉)
Department of Computer Science, University of Pretoria, Hatfield, South Afria
e-mail: mhelbig@cs.up.ac.za

© Springer International Publishing Switzerland 2016
N. Pillay et al. (eds.), *Advances in Nature and Biologically Inspired Computing*,
Advances in Intelligent Systems and Computing 419,
DOI 10.1007/978-3-319-27400-3_35

objective optimisation (MOO) is Pareto-dominance [1], where solution$_A$ dominates solution$_B$ if solution$_A$ is at least equal in quality to solution$_B$ for all objective function values, and better than solution$_B$ for at least one objective function value.

ϵ-dominance was proposed by Laumanns et al. [2] to increase convergence of a MOA and to improve the spread of the found solutions along the POF. ϵ-dominance relaxes the Pareto-dominance relation by allowing a solution with a small degradation in the objective function value to still dominate another solution. This enables an increased diversity of the POF solutions. The goal of this study is to investigate the effect of using ϵ-dominance for:

- the global best update of a sub-swarm (best position found so far by the sub-swarm), or
- managing solutions of the archive (trade-off solutions found so far), or
- both the global best update of a sub-swarm and managing solutions of the archive

The rest of the paper's layout is as follows: Background information that is required for the rest of the paper, namely MOO and dominance relations, is discussed in Sect. 2. Section 3 discusses the default dynamic vector evaluated particle swarm optimisation (DVEPSO) configuration used in this study. The experimental setup for this study is discussed in Sect. 4. The dynamic multi-objective optimisation algorithm (DMOAs) used in this study, and the benchmark functions they were evaluated on are highlighted. Furthermore, the performance measures used to measure the DMOAs performance and the statistical analysis approach used to analyse the data are also discussed. Section 5 presents and discusses the results of the study. Finally, the conclusions are presented in Sect. 6.

2 Background

This section discusses MOO and dominance relations.

2.1 Multi-objective Optimisation

The conflicts between a MOOP's objectives result in a MOOP not having a single optimum, since an improvement in one objective leads to worse values for at least one other objective. Therefore, when an algorithm solves MOOPs, it has to find a set of trade-off solutions there are as close as possible to the set of optimal trade-off solutions and that contains a diverse set of solutions. The set of optimal trade-off solutions is referred to as the POS in the decision variable space and as the POF in the objective space. These solutions are called Pareto-optimal solutions.

2.2 Dominance Relations

This section discusses dominance relations and provides definitions for the two dominance relations that are used in this study, namely Pareto-dominance and ϵ-dominance. Throughout this section minimisation is assumed.

Pareto-Dominance

Definition 1 (*Vector Domination*) Let f_k be an objective function. Then, a decision vector x_1 dominates another decision vector x_2, denoted by $x_1 \prec x_2$, if and only if

- $f_k(x_1) \leq f_k(x_2)$, $\forall k = 1, \ldots, n_k$; and
- $\exists i = 1, \ldots, n_k : f_i(x_1) < f_i(x_2)$.

The best decision vectors are referred to as being Pareto-optimal, defined as follows:

Definition 2 (*Pareto-optimal*) A decision vector x^* is Pareto-optimal if $\nexists k : f_k(x) \prec f_k(x^*)$. If x^* is Pareto-optimal, the objective vector, $f(x^*)$, is also Pareto-optimal.

ϵ-Dominance

Definition 3 (*Vector ϵ-Domination*) Let f_k be an objective function. Then, a decision vector x_1 ϵ-dominates another decision vector x_2, denoted by $x_1 \prec_\epsilon x_2$, if and only if

- $f_k(x_1) - \epsilon_k \leq f_k(x_2)$, $\forall k = 1, \ldots, n_k$; and
- $\exists i = 1, \ldots, n_k : f_i(x_1) - \epsilon_i < f_i(x_2)$.

Usage of Dominance Relations Goldberg [1] first introduced Pareto-based fitness assignment by ranking individuals based on Pareto-dominance, referred to as Pareto-ranking. With Pareto-ranking, first the non-dominated individiuals of the population are selected and assigned the highest rank. These individiuals are then removed from the population and the same process is then followed with the remaining individuals of the population. The non-dominated individuals are then selected from these remaining individiuals and assigned a lower rank than the previously selected individuals. This process is continued until all individiuals have been assigned a rank. A different approach was proposed by Fonseca and Fleming [3] where individuals were assigned a rank based on the number of individuals it dominated. Zitzler and Thiele [4] proposed a fitness assignment method based on Pareto-dominance for evolutionary algorithm (SPEA)'s external population. In SPEA an individual in the external population is assigned a fitness proportional to the number of individiuals in the current population that it dominates.

The approaches discussed above use Pareto-dominance to assign fitness values to individuals by sorting these individuals based on Pareto-dominance. When a MOA uses an archive to store the non-dominated solutions found so far in the optimisation process, typically Pareto-dominance is used to manage the archive. When a new dominated solution is found, the new solution is compared to the current solutions in the archive. If the archive contains solutions that dominate the new solution, the new solution is not added to the archive. However, if the solution dominates solutions that are currently in the archive, the new solution is added to the archive and the dominated solutions are removed from the archive.

3 Dynamic Vector-Evaluated PSO Algorithm

The vector evaluated particle swarm optimisation (VEPSO) algorithm was introduced by Parsopoulos et al. [5] to solve static MOOPs. VEPSO is a co-operative MOO particle swarm optimisation (PSO)-based algorithm, where each sub-swarm solves only one objective function. The sub-swarms share their knowledge with one another through the global best position used in the velocity update of the particles. Each particle has a local guide and a global guide that guide its search through the search space. The particle's local guide is its personal best or pbest (best position found so far by the particle) and its global guide is selected from one of the sub-swarms through a knowledge transfer strategy [6].

Greeff and Engelbrecht [7] extended VEPSO for dynamic multi-objective optimisation (DMOO). The extended algorithm, DVEPSO, works as follows: Each sub-swarm is a global best or gbest PSO. The particle's new position is selected as the particle's new pbest if its new position leads to a better objective function value than its current pbest position. When comparing positions for the pbest update, only the objective function being optimised by the sub-swarm is taken into consideration. The particle's new position is selected as the new gbest if the particle's new position dominates the current gbest position. However, if a particle's new position is non-dominated with regards to the sub-swarm's current gbest position, one of these two positions is randomly selected as the sub-swarm's new gbest. The local guide is the particle's pbest. The global guide is selected according to a random knowledge sharing topology [6].

4 Experimental Setup

This section discusses the experimental setup of the experiments conducted for this study. The DVEPSO configurations that were evaluated on the selected benchmark functions are discussed in Sect. 4.1. Sections 4.2 and 4.3 discuss the benchmark functions and performance measures used to evaluate the performance of the algorithms respectively. The approach used to analyse the performance of the various DVEPSO configurations is discussed in Sect. 4.4.

4.1 Dynamic Vector Evaluated Particle Swarm Optimisation Algorithm Configurations

The following DVEPSO configurations were used in the study:

- Default DVEPSO using gbest PSO as discussed in Sect. 3.
- DVEPSO using ϵ-dominance for the gbest update, with ϵ set to 0.003, 0.006 or 0.009, respectively.

- DVEPSO using ϵ-dominance for managing solutions in the archive, with ϵ set to 0.003, 0.006 or 0.009, respectively. ϵ-dominance is used to compare a new solution to the solutions in the archive.
- DVEPSO using ϵ-dominance for both the gbest update and for managing solutions in the archive, with ϵ set to 0.003, 0.006 or 0.009, respectively.

Each algorithm was executed on each benchmark function for 20 environments and for 30 independent runs each. All DVEPSO configurations' sub-swarm sizes were set to 20 particles. The inertia weight and c_1, c_2 were set to 0.72, 1.49, and 1.49 respectively, since these values lead to convergent behaviour [8]. The source code of DVEPSO is available in the opensource library CIlib [9].

4.2 Benchmark Functions

Based on the analysis of dynamic multi-objective optimisation problems DMOOPs in [10], nine benchmark functions were selected to compare the performance of the DVEPSO configurations. These benchmark functions are of various DMOOP types [11] and were selected based on the characteristics of their POF and POS and the specific difficulties they present to a DMOA. These functions include:

- type I DMOOPs: DIMP2 [12] (a modified version of DIMP2 [12] with a concave POF, referred to as DIMP2 in the rest of the paper), dMOP3 [13] and FDA4 [11].
- type II DMOOPs: FDA4 [11], FDA5$_{iso}$ [10] and FDA5$_{dec}$ [10].
- type III DMOOPs: HE2 [10] and HE7 [10].

A type I DMOOP's POS changes over time, while its POF remains static. A type III DMOOP is the opposite of a type I, with its POS remaining static and its POF changing over time. Both the POS and POF of a type II DMOOP change over time.

In addition, these DMOOPs have the following characteristics: Each decision variable of DIMP2 [12] has its own rate of change. dMOP2$_{dec}$'s POF changes from concave to convex and vice versa over time. Furthermore, the POF is deceptive, i.e. the POF has at least two optima, but the search space favours the deceptive local POF. The variable that controls the spread of the solutions of dMOP3 changes over time. FDA4 is a 3-objective function where the POF is a quarter sphere with radius 1. FDA5$_{iso}$ is a 3-objective DMOOP where the spread of solutions in the POF changes over time. In addition, FDA5$_{iso}$ has an isolated POF, i.e. the majority of the fitness landscape is fairly flat and no useful information is provided with regards to the location of the POF. FDA5$_{dec}$'s spread of solutions in the POF varies over time and its POF is deceptive. HE2 has a discontinuous or disconnected POF, i.e. the POF has disconnected pieces, but each piece is continuous. HE7 has a non-linear POS and each decision variable has a different POS.

4.3 Performance Measures

An analysis of performance measures in [14, 15] lead to the selection of the following
two performance measures for this study:

- the alternative accuracy measure (acc_{alt}) [16] that measures the absolute difference between the hypervolume [4] of the found or approximated POF and the
  hypervolume of the true POF. This measure is referred to as *acc* in this paper. A
  low *acc* value indicates good performance.
- stability (*stab*) [16] that quantifies the effect of changes in the environment on *acc*
  of the DMOA. It is referred to as *stab* in this paper. A low *stab* value indicates
  good performance.

The hypervolume value required for these measures was calculated according
to [17], using the worst objective function value of each dimension for the reference
vector.

4.4 Statistical Analysis

For each DMOOP, for each environment, and for each performance measure, the
following process was followed for the statistical analysis of the results [18]:

1. For each time step just before a change in the environment occurred, the average performance measure value over 30 runs was calculated for each algorithm.
   Therefore, for each algorithm and for each environment there were 30 performance measure values.
2. These performance measure values obtained by all the DVEPSO configurations
   were tested using a Kruskal-Wallis test to determine whether there was a statistical significant difference between the performance of these algorithms.
3. If the Kruskal-Wallis test indicated a statistical significant difference, a pair-wise
   Mann-Whitney U test was performed for each pair of DMOAs.
4. If the Mann-Whitney U test indicated a statistical significant difference, wins and
   losses were awarded based on the average performance measure value of each
   time step just before a change in the environment occurred as follows:

   - At each time step just before a change in the environment occurred, the average
     performance measure values of the two DMOAs were compared.
   - A win was awarded to the DMOA with the best performance measure value
     and a loss was awarded to the other DMOA.
   - The number of wins and losses were normalised for each DMOOP to prevent
     skewed results.

A confidence level of 95 % was used for all statistical tests.

5 Results

This section presents the results obtained by the various DVEPSO configurations.

5.1 Overall Performance

This section discusses the results that were obtained over all the environments, all the performance measures and all the benchmark functions. The overall wins and losses are presented in Table 1. In all tables in this section, D_d refers to the default DVEPSO configuration and Dϵ-asp-val refers to a DVEPSO configuration that uses ϵ-domination for the aspect *asp* of DVEPSO for an ϵ value of *val*. *asp* can have the value *a* for archive management, *g* for global best update or *ag* for both the archive management and global best update. *val* can have the values 0.003, 0.006 or 0.009 respectively (indicated as 3, 6 or 9 in the tables).

The results indicate that Dϵ-ag-6 performed the best, obtaining 139.95 more wins than losses. The Dϵ-g and Dϵ-ag configurations outperformed the Dϵ-a configurations. However, all ϵ-dominance configurations outperformed the D_d configuration, that performed poorly and obtained 167.3 more losses than wins. All Dϵ-a configurations performed poorly in comparison to the other Dϵ configurations, obtaining ranks 7–9, and being awarded many more losses than wins.

5.2 Performance per Measure

The wins and losses that were obtained per measure, over all benchmarks and all environments, are discussed in this section. Table 2 presents the wins and losses obtained for each performance measure. For *acc* the best performing configuration was Dϵ-g-6, that obtained 97.5 more wins than losses. Dϵ-ag-6 also obtained the best rank for *acc*. D_d performed the worst for *acc* and the second worst for *stab*. In addition to Dϵ-ag-6, both Dϵ-g-3 and Dϵ-g-6 performed well with regards to both

Table 1 Overall wins for the various DVEPSO configurations

Results	DMOO Algorithm									
	D_d	Dϵ-a-3	Dϵ-a-6	Dϵ-a-9	Dϵ-g-3	Dϵ-g-6	Dϵ-g-9	Dϵ-ag-3	Dϵ-ag-6	Dϵ-ag-9
Wins	57.8	83.85	83.4	63.9	92.25	119.85	73.3	76.6	137.45	133.3
Losses	225.1	119.45	135.5	130.5	24.9	23.9	74.7	69.4	23.5	83.95
Diff	−167.3	−35.6	−52.1	−66.6	67.35	95.95	−1.4	7.2	113.95	49.35
Rank	10.0	7.0	8.0	9.0	3.0	2.0	6.0	5.0	**1.0**	4.0

Table 2 Overall wins and losses per measure for various DVEPSO configurations

PM	Results	DMOO Algorithm									
		D_d	De-a-3	De-a-6	De-a-9	De-g-3	De-g-6	De-g-9	De-ag-3	De-ag-6	De-ag-9
acc	Wins	39.55	64.75	58.0	35.25	75.0	96.6	52.95	53.05	110.25	86.6
acc	Losses	189.45	89.25	95.0	80.75	16.0	12.4	61.05	51.95	12.75	53.4
acc	Diff	−149.9	−24.5	−37.0	−45.5	59.0	84.2	−8.1	1.1	97.5	33.2
acc	Rank	10.0	7.0	8.0	9.0	3.0	2.0	6.0	5.0	**1.0**	4.0
stab	Wins	18.25	19.1	25.4	28.65	17.25	23.25	20.35	23.55	27.2	46.7
stab	Losses	35.65	30.2	40.5	49.75	8.9	11.5	13.65	17.45	10.75	30.55
stab	Diff	−17.4	−11.1	−15.1	−21.1	8.35	11.75	6.7	6.1	16.45	16.15
stab	Rank	9.0	7.0	8.0	10.0	4.0	3.0	5.0	6.0	**1.0**	2.0

acc and *stab*. All De-a configurations performed poorly for both *acc* and *stab*, being awarded more losses than wins for both measures. In contrast, good performance was achieved by all De-g and De-ag configurations for both measures.

5.3 Performance per Environment

Table 3 presents the wins and losses over all benchmark functions and all performance measures for each of the environments.

D_d performed the worst in all environments, being awarded many more losses than wins. De-ag-6 performed the best in all environments. Both De-g-3 and De-g-6 performed well in all environments, obtaining the third and second ranks for all environments respectively. De-ag-6, De-g-3 and De-g-6 were the only algorithms that obtained more wins than losses for all environments. D_d, De-a-3 and De-a-9 performed poorly, being awarded more losses than wins for all environments.

5.4 Discussion of Results

Figure 1 illustrates the approximated POFs found by D_d and De-ag-6, from various viewpoints, for FDA5$_{dec}$ with $n_t = 10$ and $\tau_t = 10$. The figure illustrates the approximated POF found for all 30 runs.

De-ag-6 obtained more solutions that are close to the true POF and therefore obtained a higher *acc* value than D_d. Both De variations using $\epsilon = 0.006$ performed well for both *acc* and *stab*. Furthermore, configurations using ϵ-dominance only to manage the archive performed poorly compared to those that used ϵ-dominance for updating the gbest and managing the archive solutions. In addition, all ϵ-dominance DVEPSO configurations outperformed the default DVEPSO configuration.

Table 3 Overall wins and losses per environment for various DVEPSO configurations

$n_t - \tau_t$	Results	DMOO Algorithm									
		D_d	$D\epsilon$-a-3	$D\epsilon$-a-6	$D\epsilon$-a-9	$D\epsilon$-g-3	$D\epsilon$-g-6	$D\epsilon$-g-9	$D\epsilon$-ag-3	$D\epsilon$-ag-6	$D\epsilon$-ag-9
10–10	Wins	5.6	8.7	13.6	7.8	6.15	11.15	4.2	5.05	15.7	15.7
10–10	Losses	21.2	10.9	13.5	8.45	2.35	1.05	9.75	9.15	2.25	12.7
10–10	Diff	−15.6	−2.2	0.1	−0.65	3.8	10.1	−5.55	−4.1	13.45	3.0
10–10	Rank	10.0	7.0	5.0	6.0	3.0	2.0	9.0	8.0	**1.0**	4.0
10–25	Wins	15.05	17.65	16.45	16.5	26.7	27.1	21.45	21.7	31.35	34.15
10–25	Losses	51.3	37.8	40.4	32.5	6.95	5.25	15.25	13.65	8.05	15.4
10–25	Diff	−36.25	−20.15	−23.95	−16.0	19.75	21.85	6.2	8.05	23.3	18.75
10–25	Rank	10.0	8.0	9.0	7.0	3.0	2.0	6.0	5.0	**1.0**	4.0
10–5	Wins	13.65	17.7	17.65	13.2	20.4	28.1	17.35	18.65	29.6	29.15
10–5	Losses	49.6	28.65	28.3	32.95	5.6	6.0	15.3	14.7	4.9	17.45
10–5	Diff	−35.95	−10.95	−10.65	−19.75	14.8	22.1	2.05	3.95	24.7	11.7
10–5	Rank	10.0	8.0	7.0	9.0	3.0	2.0	6.0	5.0	**1.0**	4.0
1–10	Wins	11.75	19.9	17.85	13.2	19.5	26.75	15.15	15.6	30.4	27.15
1–10	Losses	51.5	21.05	26.65	28.3	5.0	5.8	17.2	15.95	4.15	19.2
1–10	Diff	−39.75	−1.15	−8.8	−15.1	14.5	20.95	−2.05	−0.35	26.25	7.95
1–10	Rank	10.0	6.0	8.0	9.0	3.0	2.0	7.0	5.0	**1.0**	4.0
20–10	Wins	11.75	19.9	17.85	13.2	19.5	26.75	15.15	15.6	30.4	27.15
20–10	Losses	51.5	21.05	26.65	28.3	5.0	5.8	17.2	15.95	4.15	19.2
20–10	Diff	−39.75	−1.15	−8.8	−15.1	14.5	20.95	−2.05	−0.35	26.25	7.95
20–10	Rank	10.0	6.0	8.0	9.0	3.0	2.0	7.0	5.0	**1.0**	4.0

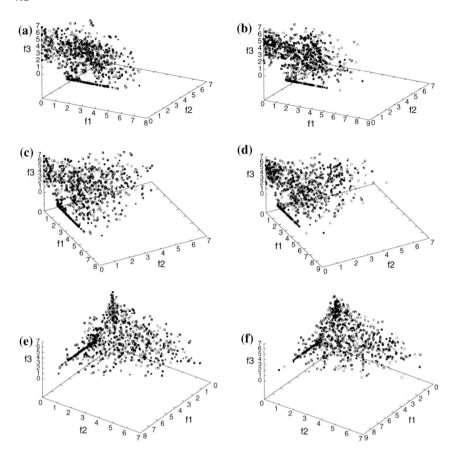

Fig. 1 Approximated POFs found by D_d and $D\epsilon$-ag-6 for FDA5$_{dec}$ with $n_t = 10$ and $\tau_t = 10$ **a** POF of D_d **b** POF of $D\epsilon$-ag-6 **c** POF of D_d from viewpoint 2 **d** POF of $D\epsilon$-ag-6 from viewpoint 2 **e** POF of D_d from viewpoint 3 **f** POF of $D\epsilon$-ag-6 from viewpoint 3

6 Conclusions

This study investigated the effect of using ϵ-dominance to update the global best and manage the archive solutions for the dynamic vector evaluated particle swarm optimisation (DVEPSO) algorithm. The results indicate that only using ϵ-dominance for managing the archive, does not lead to good performance. The best performance was obtained when ϵ-dominance was used for both the global best update and for managing the archive, setting the ϵ value to 0.006. Using ϵ-dominance only to update the global best and still using Pareto-dominance for the archive, produced good results for $\epsilon = 0.003$ and $\epsilon = 0.006$. The DVEPSO configurations that use ϵ-dominance outperformed the default DVEPSO configuration.

Future work will include investigating the performance of DVEPSO using ϵ-dominance for personal best updates of the particle swarm optimisation (PSO) sub-swarms, as well as a heterogeneous DVEPSO where some of the particles will use ϵ-dominance for either the global best or personal best updates or both, or using Pareto-dominance for either the global guide or local guide updates or both.

References

1. Goldberg, D.: Genetic Algorithms in Search, Optimization and Machine Learning. Addison-Wesly, Reading (1989)
2. Laumanns, M., Thiele, L., Deb, K., Zitzler, E.: Combining convergence and diversity in evolutionary multiobjective optimization. Evol. Comput. **10**(3), 263–282 (2002)
3. Fonseca, C., Fleming, P.: An overview of evolutionary algorithms in multiobjective optimization. Evol. Comput. **3**(1), 1–16 (1995)
4. Zitzler, E., Thiele, L.: Multiobjective evolutionary algorithms: a comparative case study and the strength pareto approach. IEEE Trans. Evol. Comput. **3**(4), 257–271 (1999)
5. Parsopoulos, K., Vrahatis, M.: Recent approaches to global optimization problems through particle swarm optimization. Nat. Comput. **1**(2–3), 235–306 (2002)
6. Greeff, M., Engelbrecht, A.: Dynamic multi-objective optimisation using PSO. In: Nedjah, N., dos Santos Coelho, L., de Macedo Mourelle, L. (eds.) Multi-Objective Swarm Intelligent Systems. Studies in Computational Intelligence, vol. 261, pp. 105–123. Springer, Berlin (2010)
7. Greeff, M., Engelbrecht, A.: Solving dynamic multi-objective problems with vector evaluated particle swarm optimisation. In: Proceedings of World Congress on Computational Intelligence (WCCI): Congress on Evoluationary Computation, pp. 2917–2924. Hong Kong (June 2008)
8. van den Bergh, F.: An analysis of particle swarm optimizers. Ph.D. thesis, Department of Computer Science University of Pretoria (2002)
9. Pampará, G., Engelbrecht, A., Cloete, T.: CIlib: a collaborative framework for computational intelligence algorithms—Part I. In: Proceedings of World Congress on Computational Intelligence, pp. 1750–1757. Hong Kong 1-8 June 2008. http://www.cilib.net, Last accessed on: 10 December 2013
10. Helbig, M., Engelbrecht, A.: Benchmarks of dynamic multi-objective optimisation algorithms. ACM Comput. Surv. **46**(3), 37 (2014)
11. Farina, M., Deb, K., Amato, P.: Dynamic multiobjective optimization problems: test cases, approximations, and applications. IEEE Trans. Evol. Comput. **8**(5), 425–442 (2004)
12. Koo, W., Goh, C., Tan, K.: A predictive gradient strategy for multiobjective evolutionary algorithms in a fast changing environment. Memet. Comput. **2**(2), 87–110 (2010)
13. Goh, C.K., Tan, K.: A competitive-cooperative coevolutionary paradigm for dynamic multi-objective optimization. IEEE Trans. Evol. Comput. **13**(1), 103–127 (2009)
14. Helbig, M., Engelbrecht, A.: Issues with performance measures for dynamic multi-objective optimisation. In: Proceedings of IEEE Symposium Series on Computational Intelligence, pp. 17–24. Singapore (April 2013)
15. Helbig, M., Engelbrecht, A.: Performance measures for dynamic multi-objective optimisation algorithms. Inf. Sci. **250**, 61–81 (2013)
16. Cámara, M., Ortega, J., de Toro, F.: A single front genetic algorithm for parallel multi-objective optimization in dynamic environments. Neurocomputing **72**(16–18), 3570–3579 (2007)
17. Fonseca, C., Paquete, L., Lopez-Ibanez, M.: An improved dimension-sweep algorithm for the hypervolume indicator. In: Proceedings of Congress on Evolutionary Computation, pp. 1157–1163 (July 2006)
18. Helbig, M., Engelbrecht, A.: Analysing the performance of dynamic multi-objective optimisation algorithms. In: Proceedings of the IEEE Congress on Evolutionary Computation, pp. 1531–1539. Cancún, Mexico (June 2013)

Detection of Zero Day Exploits Using Real-Time Social Media Streams

Dennis Kergl, Robert Roedler and Gabi Dreo Rodosek

Abstract Detection of zero day exploits is a challenging problem. Vulnerabilities that are known only by attackers but not by software vendors and neither by users have severe impact on security of systems and networks. Such vulnerabilities are exploited to intrude systems and often cause leakage of confidential data. Due to the hitherto unknown pattern of the exploitation, real-time detection is hardly possible. Hence, often an incident is detected only long time after it took place, if it is detected at all. More timely detection of attacks is necessary to trigger suitable counter-measures like reconfiguration of firewalls and sending alerts to administrators of other vulnerable targets. Therefore, to know the attributes of a novel attack's target system supports the protection of other vulnerable systems. We suggest a novel approach of post-incident intrusion detection system, to be precise—a crowd-based intrusion detection system. To accomplish this, we take advantage of social media users' postings about incidents that affect their user accounts of attacked target systems or their observations about misbehaving online services. Combining knowledge of the attacked systems and reported incidents, we should be able to recognize patterns that define the attributes of vulnerable systems. Furthermore, by matching detected attribute sets with those attributes of well-known attacks, we should be able to link attacks to already existing entries in the Common Vulnerabilities and Exposures database. If a link to an existing entry is not found, we can assume to have detected an exploitation of an unknown vulnerability, i.e., a zero day exploit or the result of an advanced persistent threat. This finding could also be used to direct efforts of examining vulnerabilities of attacked systems and simultaneously lead to faster patch deployment.

Keywords Network security · Crime data mining and network analysis · Applications on social networks · Reinforcement learning

D. Kergl (✉) · R. Roedler · G.D. Rodosek
Faculty of Computer Science, Department for Communication Systems and Network Security,
Universität der Bundeswehr München, Neubiberg, Germany
e-mail: dennis.kergl@unibw.de

© Springer International Publishing Switzerland 2016
N. Pillay et al. (eds.), *Advances in Nature and Biologically Inspired Computing*,
Advances in Intelligent Systems and Computing 419,
DOI 10.1007/978-3-319-27400-3_36

405

1   Introduction

The ongoing digital transformation of classical businesses and upcoming of new companies that rely on digitally based business models lead to more data of customers, business processes and products that are accessible via online networks. Furthermore, not only static data, e.g., in data warehouses, is connected to the Internet. Also sensor data for, e.g., predictive maintenance and even control interfaces for productive machines in industry are exposed by being accessible via networks that are connected to the Internet. Web services like online stores, communication platforms and gaming providers retain a huge amount of private data of their customers. The large number of companies that offer numerous publicly accessible web services provide a large attack surface exposing access to private data. Companies try to protect their crown jewels by combining the best possible intrusion detection systems (IDS) and hiring qualified network security staff to monitor and maintain their IT infrastructure. Today's IDS are based on recognizing signatures of well known attacks. Although a lot effort is spent on the development of more reliable detection techniques [1], educating network security staff as studied by Aoyama in [2] and improve cyber security awareness of employees, e.g., by techniques as described in [3], security incidents happen on a regular basis. These often result in identity theft or stolen customer data that includes real name, postal address, email address, credit card or bank account information.[1]

Until the vulnerable software is not identified, there neither is a pattern for intrusion detection nor is there a patch available to mitigate the vulnerability of the software. This delay makes it possible for the attackers to repeat the attack on other vulnerable targets for the whole duration of the indicated time frame.

2   Problem Description

Digitally exposed companies secure their networks with state-of-the-art intrusion detection systems. Doing this, they rely on signatures and techniques brought to them by other companies besides having their own network security staff in some cases. Publishers of signatures for malicious software investigate the techniques of exploited vulnerabilities in order to find more general signatures. To enhance the detection rate of potential intrusion attempts, patterns are more general than they need to be to describe a certain attack. The downside of more general signatures is a higher rate of false positive alarms, leading to unwanted behavior of the IDS, i.e., blocking non-malicious content. The process of monitoring network data and generating alarms when a pattern match or suspicious behavior is detected, still needs a human to assess the alarm and take proper actions. IDS solutions also often allow to specify an arbitrary threshold to adjust the sensitivity of detection. Lowering thresholds to increase the sensitivity consequently leads to decreasing the specificity of the

---

[1]https://cve.mitre.org/.

detection system and therefore leads to more false positive intrusion alarms. Dealing with too many false positive alarms is problematic for two reasons: first, the human involved in the process gets used to alarms what it makes harder for them to distinguish between false and true positive alarms and second, in a (partly) automated incident response, a false positive detection leads to unwanted behavior as locking out customers or falsely blocking access to systems and services.

The urgency for improving detection rates and specificity leads to taking individual user behavior and aggregated behavior patterns into account. A specific company is able to monitor user behavior in their own network only, so the behavior supported approach of intrusion detection is limited to the companies own network sphere. Until today there is no generally usable sensor network for monitoring user behavior in the Internet, that could be used to feed data into behavior based intrusion detection systems.

In general, recent approaches bring out sensors as close as possible to the user, monitor their behavior and, sometimes preceded by a network dependent aggregation, compare the observed behavior with the behavior that is expected. If the difference is greater than a specified threshold, a behavior based event (security alert) will be fired. Combining the classical pattern based events with new behavior based events, detection rate and specificity of IDS can be improved, especially for attacks originating in the own network. The concentration of the improvement on attacks originated in the own network clearly results from the possibility to bring out sensors. Concentrating on behavior of nodes within the monitored network does not exclude the detection of external attacks, as an internal host can be hijacked and used to perform an attack on more important and better protected systems.

Including messages from users on social media platforms that contain information about unwanted behavior of online services or even observed security incidents, can provide sensors more close to the user than the own network boundary reaches.

Therefore, knowledge about new attack patterns is hard to obtain in a short time period after the occurrence of an exploit, but it appears feasible to observe effects of new exploits, find patterns in characteristics of affected software and determine which software is vulnerable to support automated cyber defense strategies. This may include, e.g., to reconfigure firewalls or trigger deep packet inspection and verbose log analysis for affected system parts, as supposed by Seeber and Dreo in [4, 5] for software defined networks.

To reach this goal, it is important to answer the following question:

- How can user complaints about web services in social media be matched to a certain software vulnerability?

The proposed system is based on the hypothesis, that mention of a web service and knowledge about the kind and version of software that is running on the specific web service can be matched to well known vulnerabilities. If this works out, it should be feasible to detect patterns that does not match to a known vulnerability and therefore could imply the existence of a newly exploited vulnerability. To reach this goal there are some more questions to answer:

- How many complaints about unexpected behavior of a certain web service are there in social media streams and how timely is it?
- Does the information that can be gathered by profiling web services match the information available from vulnerability knowledge bases in order to match attributes between web services and vulnerabilities?
- How homogeneous is the software landscape of global services that use content delivery strategies and how has this to be integrated into our model?

We have already figured out the technical feasibility to answer these questions and first results draw a positive picture. After the system is fully implemented, we will be able to proof our hypothesis or name open problems. Many aspects of social media data, as presented in [6], have to be addressed and joined with aspects of network security to achieve social media analysis improved network security. Towards a crowd-based intrusion detection system, that is virtually a human supported vulnerability detection system.

3 Our Approach

From the entirety of Internet users we turn social media users into sensors for a novel approach of post-incident intrusion detection systems. By analyzing user posts containing incidents of interest close to their time of occurrence, one should be able to detect, e.g., complains about unusable services, stolen and fraudulent used credit card information, hijacked email or social media accounts. Sentiment analysis, as supposed by Wang et al. in [7] can support identifying these posts, as we can assume a strong negative sentiment in relevant posts. Narr et al. even propose a language independent sentiment analysis approach in [8], that should support our proposed system. The second building block of the suggested approach is a knowledge base of worlds most popular online services, in particular websites. This knowledge base contains detailed information about the services' underlying technologies as type and version of web server, scripting language engines, frameworks in use and other attributes determinable by non-invasive fingerprinting techniques. Collecting reports about events by filtering user posts from social media streams and comparing the observed patterns with patterns known from vulnerability knowledge bases should allow to identify the exploited vulnerability or at least a set of possibly exploited vulnerabilities. Figure 1 shows the components of the proposed architecture that is described in more detail below.

URL-Crawler: The starting set of URLs that will be monitored by the implemented prototype framework are gathered by extracting URLs from publicly available URL directories (Open URL-Directories). Some of these directories offer a popularity ranking, that defines our order of profiling the URLs, as we expect the more popular URLs to be mentioned more often on social media. The obtained URLs and available related metadata are stored into a URL-Database (URL-DB).

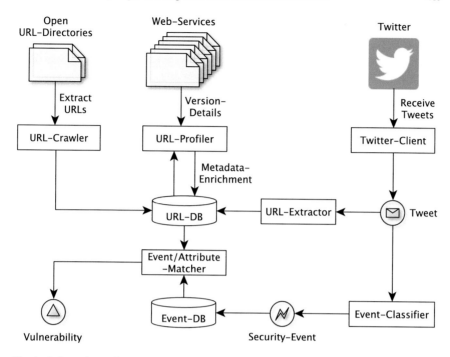

Fig. 1 Information and process flow of the proposed system

URL-Profiler: The URL-Profiler uses several fingerprinting techniques to gain information about every single web service. Techniques as proposed in [9] should be
applicable. As an attack to a web service can happen in numerous ways, we gain
as lot information as possible about the software running on the services URL.
This includes types and versions of:

- Web server
- Installed scripting engines
- Installed web frameworks
- SSH, Telnet, FTP on commonly used ports
- E-Mail server, if necessary after requesting related mail server from DNS
  URL entries in the URL-DB are enriched with the gained information. Profiling
  is repeated on a regular basis, prioritized by popularity of the specific URL.

Twitter-Client: Twitter is used as a first source of social media data. We have
defined a filter for listening to user messages in real time and alternatively use
the publicly available sample stream whose nature is described in [10]. The filter contains the most popular and already profiled URLs from the URL-DB and
keywords that are related to the expected cyber security events. This will provide
a number of 50 to 200 tweets per second. Therefore Twitter is suitable to fulfill
the part of the online social media stream source for our prototype. The received
tweets are forwarded to the next components of the system.

URL-Extractor: This part extracts URLs from tweet content in order to extend the
    URL-DB with new URLs, to build a separate ranking of mentions of all URLs on
    Twitter and to calculate an up-to-date popularity, used for prioritizing the URL-
    Profiler actions. Unexpected changes in this ranking can also be an indicator for
    some kind of event (not necessarily a security event) that is related to a URL. The
    URL-Extractor is also able to match common names of web services to existent
    URLs.

Event Classifier: Ritter et al. describe in [11] how to extract computer security
    events from Twitter. While they consider tweets in English only and have a
    clear focus on the natural language processing part of the problem, they show
    that extracting cyber security events from Twitter is possible. By adopting their
    approach and refining it with network security specific features, it should be pos-
    sible to extract events from Twitter that are mainly related to attacks on web ser-
    vices. Kinds of events that should be extracted include (in order of importance):

- Data Breach
- Site Hack with Content Replacement
- Denial of Service
- Account Hijacking
    The first and second kind of events are expected to be both mentioned on social
    media with high probability and mappable to software vulnerabilities. Services in
    some cases can still be profiled when an attack is detected and the URL has not
    been present in the URL-DB prior to the attack. Denial of Service (DOS) could
    also be mapped to a software vulnerability, but is expected to be more often caused
    by distributed attacks independent from the used software on the target machine.
    In contrast to the first two kinds, DOS-suffering services are unlikely to be pro-
    filed successfully during an attack, furthermore a scan during an on going DOS
    attack can be considered harmful and therefore we do not use mention of a DOS
    attack as a trigger for profiling. Account Hijacking in most cases is the result of
    individual security neglect like using unencrypted connections over public WiFi,
    using easy-to-guess respectively easy-to-brute-force passwords or public Internet
    terminals without essential caution. Nevertheless, a burst of such events can also
    clearly signal an exploitation of a software vulnerability. When an event that points
    to a web service outage is detected, we are able to verify this finding by imme-
    diately accessing the mentioned web service. The result of our scan can be used
    to improve the accuracy of the Event-Classifier. Similar approaches are success-
    fully implemented by Augustine et al. in [12] and by Motoyama et al. in [13].
    A detected security event is stored into the Event-DB containing kind of event,
    affected URL, timestamp of mention and hashed user name for avoiding dupli-
    cates. Hashing is performed to respect privacy as there is no need to store real
    user names. Regression analysis of occurrence rate of event-URL combinations
    might provide an indicator for the severity of an security event.

Event/Attribute-Matcher: The component providing the main logic of the pro-
    posed approach is the Event/Attribute-Matcher. Several tasks are accomplished:

- Event-DB is monitored for new clusters of URLs and/or event types
- Attributes of URLs which occur in the Event-DB are compared
- URL-DB is monitored for new versions of software

  To monitor the Event-DB is important to detect ongoing attacks to web services. Once an attack is detected, the corresponding attributes in the URL-DB are candidates for being vulnerable. If additional events provide intersecting sets of attributes, vulnerable software versions can be identified by looking at the software that is in common use. Because new software versions often provide security updates, software that is not up-to-date has a potentially higher probability of being vulnerable than software at the latest patch level. Therefore, URL-DB is monitored for new software versions and all outdated versions are marked as being more prone to attacks, resulting in a reconfiguration of the social media stream filter.

Vulnerability Matcher: A downstream component of the proposed detection system is the Vulnerability Matcher, that assesses the probability of a detected vulnerability to be one of those contained in a Common Vulnerabilities and Exposures database (CVE-DB). If it is possible to assign a detected vulnerability to an existing entry, we have found the exploitation of a well known weakness. Otherwise, if there is unambiguous indication for a vulnerability, but an assignment to a database entry is not possible, the proposed system possibly has discovered an unknown vulnerability and thus the existence of a zero day exploit for the affected software.

Monitoring the databases might help to understand the general relationship between attacks and target systems. A more specific pattern that includes, e.g., time stamp of event occurrence, links to other involved attributes, likelihood and severity of an attack, is used to reveal hidden connections between different attacks and thus lead to new insights about which kind of systems are vulnerable. Using the URL-DB to search for vulnerable systems could even predict next targets and issue warnings to these.

3.1 Example

To illustrate the objective of the proposed system we provide the following example: Assume an online service "Service A" that provides a web site by utilizing

- Apache HTTP Server,[2] version 2.4.16 and
- PHP,[3] version 5.4.45.

---

[2]http://httpd.apache.org/.

[3]https://secure.php.net/.

Beneath many others, there also is "Service B" that provides a web service utilizing

- nginx,[4] version 1.9.5 and
- PHP, version 5.4.45.

This attributes are known to us and stored in the URL-DB, because we have already profiled the corresponding URLs "service-a.example" and "service-b.example". Messages about problems with both of the mentioned services might look like:

- "Service A is down. #serviceAfail"
- "@serviceA: Cannot reach your web site!"
- "service-b.example is unreachable."

As well mentions of the service, the service URL and directed messages to service providers can be matched to an entry in our URL-DB. Now, that we know about problems with the online services of both "Service A" and "Service B", we are able to match the attributes of both services and find that both have the same version of PHP in common. This leads to the hypothesis that the identified version of PHP is vulnerable or at least unstable under certain circumstances. In contrast to this simplified example, we expect many more services with a common attribute to match, if a vulnerability is causing unexpected behavior.

4 Evaluation

The described system is composed of components with clearly defined tasks. The evaluation of the whole system should be done in three steps:

1. Measure performance and error rates of each single component
2. Determine performance and composed error rates of the combined components
3. Examine identified vulnerability candidates to obtain precision and recall rates for the whole system.

As the proposed system is still under development, we only would give a preview of the evaluation of performance and error rates of the single components:

URL-Crawler: Once most of the open available resources for URLs are crawled, the results of the different sources can be compared. URLs that exist in all datasets can be assumed to be correct, while URLs that are observed only once might by considered not to be reliable. To verify this finding, a pre-profiling can be established by sending a simple HTTP GET request and examine the result. Performance indicators are the number of total URLs gained by crawling open URL directories and the growth or, more general, change rate of these URLs per day or per week.

---

[4]http://nginx.org/.

URL-Profiler: Performance of the URL-Profiler can easily be measured by comparing the number of successfully gained attributes with the number of total attributes that are tried to be gained. As we define the list of attributes that should be monitored to detect a vulnerability, both figures are available to us and can be compared. After a while of operating the URL-Profiler, we should be able to state facts about change rate of software and versions. This change rates can support defining the interval of re-profiling the same URL. Furthermore, the rate of maximum possible profiling jobs per second will be limited by our profiling architecture. Therefore we develop a distributed profiling architecture that will be presented in future work.

Twitter-Client: As Twitter delivers about one percent of the comprehensive firehose stream, we are able to estimate the proportion of tweets that we observe by our own filters. As Sampson et al. describe in [14] it should be possible to raise the coverage rate of tweets about the monitored keywords. Their remarks mostly agree with our so far findings about Twitter stream access possibilities.

URL-Extractor: We will be able to measure features like

- number of observed URLs per second, distinguished by place of occurrence (i.e., in tweet text or in user profile).
- statistical data about redirection hops, level of target URL (e.g., top-level-domain (TLD), sub-domain, deep-link of certain level), TLD ranking.
- statistical data about occurrence of new domain names in comparison with our change rate findings from the evaluation of the URL-Crawler.

Event Classifier: This component will need to be evaluated using strategies from natural language processing. A sophisticated pipeline for detecting events in Twitter messages [15] promises a precision and recall rate that is sufficient to support our system. Measurements like the number of relevant security events per minute, in comparison with all URL mentions and can be plotted on a time line in order to lead to further insights.

Event/Attribute-Matcher: Central figures to measure will answer questions like

1. What percentage of URLs in security events has been profiled in advance?
2. How many attributes do one, two, three or more security events have in common averagely?
3. Are we able to observe anomalies like sharp raises that indicate a new combination of attributes?
   Once we successfully answered the latter two questions, we should be able to assess the capability of the proposed system as a whole.

Vulnerability Matcher: When a vulnerability candidate is identified by the previous steps, we have to match it against the CVE-DB. We expect to find attacks to well known vulnerabilities and be able to state statistics about which vulnerabilities are exploited. If no match to the CVE-DB is possible, we have to monitor the CVE-DB for new entries. When a new entry matches to a previously found vulnerability candidate, the time between our detection and the addition of the new entry will mark the benefit that is achieved by the proposed social media supported vulnerability detection system.

Referring to Fig. 1, the upper half of the process, consisting of URL-Crawler, URL-Profiler, Twitter-Client and URL-Extractor, can be evaluated by inspecting the correctness of the output of these components. E.g., by now there are more than 130.000 URLs (unique top level domain names) in the URL-DB inserted by URL-Crawler component that match a regular expression pattern for valid domain names. After integrating the Tweet-URL-Extractor and URL-Profiler, significant statistical information about how many URLs have been crawled or extracted, how many of these are mentioned on Twitter and at what frequency this happens, how many we can associate with services like SSH, Telnet, FTP, Mail services and so on and how many of these we are able to profile will be available.

The evaluation of lower half of the process diagram, consisting of Event Classifier, Event/Attribute-Matcher and Vulnerability Matcher, probably has to involve manually annotated gold standards, to make use of weakly supervised learning techniques and determine results for true positive and negatives as well as for type I and type II errors. This evaluation will be part of future work, as our current focus is on finalizing a prototype of the proposed system.

5 Conclusion and Outlook

The proposed approach is dependent on users publishing cyber security event related messages in conjunction with a specific service. Knowledge about the set of online services that a single user is consuming, would even allow to match unspecific complaints of multiple users about cyber security incidents. If there has been an attack to a web service and, e.g., credit card data was stolen, complaints about the misuse of credit card credentials can reveal a hitherto unnoticed or unpublished successful attack to a certain service. Furthermore, multiple detection of similar events could even reveal the exploitation of an unknown vulnerability. This would state a big step forward in the early detection of security incidents and thus make cyber security more transparent to users that want to be informed about security breaches and exposition of private data as soon as possible as well as to administrators that aim to protect their networks in a more proactive way.

The approach that we present can set the stage for such future developments and provide various improvements even now. The proposed system could already enhance a company's IDS and by extracting URLs from the company's own monitored network traffic and insertion of these into the URL-DB a highly adaptive system could be realized. A simple network monitor to demonstrate this capability is already launched in our test set.

The sensitivity of security event detection could be improved by including additional languages. Starting with an English set of keywords, identified security event related tweets may provide certain hashtags that mark a specific event. Furthermore, the contemporary occurrence of URLs or web service descriptors in English tweets that we have identified to describe a security event and the same URLs or descriptors in tweets in other languages can provide a new set of vocabulary to listen to in an unknown language. Providing sufficient messages with intersecting hashtags or vocabulary and related security events, one should be able to build a universal dictionary for filtering security event tweets from social media stream nearly independent from the used language.

It will be challenging to deal with content delivery networks and load balancers. Also the perspective on a specific service of a user that reports a service outage has to be considered. To deal with these aspects, it might be necessary for some services to create a map of specific world wide content delivery architectures or load balancing strategies.

Acknowledgments The author wish to thank the members of the Chair for Communication Systems and Network Security at the Universität der Bundeswehr München, headed by Prof. Dr. Gabi Dreo Rodosek, for helpful discussions and valuable comments on previous versions of this paper. This work was partly funded by FLAMINGO, a Network of Excellence project (ICT-318488) supported by the European Commission under its Seventh Framework Program.

References

1. Zuech, R., Khoshgoftaar, T.M., Wald, R.: Intrusion detection and Big Heterogeneous Data: a Survey, J. Big Data **2**(1) (2015). doi:10.1186/s40537-015-0013-4, http://www.journalofbigdata.com/content/2/1/3
2. Aoyama, T., Naruoka, H., Koshijima, I., Machii, W., Seki, K.: Control Conference (ASCC), 2015 10th Asian, pp. 1–4. IEEE (2015)
3. Yang, C.C., Tseng, S.S., Lee, T.J., Weng, J.F., Chen, K.: Proceedings of the 12th IEEE International Conference on Advanced Learning Technologies, ICALT 2012, pp. 121–123 (2012). doi:10.1109/ICALT.2012.174
4. Seeber, S., Rodosek, G.D.: 10th International Conference on Network and Service Management (CNSM), 2014, pp. 376–381. IEEE (2014)
5. Seeber, S., Rodosek, G.D.: **9122**, 134 (2015). doi:10.1007/978-3-319-20034-7, http://link.springer.com/10.1007/978-3-319-20034-7
6. Derczynski, L.R.A., Yang, B., Jensen, C.S.: Proceedings of the 16th International Conference on Extending Database Technology—EDBT'13, p. 137 (2013). doi:10.1145/2452376.2452393, http://dl.acm.org/citation.cfm?doid=2452376.2452393
7. Wang, X., Wei, F., Liu, X., Zhou, M., Zhang, M.: pp. 1031–1040 (2011)
8. Narr, S., Hulfenhaus, M., Albayrak, S.: Proceedings of KDML-2012, the 2012 Workshop on Knowledge Discovery, Data Mining and Machine Learning (2012)
9. Shamsi, Z., Nandwani, A., Leonard, D., Loguinov, D.: pp. 195–206. doi:10.1145/2591971.2591972
10. Kergl, D., Roedler, R., Seeber, S.: 2014 IEEE/ACM International Conference on Advances in Social Networks Analysis and Mining (ASONAM 2014), pp. 357–364. IEEE, Asonam (2014) doi:10.1109/ASONAM.2014.6921610, http://ieeexplore.ieee.org/lpdocs/epic03/wrapper.htm?arnumber=6921610

11. Ritter, A., Wright, E., Casey, W., Mitchell, T.: Proceedings of the 24th International Conference on World Wide Web, pp. 896–905. International World Wide Web Conferences Steering Committee (2015)
12. Augustine, E., Cushing, C.: Proceedings of the 21st international conference companion on World Wide Web, pp. 13–22 (2012). doi:10.1145/2187980.2187983, http://dl.acm.org/citation.cfm?id=2187983
13. Motoyama, M., Meeder, B., Levchenko, K., Voelker, G.M., Savage, S.: Proceedings of the 3rd conference on Online social networks (WOSN'10) (2010). http://dl.acm.org/citation.cfm?id=1863203
14. Sampson, J., Morstatter, F., Maciejewski, R., Liu, H.: Proceedings of the 26th ACM Conference on Hypertext and Social Media, pp. 237–245. ACM (2015)
15. Bontcheva, K., Derczynski, L., Funk, A., Greenwood, M.A., Maynard, D., Aswani, N.: RANLP, pp. 83 (September 2013)

Profile Matching Across Online Social Networks Based on Geo-Tags

Robert Roedler, Dennis Kergl and Gabi Dreo Rodosek

Abstract Profile matching in social networks is a challenging problem. There are many previous approaches but all of them rely on data, for which there is no ground truth. Profile information is provided by the users. Therefore, there is no certainty about the reliability of this information. For this reason, we suggest using timestamps, which are generated by the social network, and device-generated geo-tags. We take a closer look at various approaches for implementation of this profile matching algorithm. In addition, possibilities for evaluation of this algorithm are outlined afterwards.

Keywords Social networks · Profile matching · User behaviour · Geo-tags · Timestamps

1 Introduction

Social Networks are growing faster then ever, facebook[1] reached recently almost 1.5 billion users.[2] But there is not only facebook, there are many social networks, each for a different purpose. Because social networks have specialized over the years people are using a variety of different networks for pictures (e.g. Flickr[3]), videos (e.g. Youtube[4]), short messages (e.g. Twitter[5]) or geo-located services (e.g. Foursquare[6]). But in the meantime, it is the rule not the exception to have even two or three or more

---

[1]http://www.facebook.com.
[2]http://newsroom.fb.com/company-info/.
[3]http://www.flickr.com.
[4]http://www.youtube.com.
[5]http://www.twitter.com.
[6]http://www.foursquare.com.

R. Roedler (✉) · D. Kergl · G.D. Rodosek
Department for Communication Systems and Network Security, Faculty of Computer Science, Universität der Bundeswehr München, Neubiberg, Germany
e-mail: robert.roedler@unibw.de

© Springer International Publishing Switzerland 2016
N. Pillay et al. (eds.), *Advances in Nature and Biologically Inspired Computing*,
Advances in Intelligent Systems and Computing 419,
DOI 10.1007/978-3-319-27400-3_37

417

accounts in contrast to only a single account in the earlier days of social networks. People owning just one account are already a minority.

It is not sufficient to gain access to one of these networks to obtain a comprehensive profile of a user, because the information is spread across all accounts of this particular user. That means that the included information in each account on a social network can improve the profiling.

Organizations, agencies and companies are very interested in obtaining comprehensive profiles of users. It is even a business model to sell this information. Especially for marketers and advertising agencies obtaining such information can have a big business impact. Since Edward Snowden, we also know that secret services are very interested in determining every information they can.

The remainder of this paper is organized as follows. Section 2 provides an overview of the typical problems of profile matching and explains open research questions. Section 3 refers to previous work that studies or relies on profile matching to underline the large research interest. Section 4 describes the further steps how profiles are matched by overcoming the weaknesses of other approaches. We conclude this paper in Sect. 5 and provide an outlook on future research challenges.

2 Problem Description

To accomplish a full picture of a user it is necessary to match his profiles. Different approaches have been used so far, which are outlined afterwards. Examples of the following approaches are discussed in the related work section, which follows this one.

- Matching based on profile attributes like name, hobbies, town etc. This approach focuses on the accordance of these textual attributes. Because of intentional or unintentional spelling mistakes the clarity of attributes across social networks is not ensured. That's why there is a variety of distance measures to overcome different spellings.
  It is a widespread technique, nevertheless, each of these attributes is user chosen and because of that easy to fake.
- A methodology using the network structure is based on the assumption, that a user has the same friends in each social network and therefore the number of friends and their names stay almost the same.
  In cases where this assumption is not true, for example when a user tries to separate different circles of friends or a user has a private and a business social network account, this approach fails.
- Analyzing written text to derive typical features of the writing style of a user shows only little capabilities for profile matching.
  This technique only provides for longer texts and a limited number of authors practicable results.

Each of these approaches has its own weaknesses, mainly based on the data they rely on, because of ambiguity and ease to fake. In consequence, for an efficient methodology you need data which is easy to compare, like numbers, coordinates or times and dates. Moreover, numbers do not lack certainty in contrast to words with different spellings. Additionally, a smaller-greater relation is obtained hereby. Therefore, there is no need for an additional metric; the distance between two numbers can be easily calculated. In addition, attributes like timestamps and geo-coordinates are difficult to falsify. Timestamps are generated by the social network itself and geo-tags are automatically included, if enabled, without prior access for possible changes. A user can try to distort his typical usage behavior, but in contrast to text-based attributes like names, timestamps and other metadata are not on the radar of most users.

This work addresses the following hypothesis:

- Usage behavior based on timestamps and geo-tags is so unique, that it is possible to match social network profiles using just this information.
- Times and places for social network usage concerning a single user vary only rarely.

There are also some other issues that should be addressed, which are not covered under the definition of a hypothesis. These additional research questions are:

- How many timestamps and geo-tags are needed for profile matching?
- What further information can be extracted based on the evaluation? (E.g. holidays, working hours, place of residence, ...)
- What are the privacy risks thereof?
- How can this threat be reduced?

Mining user profiles based on GPS traces is already extensively considered, e.g. [1, 2]. The main challenges arising from using geo-tags instead of traces are:

- Sparse data, large temporal and spatial gaps. This results from the fact that many people just post two or three updates a day. During the time between, there is no data which could be obtained.
- Numbers of decimal places vary, geo information can be fuzzy. Depending on the mobile device which generates the geo-tags, a geo-tag contains a fixed number of decimal places. The lower the decimal places, the more area is covered.
- Number of geo-tags is possibly small. The usage of geo-tags is still rare, but the percentage is rising slowly but continuously. Nevertheless, many users do not tag all of their updates. This means, that the number of geo-tags is even smaller than the number of updates of a certain profile.
- Period of geo-tags is possibly large. To obtain sufficient data from a profile, it is necessary to crawl all updates. This results in data which extends over a long period of time, typically years.

To solve and prove these hypothesis, questions and challenges, a new technique is necessary. There is no method at the moment which is able to afford profile matching on geo-tags. Therefore, the main task and the main contribution of our work is the elaboration of a new algorithm which enables the possibility to match profiles based on geo-tags.

3 Related Work

This section discusses the related work concerning profile matching on social networks. The approaches which are typically used were already outlined in the problem description. In the following subsections, important work concerning each of these approaches is considered in more detail.

3.1 Matching Profile Attributes

Zafarani and Liu [3] match identities across different websites based only on user names. The authors show, that user names can be successfully used for this purpose. Nevertheless, in this way it is only possible to match users who want to be found due always to using the same or similar user names. Another publication matching profiles using just the user names is [4]. Amongst other things, various string matching approaches are evaluated. Additionally, countermeasures for user and web services are also discussed. In [5] various profile attributes are used for identification. Classification and different weights for the attributes are considered to detect duplicated users. An extensive consideration of similarity metrics for strings and semantics and also weights for different attributes was carried out in [6]. The paper is written with close reference to FOAF.[7] A combination of profile attributes and network structure is proposed in [7]. The main contribution is a technique to infer missing attributes. In [8] similarity measures, online services and attribute weights are used for profile matching. Depending on the type of data, personal identities, social identities and relational identities, the similarity measures and online API calls are selected. A comprehensive consideration of similarity measures for profile attributes was carried out in [9]. This includes Levenstein, Jaro Winkler, Jaccard, Soundex and some more similarity measures. A similar consideration of string similarity measures is presented in [10]. Similarity metrics are considered in more detail in [11]. Additionally, profile attributes were weighted for a more effectively matching technique. For this purpose, each attribute is considered and a subjective weighting method is applied. Further publications dealing with profile matching based on profile attributes are [12] and [13].

3.2 Network Structure

One of the most cited papers for this category is [14]. The authors justifiably claim that this publication was the first using the network structure to match profiles. It also contains an extensive summary of related techniques. However, the ground truth of

---

[7]http://www.foaf-project.org/.

the dataset is not present or only with restrictions, because it is defined by a mapping elected by the authors. Several algorithms for matching the network structures of two social networks have been evaluated in [15].

3.3 Analyzing Written Text

To a greater extent, [16] falls into this category, although one additional profile attribute, the user id is used. The authors try to identify users based on their tagging practices and their user ids. The evaluation shows good results, but the dataset is very small, within the lower four-figure sector. [17] is a substantial contribution concerning author identification. Various classification approaches are considered and evaluated. From each text 1,188 features for classification were obtained, the text itself was not used for further analysis. Novak et al. [18] used four distributions: words, misspellings, punctuation and emoticons as a feature set. They made use of clustering algorithms to obtain meaningful clusters of profiles.

3.4 Combinations

The authors in [19] consider all previous approaches and try to combine them. This includes string-based, stylometric, time profile-based and social network-based matching. Because of the complexity of these approaches, each approach is only superficially scrutinized. How user information is distributed and how it can be aggregated is examined in [20]. Profile attributes and tags assigned to images etc. were combined for this purpose.

As already outlined in the problem description, one weakness of each of these approaches is that they rely on data, for which there is no ground truth. While users fill up their profile information themselves, there is no certainty that it is truthful information. Other weaknesses of these approaches are, that most of the used datasets for evaluation contained a number of users in the lower four-figure sector. But in the big world of social networks, such results are not representative.

3.5 Approaches Similar to Our Approach

There are already very few paper with similar approaches, which are also based on geo-tags. The authors in [21] infer social ties from geographic coincidences. A grid is used to analyze which flickr users visited the same cells in a limited time range. Users with several co-occurrences were often friends. The paper [22] tries to match profiles using geo-tags and timestamps. Nevertheless the aim of this paper is to reduce the set of possible matching pairs to a size a person can manually evaluate. In a given example, there still remain 750 possible pairs for a single match. This approach is not nearly as accurate as our approach will be by considering all later mentioned matching possibilities. Additionally, small datasets were used for both papers.

4 Approach

This section describes ideas and approaches for matching profiles. As already mentioned in Sect. 2, it is important to use the right data, data which is not distorted by the users. To achieve profiling on timestamps and geo-tags, the next step will be to evaluate different correlation mechanisms and pattern mining approaches, to see to what extent these enables the possibility for unambiguous assignments. Included here are traditional correlation methods like the correlation coefficient as well as time and geolocation correlation approaches. Additionally, techniques for mining of typical usage patterns are also considered. The following list outlines aspects and approaches, which will be checked for suitability to correlate, cluster or estimate profile information for matching:

- **Usage behavior in relation to times**: many people use their social network accounts at the same times. Therefore it is possible to extract patterns of usage behavior, for an example refer to Fig. 1. These patterns can be compared on the basis of a similarity measures or correlation approaches. Additionally, it is possible to infer daily schedules including working, sleeping and free time.
- **Usage behavior in relation to places**: By counting geo-tags and their corresponding distance, it is possible to draw connections. For example, in Fig. 2 it is seen, many tags in a very short distance mark most likely the home of this user (red). A slightly wider range could be the working place (blue), where there are different restaurants for lunch in the surrounding area. Places visited during free time are scattered (green).

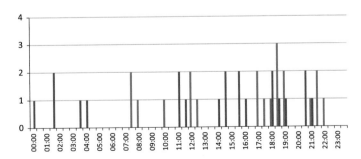

Fig. 1 Example for different usage behaviors relating to times. Each color represents a user

Fig. 2 Example for what could be inferred by analyzing geo-tags

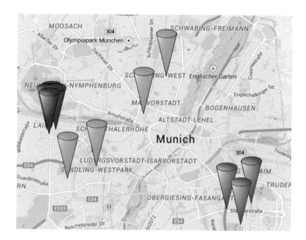

Fig. 3 Example of visiting frequencies

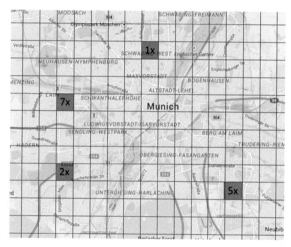

- **Visiting frequency for places**: By dividing the world into small grid cells, for each cell the visiting frequency can be calculated. Based on these frequencies, user-specific visiting patterns can be derived. Figure 3 presents an example for a user living in Munich.
- **Geo-tags related to time**: Holiday and business trips enable the possibility to distinguish between profiles. Geo-tags from different countries or with a great distance in a short sequence indicate different users. Thereby, it is possible to differentiate, even if their profiles show similar behavior most of the time. An example is shown in Fig. 4. Red and green represents two different users, but with similar usage behaviors. Just one geo-tag in a short period can decide whether it is the same user or not. The exclusion criterion for matching red and green is shown in the left upper corner.

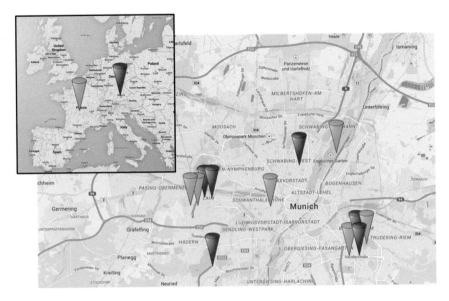

Fig. 4 Example of a geo-based exclusion criterion

- **Geo-tags accuracy**: For similar profiles with deviations relating to geo-tag accuracy, equality can be excluded using the accuracy, which depends on the generating device. Figure 5 shows an example of similar profiles but with geo-tags with different accuracies. It is important to note, however, that a user could use multiple devices. If so, it is probable that these devices are also used within each single profile. An exception to this are geo-tags in pictures, because a camera is unable to send status updates to social networks. Therefore, the accuracy of geo-tags, tagged by this camera, can only be find in the accounts that include these pictures.
- **Geo-tags sequences**: It is possible to differentiate excessive users by typical sequences of visited places. You need excessive users because one or two geo-tags a day are not enough to obtain a real sequence. The elapsed times between the geo-tags are also important. A threshold which determines, whether a geo-tag is still considered for a sequence, should be defined (Fig. 6).

These approaches will be analyzed and individually evaluated. Combinations of these are also taken into consideration. Based on the evaluation, the approaches providing the best results are merged into one algorithm. This algorithm is further optimized by weighting separate components.

Fig. 5 Example for
different geo-tag accuracies

Fig. 6 Two similar
sequences of geo-tags. The
assumption that both profiles
are related to one user is
obvious

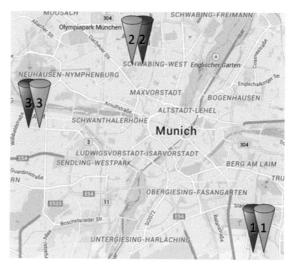

5 Evaluation

At least two datasets to match a profile are needed. As a trade-off between complexity for data gathering and the meaningfulness of the results, the aim is to use
three datasets for evaluation, e.g. Twitter, Flickr and Foursquare. These three social
networks were proposed, to show that it is not limited to networks with similar purpose and to emphasize, that geo-tags could not only be included in account updates
but also in pictures and so on. That is also the reason for choosing three instead
of just two social networks for evaluation, because otherwise it is not possible to
cover this field in the desired extent. To improve the possible results, these datasets

should be enriched by the accounts obtained by the next step. Another dataset, which is crawled amongst other sites from http://aboutme.com where people publish their various social network accounts, provides the ground truth.

Of course, there are many more account-pairs and for the intended procedure also triples of accounts belonging to single users. But to prove, whether the correct pairs are detected or not, this ground truth dataset is sufficient. If this approach works properly, these pairs and triples have to show up in the results for the evaluation datasets. Using only data from accounts included in the ground truth dataset for a further evaluation step, it is also possible to determine false positives.

The following tables and paragraphs provide some data characteristics and results. To obtain these results, we analyzed a twitter dataset, which was collected from Friday 2015-08-14 and Saturday 2015-08-15. The data was collect in 35 h and contains overall 2,588,981 geo-tagged tweets. For future datasets, we will also use our earlier findings with respect to twitter to increase the amount of received tweets [23].

Table 1 provides an overview about some user statistics. We obtained geo-tagged tweets of 827,472 distinct users. About 17 % of these users use geo-tags at least 4 times a day. This is important for sequence detection as outlined in Sect. 4. However, without examining this issue in detail, we believe that almost all user with more than 50 tweets use automatically generated messages. The maximum number of tweets we collected for a user was 5,260, which is totally implausible for manually written messages.

Table 2 contains an overview of the geo-tag accuracies that occurred. The most geo-tags provides six decimal places, nevertheless, there is enough scope to use the deviations for differentiation of different devices. Further analyzes will be part of our future work, due to space limitations of this paper.

Table 1 Amount of geo-tagged tweets per user

# tweets	# users	% of users
1	436,385	52.737
2	175,082	21.259
3	77,017	9.308
4	42,318	5.114
5	21,796	2.634
6	17,218	2.081
7	9,597	1.160
8	7,623	0.921
9	5,542	0.670
10	4,442	0.539
>10 & ≤100	29,388	3.546
>100 & ≤1000	1,078	0.130
>1000	30	0.004

Table 2 Distribution of geo-tag accuracies

# decimal places	# tweets	% of tweets
1	7,189	0.278
2	12,853	0.496
3	16,354	0.632
4	135,595	5.237
5	462,125	17.850
6	1,886,809	72.878
7	30,469	1.177
8	37,623	1.453

Nevertheless, it is obvious that it is not enough to only focus on one approach outlined in Sect. 4. A combination of all these techniques is necessary to obtain good results, because each single approach is not accurate enough to enable any reliable profile matching.

6 Conclusion and Outlook

Profile matching based on geo-tags is a work in progress at an early stage. The reasons why previous approaches were not able to find matches for users that do not want to be detected were outlined. Based on these findings, new techniques were proposed and will be outlined in more detail in future versions of this study.

There are also some countermeasures to prevent profile matching based on geo-tags, like limiting the accuracy, adding noise or discretizing to a grid. A consideration will be also part of our future work.

Acknowledgments The authors wish to thank the members of the Chair for Communication Systems and Network Security at the Universität der Bundeswehr München, headed by Prof. Dr. Gabi Dreo Rodosek, for helpful discussions and valuable comments on previous versions of this paper. This work was partly funded by FLAMINGO, a Network of Excellence project (ICT-318488) supported by the European Commission under its Seventh Framework Program.

References

1. Bayir, M.A., Demirbas, M., Eagle, N.: IEEE International Symposium on a World of Wireless, Mobile and Multimedia Networks and Workshops, 2009. WoWMoM 2009. IEEE, 2009, pp. 1–9
2. Trasarti, R., Pinelli, F., Nanni, M., Giannotti, F.: Proceedings of the 17th ACM SIGKDD International Conference on Knowledge Discovery and Data Mining, KDD'11. ACM, New York, NY, USA, 2011, pp. 1190–1198
3. Zafarani, R., Liu, H.: ICWSM, Citeseer (2009)

4. Perito, D., Castelluccia, C., Kaafar, M.A., Manils, P.: Privacy Enhancing Technologies. Springer, New York, pp. 1–17 (2011)
5. Vosecky, J., Hong, D., Shen, V.Y.: First International Conference on Networked Digital Technologies, 2009. NDT'09. IEEE, 2009, pp. 360–365
6. Raad, E., Chbeir, R., Dipanda, A.: 2010 13th International Conference on Network-Based Information Systems (NBiS). IEEE, 2010, pp. 297–304
7. Akcora, C.G., Carminati, B., Ferrari, E.: 2011 IEEE International Conference on Information Reuse and Integration (IRI). IEEE, 2011, pp. 292–298
8. Soltani, R., Abhari, A.: 2013 International Symposium on Performance Evaluation of Computer and Telecommunication Systems (SPECTS). IEEE, 2013, pp. 64–70
9. Peled, O., Fire, M., Rokach, L., Elovici, Y.: 2013 International Conference on Social Computing (SocialCom). IEEE, 2013, pp. 339–344
10. Cohen, W., Ravikumar, P., Fienberg, S.: KDD Workshop on Data Cleaning and Object Consolidation, vol. 3, pp. 73–78 (2003)
11. Na, Y., Yinliang, Z., Lili, D., Genqing, B., Liu, E., Clapworthy, G.J.: Communications. China **10**(12), 37 (2013)
12. Carmagnola, F., Osborne, F., Torre, I.: Proceedings of the 1st International Workshop on Information Heterogeneity and Fusion in Recommender Systems. ACM, 2010, pp. 9–15
13. Motoyama, M., Varghese, G.: Proceedings of the Eleventh International Workshop on Web Information and Data Management. ACM, 2009, pp. 67–75
14. Narayanan, A., Shmatikov, V.: 2009 30th IEEE Symposium on Security and Privacy. IEEE, 2009, pp. 173–187
15. Chen, D., Hu, B., Xie, S.: CS224W Course Project Writup (2012)
16. Iofciu, T., Fankhauser, P., Abel, F., Bischoff, K.: ICWSM (2011)
17. Narayanan, A., Paskov, H., Gong, N.Z., Bethencourt, J., Stefanov, E., Shin, E.C.R., Song, D.: 2012 IEEE Symposium on Security and Privacy (SP). IEEE, 2012, pp. 300–314
18. Novak, J., Raghavan, P., Tomkins, A.: Proceedings of the 13th International Conference on World Wide Web. ACM, 2004, pp. 30–39
19. Johansson, F., Kaati, L., Shrestha, A.: Proceedings of the 2013 IEEE/ACM International Conference on Advances in Social Networks Analysis and Mining. ACM, 2013, pp. 1004–1011
20. Abel, F., Henze, N., Herder, E., Krause, D.: User Modeling, Adaptation, and Personalization. Springer, New York, pp. 16–27 (2010)
21. Crandall, D.J., Backstrom, L., Cosley, D., Suri, S., Huttenlocher, D., Kleinberg, J.: Proc. Natl. Acad. Sci. **107**(52), 22436 (2010)
22. Goga, O., Lei, H., Parthasarathi, S.H.K., Friedland, G., Sommer, R., Teixeira, R.: Proceedings of the 22nd International Conference on World Wide Web. International World Wide Web Conferences Steering Committee, 2013, WWW '13, pp. 447–458
23. Kergl, D., Roedler, R., Seeber, S.: 2014 IEEE/ACM International Conference on Advances in Social Networks Analysis and Mining (ASONAM). IEEE, 2014, pp. 357–364

Detection of Criminally Convicted Email Users by Behavioral Dissimilarity

Maqsood Mahmud, Pranavkumar Pathak, Vaishalibahen Pathak and Zahra Afridi

Abstract The phenomenon of social interactions is prevailing charismatically like a spider net in nowadays society despite the people busy lives. In this fashion, people willingly supply their private or public data without sensing the threat of any information theft. These kinds of information could be easily misused and could be analyzed by any third party for malicious or non-malicious purposes. In this paper, detection of irregular or anomalous individual are focused. Individual with behavioral dissimilarity are discovered and validated with the real denounced victims. An affluent feature set of 15 characteristics is anticipated for deviation detection. The kth nearest neighbour technique is applied on the Enron dataset for finding accused email users. Noteworthy outputs are achieved by implication of the KNN method.

Keywords Outliers · Early warnings · Social communication networks · Anomaly detection · Behavioral dissimilarity

M. Mahmud (✉)
Department of Computer Sciences, ALMAAREFA Colleges of Science
and Technology (MCST), Riyadh, Kingdom of Saudi Arabia
e-mail: mmahmud@mcst.edu.sa

P. Pathak · V. Pathak
Department of Computer Sciences, Sardar Patel University, Anand, India
e-mail: pranavpp@gmail.com

V. Pathak
e-mail: vbhatt_85@yahoo.co.in

Z. Afridi
Department of Public Administration, Kenshu Institute of VU, Wah Campus,
Virtual University of Pakistan, Wah, Pakistan
e-mail: zahra.afridi@yahoo.com

© Springer International Publishing Switzerland 2016
N. Pillay et al. (eds.), *Advances in Nature and Biologically Inspired Computing*,
Advances in Intelligent Systems and Computing 419,
DOI 10.1007/978-3-319-27400-3_38

429

1 Introduction

Nowadays a trend is found that Internet is being used by e-criminals, fraudulent persons and scammers for execution of their malicious plans. The early warning systems depends on the philosophy of recovering and digging out data obtainable online or offline in the shape of emails on organization's server, news stuff or any data supplied by investigators. The thought at the back such warning architectures is primarily rate of recurrence or frequency depending [1].

The study of associations among entities like people, groups, or websites is systematically studied in social network analysis. Sometime social changes change the user's acts and behaviours regularly to distract detection. At a specific time an irregular entity residue sufficiently dissimilar from the "expected" or "normal" behaviour as can be seen in most of the entities visualized by the virtue of statistical analysis. The non-suspicious entities may be discarded by using the information in combination with other accessible meta-data. In an email network, it is usually observed that a person who directs or order a cluster or group of people is replicated in the majority of the emails that stream within the cluster. Consequently, this user may be noticed as an exceptional case due to very elevated interaction with their friends, but this user can be eradicated using some additional information [2].

Nowadays anomaly or irregularity detection is highly developed field of research. It can be efficiently practiced to different areas like credit-card fraud revealing; network intrusion detection, email depending network analysis etc. "Normal behavior is more pre-dominant than abnormal behavior" is the proposition on which the anomaly detection algorithms works. This phenomenon is shown by many network entities. For designing the proposed system, The Enron dataset was chosen for testing purposes. A number of features were selected out of the dataset to find out outliers in nodes. These nodes were deemed as malicious nodes as our features concentrated on the deviated behaviours. These features are discussed in details in later sections. Basically three main files were created for extracting the information from dataset. First file for (i) user information, the second for (ii) message information and third for (iii) interaction details information between users and messages. A runtime matrix was also generated to measure further processed information from these three basic files based on the fifteen selected features. After populating the Runtime Matrix, the square of Euclidean distance is calculated between a specific node/point with every node/point. The square of distance is taken for the sake of convenience. This process is perfumed with every node. Minimum distance is identified for every point. Then these minimum distances are listed in descending order. The top greatest points are considered as outliers. These outliers give clue of malicious nodes/users.

The rest of the paper is organized as follows. "Related Work" is elaborated in Sect. 2, "Proposed System" in Sect. 3, Sect. 4 describes "MUDS Architecture", "Results and Discussions" in Sect. 5 and "Conclusion and Future Work" in Sect. 6. Acknowledgments and References is followed by Sect. 6.

2 Related Work

In [2], Gupta and Dey proposed a proficient algorithm for irregularity discovery from social networks. Irregular individuals are discovered depending on their behavioural differences from other individuals. In [3], Ramaswamy et al. have proposed a new technique for distance depending outliers that depends on the distance of a point from its neighbour. In [4], Nithi and Dey proposed an efficient algorithm for anomaly detection from call data records. Anomalous users were detected based on fuzzy attribute values derived from their communication patterns. The kth closest neighbour algorithm [3] was implemented to find top n outliers. Authors showed experimental results which were better than Kojak because proposed algorithm showed 80 % results while Kojak showed 60 % results. The interesting thing was that it dealt with real life dataset and found extremely well results. In [5], authors have introduced the idea of multi criteria weighted graph similarity techniques and its application to observe some characteristics of social networks is brought under observation. In [6], Weinstein et al. described initial results on modelling, detection, and tracking of terrorist groups and their intents based on multimedia data. In [7], authors have presented a new network irregularity detection method depending on wavelet analysis, estimated auto regressive and outlier discovery methods. To differentiate network traffic behaviours, authors have presented 15 characteristics and have implemented these features as the put in signals in wavelet-based method. Kurkovsky et al. [8] have presented a multi-modal social networking architecture for the purpose of contributing geographic data amongst neighbouring individuals. Larsen and Vejin [9] used centrality measures. This methods was used to observe the destabization of network from multifaceted networks. They recently presented algorithms for building chain of command of the secret networks. In [10] authors have provided a detailed study of the research relevant to the study of vibrant modelling and link forecasting of SCN. In [11], authors have presented a framework, which consists of components of information extraction, blog spider, visualization of network and its analysis. In [12], de Lin and Chalupskyin described a novel unsupervised framework to identify abnormal instances. In the second part of the paper, authors have described an explanation mechanism to automatically generate human-understandable explanations for the discovered results. In [13], Bhatia and Gaur proposed a novel algorithm breadth first clustering. This clustering used arithmetical method for population mining in community networks. This method is straightforward, robust and can be configured without difficulty for huge community networks. Hui-Yi and Hung-Yuan [14], highlighted about the current, modern and advanced social networks and its prevailing impact on the society i.e. Facebook. Carrier and Spafford [15] described the process of digital investigation. In this paper they highlights that digital inspection have currently turned out to be more widespread, physical inspection subsisted for long age and the physical practices are practiced to the digital investigation. Numerous definitions subsist for each of the vocabulary and these were selected since they most precisely reproduce to the author's perspective of the issue [16].

3 Proposed System

The system that is going to be proposed will find out outliers and ultimately detect malicious nodes in the dataset. It was assumed earlier that users could be characterized by the number of emails sent or received, number of contacts, servers used for emails interactions etc. Statistical properties like average number of emails sent per day, average number of contacts, etc. have also been used. However, a user's actions cannot be inspected by summation values or mean values. Individual interaction patterns can better pigeon-hole a user in the perspective of global behavioral features. For this purpose we considered incoming features, out-going features and global features in detail. The rule of outlier discovery describes the irregularity or anomaly discovery algorithm. An outlier is termed as an observation, that diverges to a great extent from other observations that a person feel that this observation stimulate doubts that it was produced by a unlike methods. Unsupervised methods to recognize outliers are helpful in finding irregularities or uncharacteristic entities from huge datasets. An outlier is illustrated as follows [3]: Definition: Outliers of a set are the peak "n" data elements that are extreme from their kth adjacent neighbors. This distance measure is very proficient for examining spatial datasets. The following procedure or algorithm finds top n outliers or irregularities. The values "n" and "k" are given as inputs. This method finds all those entities, that are very dissimilar from their neighbors and that's why entitled to be outliers. The complexity of the below procedure or algorithm is O (N2). N is the total quantity of nodes. The procedure is stated below [2]:

Step 1: Select a value for k.

Step 2: For each node determine the distance from its kth closest neighbor using Euclidean distance. Step 3: Organize the data points in descending order of the distance obtained in step 2. Step 4: Select top n points as outliers. The initial design of the system depends upon the three relational schemas (i) UserFile (ii) MessageFile (iii) UserInteractionFile. These are given below with the fields. *UserFile (Uid, Name, EmailId)* UserFile Schema is the User information extracted from dataset. It contains User ID (UId), Name of Nodes as (Names) and Email address of users as (EmailId) of all users in dataset. *MessageFile (No, Mid, Subject, Size, Date, Time, Content)* Message File Schema is about the message details. It contains Message ID as (Mid), subject of a message as (Subject), size of an email as (Size (KB)), Date, Time and Contents as well. Content Analysis will be discussed in future research work. *UserInteractionFile (SourceUser, DestUser, MsgId, CommType, Attachements)* UserInteractionFile schema is about the interaction details of users with the corresponding emails sent and received. The first column is about the Source User who sends an email; the second column is about the Destination User who receives the message, augmented with message ID from Message detail file. Communication type is recorded to know the whether the communication is in the form of "TO" or "CC" or "BCC". Attachment information is also noted to see the frequent or infrequent attachment emails. It should be noted

Table 1 MUDS prototype results

	P1	P2	Distance
1			
2	jeff.dasovich@enron.com	veronica.espinoza@enror	9634.455078
3	sally.beck@enron.com	taffy.milligan@enron.con	4513.056641
4	richard.shapiro@enron.com	james.steffes@enron.cor	2641.174561
5	janette.elbertson@enron.com	david.forster@enron.corr	2272.520752
6	david.forster@enron.com	kay.chapman@enron.con	2241.006348
7	kay.mann@enron.com	tana.jones@enron.com	2103.902588
8	tana.jones@enron.com	kay.mann@enron.com	2103.902588
9	monika.causholli@enron.com	paul.kaufman@enron.cor	1437.817627
10	kay.chapman@enron.com	mary.hain@enron.com	1248.970215
11	mary.hain@enron.com	kay.chapman@enron.con	1248.970215
12	matthew.lenhart@enron.com	mary.cook@enron.com	1075.147217
13	mark.guzman@enron.com	geir.solberg@enron.com	948.7890625
14	stephanie.panus@enron.com	christi.nicolay@enron.cor	923.8248291
15	susan.scott@enron.com	liz.taylor@enron.com	896.3186035
16	steven.merris@enron.com	michael.mier@enron.con	865
17	craig.dean@enron.com	leaf.harasin@enron.com	828
18	rosalee.fleming@enron.com	taffy.milligan@enron.con	694.0465698
19	john.lavorato@enron.com	bill.williams@enron.com	689.1584473
20	liz.taylor@enron.com	drew.fossum@enron.con	673.3147583
21	drew.fossum@enron.com	liz.taylor@enron.com	673.3147583

that we used Square of Euclidean distance for our convenience. In general, the distance between two points x and y in a Euclidean space $\mathfrak{R}^n$ is given by Eq. 1.

$$d = \|x - y\| = \sqrt{\sum_{i=1}^{n} (x_i - y_i)^2} \qquad (1)$$

The thoughtful features sets defined for our experimentation on the dataset are listed below in Table 1. These features are about fifteen in number. These features are based on the behavioral aspect as well as self-identifiers. Table 1 is about the listing of features used for the experimentation. The details of these features are discussed below.

3.1 Feature Selection

In order to extract discriminative features for malicious entity identification, the e-mail communication network is modelled as a directed graph $G = (V, E)$ where, V is the set of nodes representing users and $E \subseteq V \times V$ is the set of directed edges representing communications between them. A directed edge originating from a node u and terminating to a node v is represented as $\langle u, v \rangle$. Once the communication graph is created, we have identified a set of 15 graph-based features that are explained in the following paragraphs.

Out-degree: Out-degree of a node is defined as the number of originating edges from it. For a node u, the out-degree is represented as $deg^+ (u)$ and it can be calculated using Eq. 2.

$$deg^+(u) = |\{\langle u, v\rangle | \forall v \in V \wedge \langle u, v\rangle \in E\}| \qquad (2)$$

In-degree: In-degree of a node is defined as the number of terminating edges to it. For a node u, the in-degree is represented as $deg^-(u)$ and it can be calculated using Eq. 3.

$$deg^-(u) = |\{\langle v, u\rangle | \forall v \in V \wedge \langle v, u\rangle \in E\}| \qquad (3)$$

In-Out Ratio (R_{IO}): This feature shows the ratio of in-degree and out-degree of a user u. The R_{IO} shows the variation in the flow of $deg^- v$ and $deg^+ v$ which shows In-Out degree behavior of a user mutually. Mathematically it can be interpreted as:

$$R_{IO}(u) = \begin{cases} \frac{deg^-(u)}{deg^+(u)}, & deg^+(u) \neq 0 \\ \propto, & otherwise \end{cases} \qquad (4)$$

Out-In Ratio (R_{OI}): This feature shows the ratio of out-degree and in-degree of a user u. The $R_{OI}(u)$ ratio shows the variation in the flow of $deg^+ v$ and $deg^- v$ which shows out-in degree behavior of a user mutually. Mathematically it can be represented as:

$$R_{OI}(u) = \begin{cases} \frac{deg^+(u)}{deg^-(u)}, & deg^-(u) \neq 0 \\ \propto, & otherwise \end{cases} \qquad (5)$$

Ratio Variance (V_R): This feature is for knowing the difference between the two ratio's obtained from the above two features. Mathematically it can be shown as:

$$V_R(u) = |R_{IO}(u) - R_{OI}(u)| \qquad (6)$$

Message Sent Time (T_S): This feature is considered to model the message sent time pattern of a user. For this purpose, all those messages sent during off-office time are assigned a weight 1 and remaining as zero.

$$T_S(u) = \frac{\sum_{i=1}^{deg^+(u)} t(u, v_i)}{\sum_{i=1}^{deg^+(u)} f(u, v_i)} \qquad (7)$$

Message Received Time (T_R): This feature is considered to model the message received time pattern of a user. For this purpose, all those messages received during off-office time are assigned a weight 1 and remaining as zero.

$$T_R(u) = \frac{\sum_{i=1}^{deg^-(u)} t(v_i, u)}{\sum_{i=1}^{deg^-(u)} f(v_i, u)} \tag{8}$$

Message Sent Day (D_S): This feature is deemed to model the message sent day pattern of a user. For this purpose, all those messages sent on off-days i.e. (Saturday and Sunday) are assigned a weight 1 and remaining as zero.

$$D_S(u) = \frac{\sum_{i=1}^{deg^+(u)} d(u, v_i)}{\sum_{i=1}^{deg^+(u)} f(u, v_i)} \tag{9}$$

Message Received Day (D_R): This feature is perceived to model the message received time pattern of a user. For this purpose, all those messages received during off-days are assigned a weight 1 and remaining as zero.

$$D_R(u) = \frac{\sum_{i=1}^{deg^-(u)} d(v_i, u)}{\sum_{i=1}^{deg^-(u)} f(v_i, u)} \tag{10}$$

Sent Messages with Attachments (A_S): This feature is deemed to model the message sent with attachments of a user. For this purpose, all those messages sent with attachments are assigned a weight 1 and remaining as zero.

$$A_S(u) = \frac{\sum_{i=1}^{deg^+(u)} a(u, v_i)}{\sum_{i=1}^{deg^+(u)} f(u, v_i)} \tag{11}$$

Received Messages with Attachments (A_R): This feature is perceived to model the message received with attachments of a user. For this purpose, all those messages received with attachments are assigned a weight 1 and remaining as zero.

$$A_R(u) = \frac{\sum_{i=1}^{deg^-(u)} a(v_i, u)}{\sum_{i=1}^{deg^-(u)} f(v_i, u)} \tag{12}$$

Sent Message Size Mean (M_S): This feature models the mean of sent message's sizes of a particular user. For this purpose, all messages sent are assigned a weight according to their sizes.

$$M_s(u) = \frac{\sum\limits_{i=1}^{deg^+(u)} s(u, v_i)}{\sum\limits_{i=1}^{deg^+(u)} f(u, v_i)} \tag{13}$$

Received Message Size Mean (M_R): This feature models the mean of received message's sizes of a particular user. For this purpose, all messages received are assigned a weight according to their sizes.

$$M_R(u) = \frac{\sum\limits_{i=1}^{deg^-(u)} s(v_i, u)}{\sum\limits_{i=1}^{deg^-(u)} f(v_i, u)} \tag{14}$$

Least Contacted User (LC_dU): The feature shows directional behavior of communication for user u in an email network. The underlying relational is to find abnormality in the communication by profiling the behavior of a specific malicious node while communicating with users v_i. The result is then used in k-distance calculation for final classification. Mathematically we define LC_dU as:

$$LC_dU(u) = \frac{\sum\limits_{i=1}^{deg^-(u)} f(v_i, u)}{\sum\limits_{i=1}^{deg^+(u)} f(u, v_i)}, where\ f(v_i, u) = \begin{cases} 0, & if,\ f(u, v_i) = 0 \\ f(v_i, u), & otherwise \end{cases} \tag{15}$$

Least Contacting User (LC_gU): The feature shows directional behavior of communication for user u in an email network. The underlying relational is to find abnormality in the communication by profiling the behavior of a specific malicious node while communicating with users v_i.

$$LC_gU(u) = \frac{\sum\limits_{i=1}^{deg^+(u)} f(u, v_i)}{\sum\limits_{i=1}^{deg^-(u)} f(v_i, u)}, where\ f(u, v_i) = \begin{cases} 0 & if\ f(v_i, u) = 0 \\ f(u, v_i) & otherwise \end{cases} \tag{16}$$

4   MUDS System Architecture

Figure 1 shows the proposed prototype MUDS (Malicious Users Detection System) design. In this prototype design, "Mailboxes" are extracted from Internet by the help of POP and IMAP. The "Mailbox Extraction" module is further connected with "Email Repository" in a unidirectional fashion while connected bidirectional with "User Interface". In the diagram "Transformation Sub System" module transform all emails in desired way to help in outlier detection. The "Intermediate Representation (IR)" module takes information from "Transformation Sub System" and stores in the next module called "Intermediate Representation" module. In IR, four relations are extracted (i) User Table (ii) Message Table (iii) Interaction Table (iv) Results Table. The IR module helps kth nearest neighborhood algorithm module to perform algorithm and extract results using Euclidean distances based on the above selected fifteen features. The "Computation" module has bidirectional relation with IR. "Outlier Ranking" module is attached with IR in a unidirectional way. It computes the malicious user and rank them to find out the accused malicious users. "User interface" is connected with "Outlier Ranking" module to interact and represent information to operator or intelligence agencies.

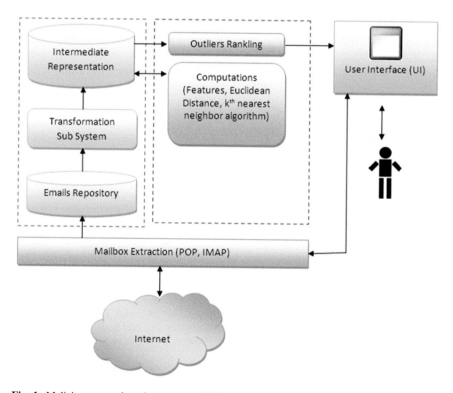

Fig. 1 Malicious users detections system (MUDS) system architecture

5 Results and Discussion

The results obtained from the MUDS are based on the feature presented in Sect. 3.1. A variant of KNN method is simulated to gain desired results. The desired result in ENRON dataset are the detection of three malicious users in top twenty transactions. As it is observed in Table 1. Mr. Sally beck and Mr. Richard Shapiro are flittered in the MUDS systems. Since these victims were found guilty for Enron fraud scandal in Oct 2001. The third accused user Mr. John Lavorato is shown on 19th position. These simulated results show the matching of victims in real world. The algorithm selection for the dataset Enron for finding the malicious user needs careful and thorough testing of dataset. In this case kth nearest neighborhood with selected fifteen features gives a perfect match for finding the malicious users rather than other classification methods.

6 Conclusions and Future Work

Outliers were successfully found out by using kth nearest neighbor algorithm. Features were thoughtfully chosen to meet the requirements of outlier detection and hence finding malicious users. These results can be used further for the analysis of more complex social networks nowadays e.g. Facebook, Twitter, LinkedIn or Orkut etc. Partition based algorithm [3] can also be used in case of very large social networks. Contents analysis of email body may also addressed in future to give more strengthened results of malicious users with proof from text rather than behavior analysis.

References

1. Qureshi, P.A.R., Memon. N., Wiil, U.K.: EWaS: novel approach for generating early warnings to prevent terrorist attacks 2010. In: Second International Conference on Computer Engineering and Applications, pp. 410–414 (2010)
2. Gupta, N., Dey, L.: Detection and characterization of anomalous entities in social communication networks. In: 20th International Conference on Pattern Recognition, pp. 738–741, Aug 2010
3. Ramaswamy, S., Rastogi, R., Shim, K.: Efficient algorithms for mining outliers from large data sets. In: Proceedings of the 2000 ACM SIGMOD International Conference on Management of Data, Texas, United States (2000)
4. Nithi, Dey, L.: Anomaly detection from call data records. Soc. Netw. 237–242 (2009)
5. Tarapata, Z., Kasprzyk, R.: An application of multicriteria weighted graph similarity method to social networks analyzing. In: International Conference on Advances in Social Network Analysis and Mining, vol. 1, pp. 366–368, Jul 2009
6. Weinstein, C., Campbell, W., Delaney, B., Leary, G.O., Street, W.: Modeling and Detection Techniques for Counter-Terror Social Network Analysis and Intent Recognition (2009)

7. Lu, W., Tavallaee, M., Ghorbani, A.A.: Detecting network anomalies using different wavelet basis functions. In: 6th Annual Communication Networks and Services Research Conference (cnsr 2008), pp. 149–156, May 2008
8. Kurkovsky, S., Strimple, D., Nuzzi, E., Verdecchia, K.: Mobile voice access in social networking systems. In: Fifth International Conference on Information Technology: New Generations (itng 2008), pp. 982–987, Apr 2008
9. Larsen, H.L., Vejin, N.B.: Practical Approaches for Analysis, Visualization and Destabilizing Terrorist Networks (2006)
10. Negnevitsky, M., huey Lim, M.J., Hartnett, J., Reznik, L.: Email communications analysis: how to use computational intelligence methods and tools? System 16–23 (2005)
11. Mining, U.W.E.B., S. Network, A. To, S. The, E. Of, C. Communities, and I.N. Blogs.: Using web mining and social network analysis to study the emergence of cyber communities in blogs. Computer
12. de Lin, S., Chalupsky, H.: Discovering and explaining abnormal nodes in semantic graphs. IEEE Trans. Knowl. Data Eng. **20**, 1039–1052 (2008)
13. Bhatia, M.P.S, Gaur, P.: Statistical approach for community mining in social networks. In: IEEE International Conference on Service Operations and Logitics, and Informatics, pp. 207–211, Oct 2008
14. Hui-Yi, H., Hung-Yuan, P.: Use behaviors and website experiences of Facebook community. Statistics **1**, 379–383, 2010
15. Carrier, B., Spafford, E.H.: Getting physical with the digital investigation process. Int. J. **2**, 1–20 (2003)
16. Saferstein, R.: Criminalistics: An Introduction to Forensic Science, 7th edn. Pearson, London (2000)

Cognitive Radio Networks: A Social Network Perspective

Efe F. Orumwense, Thomas J. Afullo and Viranjay M. Srivastava

Abstract Cognitive radio has received various research attention due to its capability of addressing inefficient spectrum usage and spectrum scarcity problems in wireless communications. In this article, cognitive radio networks in respect to social networking is studied and the trust between cognitive nodes in the network is evaluated. Interaction and cooperation amongst cognitive nodes is seen to improve sensing reliability, increased learning and better energy efficiency which in turn enhances the overall network throughput.

Keywords Cognitive radio networks · Social network · Primary user · Secondary user · Spectrum band · Wireless communication

1 Introduction

Wireless communication has indeed been one of the fastest developing sector of the communications industry in recent years due to the fact that wireless applications has steadily been on the increase. As a result, various wireless applications and systems operating in unlicensed spectrum bands have gradually led to the over-crowding of the spectral bands making them scarce and unavailable. However, investigation into the spectrum scarcity problems by numerous regulatory bodies around the world, including the United States Federal Communication Commission (FCC) and the Independent Regulator and Competition Authority (OfCom) in the

E.F. Orumwense (✉) · T.J. Afullo · V.M. Srivastava
School of Electrical, Electronics and Computer Engineering, Howard College,
University of KwaZulu-Natal, Durban 4041, South Africa
e-mail: efe.orumwense@gmail.com

T.J. Afullo
e-mail: afullot@ukzn.ac.za

V.M. Srivastava
e-mail: viranjay@ieee.org

© Springer International Publishing Switzerland 2016
N. Pillay et al. (eds.), *Advances in Nature and Biologically Inspired Computing*,
Advances in Intelligent Systems and Computing 419,
DOI 10.1007/978-3-319-27400-3_39

United Kingdom, have reported that although the demand for spectrum will significantly increase in the near future the major problem is not the spectrum scarcity but the inefficiency in spectrum usage [1–3].

Hence to address the inefficient spectrum usage and spectrum scarcity problems, a new approach for spectrum management is required. This approach should be capable of providing wireless access to unlicensed cognitive radio users, also known as secondary users (SUs), by allowing them to opportunistically gain access to unoccupied licensed spectrum while simultaneously guarantying the rights of incumbent users, also known as primary users (PUs) who possesses a *"first class"* access or legacy rights across the spectrum [4]. This implies that a licensed spectrum band can be accessed by a secondary user only if not in use by a primary user. This new approach is referred to as Dynamic Spectrum Access (DSA) [5].

The cognitive radio technology [6–8], plays an important role in ensuring the realization of this DSA paradigm. The concept of cognitive radio was first proposed by Joseph Mitola [6] where cognitive radio was described as software defined radio (SDR) [9] which possesses a more flexible approach to wireless communication. A cognitive radio has the ability to learn from its environment and intelligently adjust its parameters based on what has been learned. So in DSA, a cognitive radio can learn about the spectrum usage status of a band and automatically decides if the band is occupied by the primary user or not.

Since cognitive radio was first presented, several opinions and results have been received and since then, the term cognitive radio has become overloaded with many potential meanings. The definition adopted by the Federal Communication Commission (FCC) [10], a body set up to regulate spectrum usage in the US is

> *Cognitive Radio: A radio or system that senses its operational electromagnetic environment and can dynamically and autonomously adjust its radio operating parameters to modify system operation, such as maximize throughput, mitigate interference, facilitate interoperability, access secondary markets* [10].

When CRs are interconnected, they form Cognitive Radio Networks (CRNs). The basic components of a cognitive radio network are mobile station/cognitive radio user, base station/access point or fusion center and backbone/core networks. These three basic components compose three kinds of networks architectures in the CRNs.

In this paper, a brief overview of cognitive radio from a social network point of view is discussed and how a cognitive radio user or human behaviour can impact on cognitive radio networks is also analyzed. The paper is organized as follows. In Sect. 1.1, cognitive radio architecture is discussed. Section 2 presents a social network analysis for cognitive radio networks. Section 3 studies the cognitive radio user behaviour while Sect. 4 concludes the paper.

1.1 Cognitive Radio Architecture

A cognitive radio network (CRN) is not just a network of interconnected cognitive radios but CRN are composed of various kinds of communication systems and networks that can be viewed as a sort of heterogeneous network. Cognitive radios in a CRN, has the ability to sense available networks and communication systems around it. A typical CRN environment also consists of a primary user or a number of primary radio networks that coexist within the same geographical location of a cognitive radio network. A primary network is an existing network that is licensed to operate in a certain spectrum band. Hence, a primary network is also referred to as a licensed network. The design of cognitive radio network architecture has the objective of optimizing the entire network utilization, rather than only maximizing spectral efficiency.

CRNs can be deployed in centralized, distributed, ad-hoc or mesh architectures, and serve the needs of both licensed and unlicensed user applications. The basic components of CRNs are cognitive users, the primary user, base stations and core networks. These four basic components compose the three kinds of network architectures in CRNs which are infrastructure, ad hoc and mesh architectures [11].

The infrastructure base architecture as shown in Fig. 1, operates in a manner that the cognitive radio base station controls and coordinates the transmission activities of the secondary cognitive radio users. In the ad hoc based infrastructure as shown in Fig. 2, there is no infrastructural support. The CR users communicate directly with each other in an ad hoc manner and information is shared between the cognitive radio users who fall within this communication range. While the mesh infrastructure illustrated in Fig. 3 combines both the infrastructure and ad hoc based architectures.

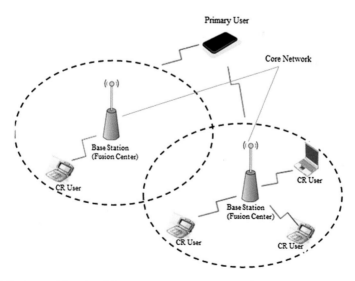

Fig. 1 Infrastructural based architecture

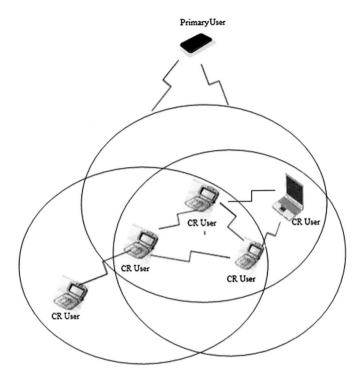

Fig. 2 Ad-Hoc infrastructure

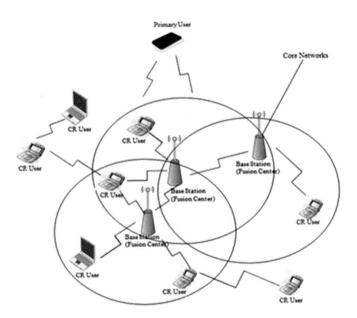

Fig. 3 Mesh infrastructure

2 Social Network Analysis for Cognitive Radio Networks

A social network accesses a network as a group of nodes with their interrelations (e.g., physical distance, contact frequencies) to benefit from these structural and social ties for mutual benefits and greater efficiency. A CRN is undoubtedly a social network in which the CRs may have various ties with each other depending on their spatial, intrinsic and social properties. Hence, discovery the interconnections among CRs and designing protocols accordingly can substantially improve a CRN performance.

In Fig. 4, an illustration of a social network consisting of CRs in which the nodes have different characteristics and a diverse view of the same network according to their social ties is shown. Social graphs are also important in keeping track of interactions and social ties among cognitive user. Interactions are very useful in assessing the structure of the network, an example is the connectivity and proximity of two cognitive users, whereas social ties may help in identifying the trust amongst the nodes and the influential nodes in the network. This information can be exploited in the design and implementation of energy-efficient cooperative protocols. Previous works on cooperative spectrum sensing [12–14], usually encourage cooperation amongst individual CRs and tacitly assume that each CR node is cooperative. However, such cooperative behavior may not be applicable in practical CRNs. For example, a CR node that is not having any active communication will run out of battery just because it constantly receives spectrum sensing information

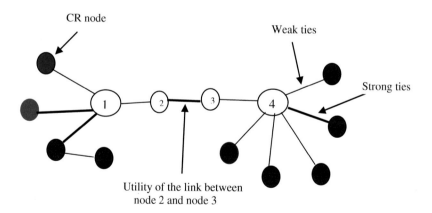

Fig. 4 A CRN as a social network with different social ties

from other CR nodes and consumes much of its battery on sensing. Instead, we visualize a more realistic operational scheme in which the cooperation willingness of CR nodes depends on the social ties amongst the users of these CR devices.

In ref. [15], it is shown that the CR nodes benefit from such a social-aware cooperative sensing scheme, where each CR node selects a cooperation set based on its friendship ties together with the historical sensing performance of the cooperating nodes. Moreover, the CR nodes can also make use of other CR nodes recommendations on the availability of primary user channels [16]. Thus, CR nodes are expected to spend less effort in spectrum sensing and learning that will eventually lead to higher energy efficiency in the network.

Environmental awareness improves the CR performance by letting the CR proactively take the best action at the expense of increased energy consumption due to constant environment monitoring. On the other hand, the CRs can share their experiences in order to reduce this burden. However, this solution raises the following question: To which extent can a CR trust other CR's reports, and what if the recommendations are inaccurate? These questions motivates us to examine trust in cognitive radio social networks.

2.1 Social Network Trust

Trust is a key factor that cuts across many facets of disciplines and based upon it many relationships are formed. Whether it is used for security on recommended systems, the issue of trust will help in successful message transmission amongst network entities. Although trust can be rather sophisticated in different contexts, we can model trust among CRs based on their social ties (e.g., friendship and community) and dynamically adapt the trust between CRs based on their interactions and feedback. In another sense, CR nodes of a cognitive radio network can form a social relationship between themselves to help build trust in the network. A set of CR nodes can form a sub group and give positive or negative rating on each other based on their previous encounters in order to determine and assess the trust rating of each other.

In this way, untrustworthy nodes can be detected cooperatively because of their low trust rating. Therefore, the information can be diverted from these untrustworthy nodes or the information originating from these nodes can either be ignored or disregarded before the fusion center makes a final decision.

Also in a cooperative scenario, a node can change its association with a neighboring node when it finds out that the level of trust value of that node has drastically been reduced thus ensuring the network operates in a trustworthy manner. We expect such a cooperative learning technique will tighten the security in the network against untrustworthy users and attacks and also improve the energy efficiency of the network.

All of these discussed CR tasks can benefit from social awareness and networking to achieve higher performance at lower energy cost, and thereby providing an energy-efficient operation.

2.2 Social Awareness in Cognitive Radio Networks

Cognitive radio is usually employed by secondary users in a wireless communication environment to opportunistically use an unused spectrum so as to maximize spectral efficiency. In these wireless communication environment, the users are human-centered, and human's behaviour will exert an influence on the network performance. Therefore, it is imperative for the development of wireless communications to study the social patterns of the humans involved in wireless communications. Generally, nodes of a cognitive radio network follows a *store-carry-forward* fashion to transmit data. Connections between the nodes are not always reliable and secure so each node has to store the data it has received, carries the data when moving and forwards them to the other nodes which are possible to reach the destination. In other words, the nodes do not get connected continuously which plays a vital role in the energy efficiency of the network [17].

Since the social relation of cognitive radio users have an impact on their behaviour, it is significant to exploit the social factors that relates to the workability of the network. These is classified into two categories: the positive social characteristics and the negative social characteristics. The positive social properties include friendship, community and centrality. The friendship of two nodes reflects their social strength. Based on homophily phenomenon, individuals are likely to be friends with people who have the same interests [18]. Nodes in good friendship are more possible to get in touch. Community is a significant concept in sociology, it basically means a group of people living in the same location [19]. In addition, a community can refer to a university, company or even nationality. It is reasonable that people in the same community have a higher sense of identity for each other. For instance, it is ordinary that people have a favourable impression of the graduates from their own university, even if they never met each other before. The centrality reflects the importance of a node in the network, in terms of degree and closeness. The nodes of high centralities, that is nodes of influence, deserve to be concerned for relay selection [20]. The major negative social property is selfishness (malicious nodes) [21]. A selfish node aims at maximizing its own utility regardless of the global performance of the network. In specific, the selfish node [22], drops the message of others, and always replicates its own data to enhance the delivery of its own message.

From Fig. 5, we can see the upper social connectivity layer displays the social ties among cognitive radios in a cognitive radio network. More precisely, it is the social relationship between the users who carry cognitive radio devices. A link in the social connectivity layer means the two users are friends. The wireless connectivity layer reflects the wireless communication of cognitive radios.

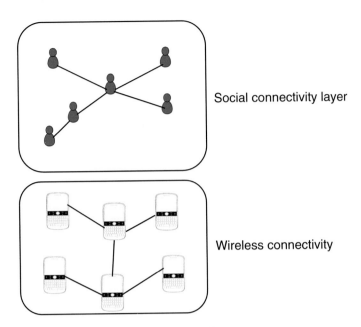

Fig. 5 Connectivity layer of a cognitive radio network

A connection between two devices in wireless connectivity layer suggests that they are inside the wireless transmission range of each other.

Device to device communications is an important element of next generation cellular network, which allows data to be transmitted directly between cognitive radio users instead of a base infrastructure. Since it is human beings taking these wireless devices, interaction of these devices are largely influenced by the social structure of their users. So therefore, the social properties of community and centrality is employed to help assign the limited spectrum resources among users in the network. They are also used to maintain a stable transmission link between devices.

3 Cognitive Radio User Behavior

We discuss cognitive radio user behavior as it relates social network since the most important stakeholder in any communication network is the user. A CR is a learning entity that autonomously senses and makes decisions based on the environment in which it operates. However, user–device interaction is usually overlooked. We discourse an example from modern cellular devices about how user–device interaction can save energy, but the arguments also apply to CR devices. Almost all up-to-date cellular devices or smartphones have several wireless interfaces: Bluetooth, WiFi, and GPS units in addition to third/fourth generation (3G/4G) cellular. Moreover, these devices have all of these circuits switched on in their factory setting. On the other hand, an average user does not frequently use these protocols, especially Bluetooth and GPS. The critical point is that the user does not care about the energy consumption of these circuits unless the device has a low battery. In addition, some users do not know how to turn them off to save energy. Thus, these protocols periodically seek some pairing or association all the time.

A CR device can learn and analyze user behavior such as when and/or where the user utilizes these types of additional communication units, and turn them off when it predicts that they are not needed. In addition to the end-user side, user behavior modeling can also let the operator and network designer make their short- and long-term plans with a view to increasing energy efficiency. These plans include elements such as frequency allocation, radio access network design, and operational time table for network equipment. For instance, the operator may switch the backbone equipment to low-power mode if its prediction based on the users' behavior indicates that the network traffic will be minimal for a specific location and time period.

4 Conclusion

In this article, we have studied cognitive radio networks in a social networking point of view. Trust in social networking and cognitive radio user behavior is also examined. Social networking approach is very vital in cognitive radio networks due to the central role it plays in the interaction and cooperation amongst cognitive radios. In this regard, cognitive radio devices can now benefit in terms of improved sensing reliability and increased energy efficiency. The improvement in the cognitive user's learning capabilities will also better its behavior and help act accordingly and also detect untrustworthy users.

References

1. E.C.N. Federal Communications Commission (FCC): "03-222", Notice of Proposed Rule Making and Order. In: Implementation of the Final ACTS OF THE World Radio Communication Conference (WRC-07), Geneva, August (2003)
2. McHenry, M.: Spectrum white space measurements. Presented to New America Foundation Broadband Forum, Shared Spectrum Company, Technical report, June 2003
3. US Federal Communications Commission: "Spectral policy task force report. Report of the Unlicensed Devices and Experimental Licenses Working Group. Technical report ET Docket 02-155, November 2002
4. Cabric, D., Mishra, S., Willkomm, D., Brodersen, R., Wolisz, A.: A cognitive radio approach for usage of virtual unlicensed spectrum. In: Proceedings of the 14th IST Mobile and Wireless Communications Summit, Dresden, Germany, pp. 1–5, June 2005
5. Zhao, Q., Sadler, B.M.: A survey of dynamic spectrum access. IEEE Sig. Process. Mag. 24(3), 79–89 (2007)

6. Mitola, J., Maguire, G.Q.: Cognitive radio: Making software radios more personal. IEEE Commun. Mag. **6**(4), 13–18 (1999)
7. Chen, K.C., Peng, Y.C., Prasad, N., Liang, Y.C., Sun, S.: Cognitive radio network architecture; Part 1—General structure. In: Proceedings of the ACM International Conference on Ubiquitous Information Management and Communication, Seoul, pp. 114–119, February 2008
8. Haykin, S.: Cognitive radio: brain-empowered wireless communications. IEEE J. Sel. Areas Commun. **23**(2), 2015–2020 (2005)
9. Mitola III, J.: Software radio architecture: a mathematical perspective. IEEE J. Sel. Areas Commun. **17**(4), 514–538 (1999)
10. Federal Communications Commission: Notice of proposed rulemaking. First report and order, in the matter of unlicensed operation in the TV broadcast bands, ET Docket No. 04-186 (FCC 04-113), May 2004
11. Chen, K.C., Peng, Y.C., Prasad, N., Liang, Y.C., Sun, S.: Cognitive radio network architecture; Part 1—General structure. In: Proceedings of the ACM International Conference on Ubiquitous Information Management and Communication, Seoul, pp. 114–119, February 2008
12. Ganesan, G., Li, Y.G.: Cooperative spectrum sensing in cognitive radio networks. In: Proceedings IEEE Symposium. New Frontiers in Dynamic Spectrum Access Networks (DySPAN 2005), Baltimore, USA, pp. 137–143, November 2005
13. Mishra, S.M., Sahai, A., Brodersen, R.: Cooperative sensing among cognitive radios. In: Proceedings IEEE International Conference on Communications, Turkey, vol. 4, pp. 1658–1663, June 2006
14. Ghasemi, A., Sousa, E.S.: Collaborative spectrum sensing for opportunistic access in fading environments. In: Proceedings IEEE Symposium. New Frontiers in Dynamic Spectrum Access Networks (DySPAN 2005), Baltimore, USA, pp. 131–136, November 2005
15. Güven, C., Bayhan, S., Alagöz, F.: Effect of social relations on cooperative sensing in cognitive radio networks. In: 1st International Black Sea Conference Communication and Networking, pp. 247–51, July 2013
16. Li, H.: Social behaviour in cognitive radio. In: Cognitive Communications: Distributed Artificial Intelligence (DAI), Regulatory Policy, Economics, Implementation. Wiley (2012)
17. Orumwense, E.F., Afullo, T.J., Srivastava, V.M.: Secondary user energy consumption in cognitive radio networks. In: Proceedings of the IEEE Africa Conference (Africon 2015), Addis Ababa, Ethiopia, September 2015
18. McPherson, M., Smith-Lovin, L., Cook, J.M.: Birds of a feather: homophily in social networks. Ann. Rev. Soc. **27**(10), 415–444 (2001)
19. McMillan, D., Chavis, D.: Sense of community: a definition and theory. J. Commun. Psychol. **14**(1), 6–23 (1986)
20. Marsden, P.V.: Egocentric and sociocentric measures of network centrality. Soc. Netw. **24**(4), 407–422 (2002)
21. Orumwense, E.F., Afullo, T.J., Srivastava, V.M.: Effects of malicious users on the energy efficiency of cognitive radio networks. In: Proceedings of the Southern Africa Telecommunications Networks and Applications Conference (SATNAC), Hermanus, South Africa, pp. 431– 435, September 2015
22. Orumwense, E.F., Oyerinde, O.O., Mneney, S.H.: Impact of primary user emulation attacks on cognitive radio networks. Int. J. Commun. Antenna Propag. **4**, 19–26 (2014)

Author Index

A

Abayomi, Abdultaofeek, 271
Abera, Eyob Shiferaw, 211
Abraham, Ajith, 27, 305, 223, 211, 117, 97
Adetiba, Emmanuel, 271, 281
Afridi, Zahra, 429
Afullo, Thomas J., 441
Ahmed, Nada, 27
Ali, Ahmed Fouad, 359
Analco, Martín Estrada, 189
Atkin, Jason, 247
Ayala-Rincón, Mauricio, 73

B

Belay, Ayalew, 211
Bergerhoff, Leif, 261
Bernábe-Loranca, María Beatríz, 189

C

Cherif, Nadjib Fodil, 37
Cilliers, Michael, 293
Coulter, Duncan, 293, 327

D

da Silveira, Lucas A., 73
Dash, Sujata, 1
de Lima, Thaynara A., 73
Dlodlo, Mqhele E., 347
Dombo, Dereck, 315
Downs, Kevin, 327
Drias, Habiba, 37
du Plessis, Mathys C., 129
Du, Xiaorong, 61

E

Ehlers, Elizabeth Marie, 179
El-aal, Shereen A., 105
Engelbrecht, Andries P., 381
Esmat, Saleh, 359

Ezzedine, Anna Bou, 235

F

Folly, Komla Agbenyo, 315
Frankland, Clive, 165

G

Ghali, Neveen I., 105
Greyling, Jean H., 129
Grmanová, Gabriela, 235

H

Habtie, Ayalew Belay, 223, 305
Hassan, Ahmed, 201
Hassan, Amira Kamil Ibrahim, 117
Hassanien, Aboul Ella, 359
Helbig, Mardé, 381, 393
Heukelman, Delene, 271
Heydenrych, Mark, 179

I

Igwe, Kevin, 151
Ismail, Fatma Helmy, 359

K

Kala, Jules R., 141
Kechid, Amine, 37
Kergl, Dennis, 405, 417

L

Lacko, Peter, 235
Lamine, Elyes, 49
Landa-Silva, Dario, 247
Laurinec, Peter, 235
Liu, Shanshan, 97
Lóderer, Marek, 235
Louwrens, Michael W., 129
Lucká, Mária, 235
Lusilao-Zodi, Guy-Alain, 347

M
Mahmud, Maqsood, 429
Midekso, Dida, 223, 305
Moodley, Deshendran, 141

N
Naeem, Usman, 49

O
Ojha, Varun Kumar, 27
Okoye, Kingsley, 49
Olugbara, Oludayo O., 271, 281
Orumwense, Efe F., 441

R
Pathak, Pranavkumar, 429
Pathak, Vaishalibahen, 429
Penna, Alejandro Fuentes, 189
Pillay, Nelishia, 151, 337, 201, 165
Pinheiro, Rodrigo Lankaites, 247
Poonia, Ajeet Singh, 87

R
Rajpurohit, Jitendra, 87
Ramadan, Rabie A., 105
Rawat, Sanyog, 369
Ray, Kanad, 369
Rodosek, Gabi Dreo, 405, 417
Roedler, Robert, 405, 417
Rozinajová, Viera, 235
Ruíz-Vanoye, Jorge, 189

S
Sánchez, Abraham, 189
Sharma, Shweta, 87
Sharma, Tarun Kumar, 87
Singh, Pushpendra, 369
Soncco-Álvarez, José L., 73
Sopov, Evgenii, 15
Srivastava, Viranjay M., 441
Sun, Yunlin, 61

T
Taiwo, Tunmike B., 281
Tawil, Abdel-Rahman H., 49
Tshiningayamwe, Loini, 347

V
Velazquez, Rogelio González, 189
Viriri, Serestina, 141
Vrablecová, Petra, 235

W
Wang, Lin, 97
Wang, Xiaoyang, 61
Weickert, Joachim, 261

Y
Yang, Bo, 97

Z
Zhang, Lei, 61

Printed in the United States
By Bookmasters